LONDON MATHEMATICAL SOCIETY LECTURE NOTE SERIES

Managing Editor: Professor J.W.S. Cassels, Department of Pure Mathematics and Mathematical Statistics, University of Cambridge, 16 Mill Lane, Cambridge CB2 1SB, England

The books in the series listed below are available from booksellers, or, in case of difficulty, from Cambridge University Press.

London Mathematical Society Lecture Note Series. 185

Representations of Solvable Groups

Olaf Manz
Universität Heidelberg
and
UCI Utility Consultants International, Frankfurt

and

Thomas R. Wolf
Ohio University

CAMBRIDGE
UNIVERSITY PRESS

Published by the Press Syndicate of the University of Cambridge
The Pitt Building, Trumpington Street, Cambridge CB2 1RP
40 West 20th Street, New York, NY 10011-4211, USA
10 Stamford Road, Oakleigh, Melbourne 3166, Australia

First published 1993

Printed in Great Britain at the University Press, Cambridge

Library of Congress cataloguing in publication data available

British Library cataloguing in publication data available

ISBN 0 521 39739 1

CONTENTS

PREFACE

Representation theory has very strong interplay with group structure. This is particularly true for finite solvable groups G, because their chief factors are irreducible modules for G over fields of prime order. In this monograph, we present some topics and problems arising in the representation theory of solvable groups. In particular, we study modules over finite fields, yet give applications to ordinary and Brauer characters of solvable groups.

It is not our intent to develop representation theory from scratch, but rather to discuss techniques and problems in current research. On the other hand, we wish that the manuscript be accessible to a reasonably wide group of people, including advanced graduate students, working in group theory. We refer to two basic references, namely:

[Hu] B. Huppert, "Endliche Gruppen I" and
[Is] I. M. Isaacs, "Character Theory of Finite Groups".

We believe that readers fairly familiar with these texts should have little problem reading the manuscript. We do also quote some material from the first chapter appearing in the sequel to [Hu], namely Chapter VII of "Finite Groups II" by B. Huppert and N. Blackburn [HB]. That chapter is entitled "Elements of General Representation Theory". Many of the results from these sources for which we have frequent use are presented (generally without proof) in Chapter 0, "Preliminaries". Since we present some applications to block theory, we state and/or prove several related results in Chapter 0. To this end, we have quoted some material here from "Representations of Finite Groups", by H. Nagao and Y. Tsushima [NT], although many of the quoted results also appear in the sketchy introduction to block theory that appears in the last chapter of [Is]. In our preliminary chapter, we do include proofs of Fong reduction and the Fong–Swan Theorem.

Of course, module (and character) induction is a powerful tool in representation theory, particularly when paired with Clifford's Theorem. Consequently, we need to study "quasi-primitive" linear groups, where those techniques do not apply. For solvable groups, the condition of quasi-primitivity imposes strong restrictions on the normal structure of the group. We study this extensively in Section 1, without restriction on the underlying field. An important class of solvable (quasi-primitive) linear groups over finite fields are the "semi-linear" groups. We study these in Section 2 along with conditions that force a linear group to indeed be a semi-linear group. Section 3 gives bounds for orders and derived lengths of solvable linear groups and permutation groups.

Much of Chapters II and III (Sections 4 through 11) deals with orbits of solvable linear groups or, as in Section 5, orbits of permutation groups. Of course, for solvable groups, orbit sizes of linear groups and those of permutation groups are closely related. This becomes clear in Section 6, where we give a new proof of Huppert's classification of doubly transitive solvable permutation groups. Many of the questions about orbit sizes of linear groups are related to the existence (or non-existence) of "regular" orbits. Our emphasis here again is on finite fields, because otherwise regular orbits always exist. The main feature of Chapter III, which is critical for Chapters IV and V, is the study of linear groups with "Sylow centralizers".

Chapters IV and V deal with ordinary and modular characters and their degrees. In Section 12, we prove Brauer's height-zero conjecture for solvable G, using material from Sections 5, 6, 9 and 10. In Section 15, we give a character-counting argument and use it to prove the Alperin–McKay conjecture for p-solvable groups. In Section 16, we discuss the derived length and the number of character degrees of a solvable group. This partially relies on a theorem of Berger, presented in Section 8, which unlike other orbit theorems gives the existence of small orbits.

The final chapter introduces the theory of "π-special" characters and gives some applications thereof. Also included is Isaacs' canonical lift of Brauer characters (for $p > 2$).

Olaf Manz
Heidelberg and Frankfurt
Germany

T. R. Wolf
Athens, Ohio
USA

ACKNOWLEDGEMENTS

The writing of this book began in Autumn 1988 as part of the project *Darstellungstheorie endlicher Gruppen und endlich-dimensionaler Algebren*, sponsored by the Deutsche Forschungsgemeinschaft (DFG). We thank the DFG for its generous support. We also thank the National Science Foundation, the Mathematical Sciences Research Institute (Berkeley) and the Ohio University Research Council for assistance. We also thank the following universities for their assistance and resources: Johannes Gutenberg Universität, Mainz (in particular, Fachbereich Mathematik), Universität Heidelberg (IWR), and Ohio University (Department of Mathematics).

We thank editors Roger Astley and David Tranah of Cambridge University Press for their assistance. Mei Lan Jin (Mathematics Typing Studio, Marion, Ohio) has done a splendid job of preparing a camera-ready manuscript using TEX. Finally, we thank numerous mathematicians whose work has influenced these pages and/or with whom we had valuable discussions.

Chapter 0

PRELIMINARIES

For this manuscript, all groups will be assumed finite. If G is a group and $\mathcal{F}$ an (arbitrary) field, an $\mathcal{F}[G]$-module V will mean that V is a right $\mathcal{F}[G]$-module and that V is finite dimensional over $\mathcal{F}$. Recall that V is *completely reducible* if V is the sum of simple $\mathcal{F}[G]$-modules. In this case, V is actually a direct sum of simple modules. Indeed, if $V \neq 0$ is completely reducible, then $V = V_1 \oplus \cdots \oplus V_l$ where $V_i \neq 0$ is the direct sum of simple isomorphic $\mathcal{F}[G]$-modules and if W_i and W_j are simple submodules of V_i and V_j (resp.), then $W_i \cong W_j$ (as $\mathcal{F}[G]$-modules) if and only if $i = j$. Then V_i are called the *homogeneous components* of V and are unique (not merely up to isomorphism, but the V_i are unique submodules). Now $V_1 = U_1 \oplus \cdots \oplus U_t$ for isomorphic $\mathcal{F}[G]$-modules U_i. While t is unique, the U_i are unique only up to isomorphism.

Because solvable groups have an abundance of normal subgroups, we begin by recalling Clifford's Theorem:

0.1 Theorem. *Suppose that V is an irreducible $\mathcal{F}[G]$-module and $N \trianglelefteq G$. Then*

(a) *V_N is completely reducible and so $V_N = V_1 \oplus \cdots \oplus V_l$ where the V_i are the homogeneous components of V_N;*

(b) *G/N transitively permutes the V_i by right multiplication;*

(c) *If W_i and W_j are irreducible N-submodules of V_i and V_j (resp.), then $\dim(W_i) = \dim(W_j)$ for all i, j; and*

(d) *If $I = \{g \in G \mid V_1 g = V_1\}$ is the inertia group in G of V_1, then V_1 is an irreducible I-module and $V \cong V_1^G$ (induced from I to G).*

Proof. This is Hauptsatz V, 17.3 of [Hu].   □

0.2 Proposition. *Suppose that V is an irreducible $\mathcal{F}[G]$-module, $N \trianglelefteq G$ and V_N is not homogeneous.*

(i) *If $C \trianglelefteq G$ is maximal such that V_C is not homogeneous, then G/C faithfully and primitively permutes the homogeneous components of V_C.*

(ii) *There exists $N \leq D \triangleleft G$ such that $V_D = W_1 \oplus \cdots \oplus W_s$ for D-invariant W_i that are faithfully and primitively permuted by G/D ($s > 1$). Furthermore, whenever $N \leq L \leq D$ with $L \trianglelefteq G$, V_L is not homogeneous and each W_i is a sum of homogeneous components of V_L.*

Proof. Write $V_N = V_1 \oplus \cdots \oplus V_t$ where the V_i are the homogeneous components of V_N. Suppose that $N \leq M \trianglelefteq G$ and $W = V_1 \oplus \cdots \oplus V_s$ is M-invariant. We claim that W is a direct sum of homogeneous components of V_M. To see this, let X and Y be isomorphic irreducible M-submodules of V with $X \leq W$. Now X_N and Y_N have isomorphic irreducible submodules X_0 and Y_0 (resp.). Since the V_i are homogeneous components of V_N, X_0 and Y_0 are contained in the same V_i. Thus $Y_0 \leq W$ and $Y \cap W \neq 0$. Then $Y \leq W$, establishing the claim.

(i) Now G transitively permutes the homogeneous components of V_C. Let K be the kernel of this permutation action, so that $C \leq K \trianglelefteq G$. Applying the last paragraph to V_C, each homogeneous component of V_C is a direct sum of homogeneous components of V_K. By maximality of C, $C = K$, proving that G/C acts faithfully on the homogeneous components of V_C. This action is primitive by the first paragraph and choice of C.

(ii) Since G transitively permutes $\Omega = \{V_1, \ldots, V_t\}$, we may write $\Omega = \Delta_1 \dot{\cup} \cdots \dot{\cup} \Delta_s$ with $s > 1$ and G primitively permuting $\{\Delta_1, \ldots, \Delta_s\}$. In other words, $V_N = W_1 \oplus \cdots \oplus W_s$ where $s > 1$ and each W_i is a sum of some homogeneous components of V_N and such that G primitively permutes the

W_i. Let D be the kernel of the permutation action of G on $\{W_1, \ldots, W_s\}$. Then $V_D = W_1 \oplus \cdots \oplus W_s$ for D-invariant W_i that are faithfully and primitively permuted by G/D. Furthermore, whenever $N \leq L \leq D$ with $L \trianglelefteq G$, each $(W_i)_L$ is a sum of homogeneous components of V_L, by the first paragraph. Since $s > 1$, V_L is not homogeneous. $\quad\square$

The structure of solvable primitive permutation groups is well-known and discussed below in Section 2. In particular, a nilpotent and primitive permutation group has prime order (see [Hu, Satz II, 3.2]).

0.3 Corollary. *Suppose that V is an irreducible G-module, $N \trianglelefteq G$ and V_N is not homogeneous. If G/N is nilpotent, there exists $N \leq C \triangleleft G$ with $|G : C| = p$, a prime such that $V_C = V_1 \oplus \cdots \oplus V_p$ for homogeneous components V_i of V_C.*

0.4 Proposition. *Suppose that V is an irreducible $\mathcal{F}[G]$-module and that $\mathcal{K}$ is an extension field of $\mathcal{F}$.*

   (i) *If $\operatorname{char}(\mathcal{F}) \neq 0$, then $V \otimes_{\mathcal{F}} \mathcal{K} = W_1 \oplus \cdots \oplus W_l$ for non-isomorphic irreducible $\mathcal{K}[G]$-modules W_i.*

   (ii) *If $\mathcal{K}$ is a Galois extension of $\mathcal{F}$, then $V \otimes_{\mathcal{F}} \mathcal{K} \cong e(V_1 \oplus \cdots \oplus V_t)$ for a positive integer e and non-isomorphic irreducible $\mathcal{K}[G]$-modules V_i. Furthermore the V_i are afforded by representations X_i that are conjugate under $\operatorname{Gal}(\mathcal{K} : \mathcal{F})$. Indeed $\{X_1, \ldots, X_t\}$ is a single orbit under $\operatorname{Gal}(\mathcal{K} : \mathcal{F})$.*

Proof. See [HB, Theorems VII, 1.15 and VII, 1.18 (b)]. The $\mathcal{K}[G]$-module $V \otimes_{\mathcal{F}} \mathcal{K}$ is denoted by $V_{\mathcal{K}}$ in [HB] and by $V^{\mathcal{K}}$ in [Is]. $\quad\square$

Suppose V is a faithful irreducible $\mathcal{F}[G]$-module for some field $\mathcal{F}$. If $\mathcal{K}$ is an extension field of $\mathcal{F}$, then G has a faithful irreducible $\mathcal{K}[G]$-module W by Proposition 0.4. By choosing $\mathcal{K}$ to be algebraically closed, G has a faithful absolutely irreducible representation $X : G \to M_n(\mathcal{K})$ for some n. Then the centralizer in $M_n(\mathcal{K})$ of $X(G)$ consists of scalar matrices. If G is abelian, then G must be cyclic and $n = 1$. We thus have the following well-known result which is of particular importance to the structure of quasi-primitive linear groups.

0.5 Lemma. *If an abelian group A has a faithful irreducible module W (over an arbitrary field $\mathcal{F}$), then A is cyclic. If furthermore W is absolutely irreducible, then $\dim_{\mathcal{F}}(W) = 1$.*

The following lemma is sometimes referred to as Fitting's lemma, although [Hu] credits Zassenhaus.

0.6 Lemma. *Suppose G acts on an abelian group A by automorphisms and $(|G|, |A|) = 1$. Then $A = [G, A] \times \mathbf{C}_A(G)$.*

Proof. See [Hu, Satz III, 13.4]. $\qquad\qquad\qquad\qquad\qquad\qquad\qquad\qquad\square$

We use $\mathrm{Irr}\,(G)$ to denote the set of the ordinary (i.e. complex) irreducible characters of the group G and let $\mathrm{char}\,(G)$ denote the set of all ordinary characters of G. Of course, $\mathrm{char}\,(G) \subseteq \mathrm{cf}\,(G)$, the set of class functions of G, and we let $[\chi, \theta]$ denote the inner product of χ, $\theta \in \mathrm{cf}\,(G)$. For $N \trianglelefteq G$ and $\theta \in \mathrm{Irr}\,(N)$, we let $\mathrm{Irr}\,(G|\theta) = \{\chi \in \mathrm{Irr}\,(G) \mid [\chi_N, \theta] \neq 0\}$. By Frobenius reciprocity, $\mathrm{Irr}\,(G|\theta)$ is the set of irreducible constituents of the induced character θ^G.

Let $\mathcal{F}$ be a field of characteristic p such that $\mathcal{F}$ contains a $|G|$-th root of unity. Then $\mathcal{F}$ is a splitting field for all subgroups of G (i.e. every irreducible $\mathcal{F}$-representation of every subgroup of G is absolutely irreducible). It is customary to choose $\mathcal{F}$ so that $\mathcal{F}$ is a quotient ring of an integral domain of characteristic zero. This is often done via p-modular systems, as in Section 3.6 of [NT]. A slightly different approach is given in Chapter 15 of [Is]. We should point out here that Chapter 15 of [Is] is only intended as an introduction to modular theory and as such is not complete. Recall that each $g \in G$

has a unique factorization $g = g_p g_{p'} = g_{p'} g_p$ where g_p is a p-element and $g_{p'}$ is p-regular (i.e. $p \nmid o(g_{p'})$). Each irreducible $\mathcal{F}$-character χ of G can then be lifted to a complex-valued function φ, defined on p-regular elements of G. Now φ is called an *irreducible Brauer character* of G, the set of which is denoted $\mathrm{IBr}_p(G)$. (Actually there are some choices involved in this procedure, but it is usual to do this simultaneously for all irreducible representations of all subgroups of G to avoid complications). Because $\chi(g) = \chi(g_{p'})$ for all $g \in G$, defining the lift $\varphi \in \mathrm{IBr}_p(G)$ only on p-regular elements loses no information and avoids technical difficulties. Now there is a 1–1 correspondence between $\mathrm{IBr}_p(G)$ and the irreducible $\mathcal{F}$-representations. Indeed, if $\varphi \in \mathrm{IBr}_p(G)$ corresponds to the $\mathcal{F}$-representation afforded by an $\mathcal{F}[G]$-module V, then $\varphi(1) = \dim(V)$. Also $\mathrm{IBr}_p(G)$ is linearly independent over $\mathbb{C}$ and $|\mathrm{IBr}_p(G)|$ is the number of p-regular classes of G.

Let $N \trianglelefteq G$ and $\varphi \in \mathrm{IBr}_p(N)$. We write $\mathrm{IBr}_p(G|\varphi) = \{\beta \in \mathrm{IBr}_p(G)| \ \varphi$ is a constituent of $\beta_N\}$. Now the induced character φ^G is a positive $\mathbb{Z}$-linear sum $\sum a_i \mu_i$ of irreducible Brauer characters μ_i, even though the corresponding induced module may not be completely reducible. By Nakayama reciprocity [HB, Theorem VII, 4.13 (a)] and Clifford's Theorem 0.1, each $\chi \in \mathrm{IBr}_p(G|\varphi)$ is a constituent of φ^G. When G/N is a p'-group, we get the converse and more.

0.7 Proposition. *Suppose that G/N is a p'-group, that $\varphi \in \mathrm{IBr}_p(N)$ and $\theta \in \mathrm{IBr}_p(G)$. Then the multiplicity of φ in θ_N equals the multiplicity of θ in φ^G.*

Proof. Let $\mathcal{F}$ be a splitting field for N and G in characteristic p. Let V be an (irreducible) $\mathcal{F}(G)$-module affording θ, and W an (irreducible) $\mathcal{F}(N)$-module affording φ. Now V_N is completely reducible by Clifford's Theorem. Since G/N is a p'-group and W an irreducible N-module, indeed W^G is completely reducible (see [HB, VII, 9.4]). With both V_N and W^G completely reducible and $\mathcal{F}$ a splitting field for N and G, it follows from Nakayama reciprocity ([HB, VII, 4.13]) that the multiplicity of W as a composition factor of V_N equals the multiplicity of V as a composition factor of W^G.

The proposition now follows. $\square$

0.8 Theorem. *Let* $N \trianglelefteq G$ *and* $\varphi \in \mathrm{IBr}_p(N)$. *If* $I = I_G(\varphi)$, *then* $\psi \to \psi^G$ *is a bijection from* $\mathrm{IBr}_p(I|\varphi)$ *onto* $\mathrm{IBr}_p(G|\varphi)$.

Proof. For ordinary characters, this is Theorem 6.11 of [Is]. More generally, a similar proof works here. Let $\chi \in \mathrm{IBr}_p(G|\varphi)$. Clifford's Theorem 0.1 (d, a) shows that $\chi = \mu^G$ for some $\mu \in \mathrm{IBr}_p(I|\varphi)$ and that $\chi_I = \mu + \Lambda$ for a (possibly zero) Brauer character Λ of I with no irreducible constituent of Λ lying in $\mathrm{IBr}_p(I|\varphi)$.

Let $\psi \in \mathrm{IBr}_p(I|\varphi)$. By Nakayama reciprocity [HB, VII, 4.13 (a)], there exists $\gamma \in \mathrm{IBr}_p(G)$ such that ψ is a constituent of γ_I. But then $\gamma \in \mathrm{IBr}_p(G|\varphi)$ and the last paragraph implies that ψ is the unique irreducible constituent of γ_I lying in $\mathrm{IBr}_p(I|\varphi)$ and $\psi^G = \gamma$. So $\psi \to \psi^G$ is a 1–1 and onto map from $\mathrm{IBr}_p(I|\varphi)$ onto $\mathrm{IBr}_p(G|\varphi)$. $\square$

Theorem 0.8 applies to ordinary characters too; just choose p so that $p \nmid |G|$.

0.9 Lemma. *Suppose that* $N \trianglelefteq G$, $\varphi \in \mathrm{IBr}_p(G)$ *and* φ_N *is irreducible. Then* $\alpha \to \alpha\varphi$ *is a one-to-one map from* $\mathrm{IBr}_p(G/N)$ *onto* $\mathrm{IBr}_p(G|\varphi_N)$.

Proof. By [HB, Theorem VII, 9.12 (b,c)], note $\alpha\varphi \in \mathrm{IBr}_p(G)$ for each $\alpha \in \mathrm{IBr}_p(G/N)$ and the mapping $\alpha \to \alpha\varphi$ is one-to-one. Let $\mu \in \mathrm{IBr}_p(G|\varphi_N)$. It suffices to show $\mu = \beta\varphi$ for some $\beta \in \mathrm{IBr}_p(G/N)$. We mimic the proof of [HB, Corollary VII, 9.13].

Let $\mathcal{F}$ be an algebraically closed field of characteristic p and V an irreducible $\mathcal{F}[G]$-module affording φ. Since $\mu \in \mathrm{IBr}_p(G|\varphi_N)$, Nakayama reciprocity implies that μ is a constituent of $\varphi_N{}^G$ (see comments preceding Proposition 0.7). Thus μ is afforded by a composition factor of $V_N{}^G \cong V \otimes_{\mathcal{F}} \mathcal{F}(G/N)$ (see [HB, VII, 4.15(b)]). If $0 = U_0 < U_1 \cdots < U_m =$

$\mathcal{F}(G/N)$ is a composition series of the $\mathcal{F}(G)$-module $\mathcal{F}(G/N)$, then for each i
$V \otimes_{\mathcal{F}} U_i / V \otimes_{\mathcal{F}} U_{i-1} \cong V \otimes U_i / U_{i-1}$ is irreducible, again by [HB, VII, 9.12(b)].
Thus, by the Jordan–Hölder theorem, μ is afforded by $V \otimes_{\mathcal{F}} W$ for an irre-
ducible $\mathcal{F}(G/N)$-module W. So $\mu = \beta\varphi$ for some $\beta \in \mathrm{IBr}_p(G/N)$. $\square$

In presenting Gallagher's Theorem (Lemma 0.9 for ordinary characters),
Issacs [Is, Theorem 6.16] first proves the following stronger result under the
assumption that φ is G-invariant.

0.10 Lemma. *Suppose that $N \unlhd G$, that $\varphi, \theta \in \mathrm{Irr}(N)$ and $\theta = \chi_N$ for
some $\chi \in \mathrm{Irr}(G)$. Assume also that $\varphi\theta \in \mathrm{Irr}(N)$ and $I_G(\varphi) = I_G(\varphi\theta)$. Then
$\sigma \to \sigma\chi$ is a bijection from $\mathrm{Irr}(G|\varphi)$ onto $\mathrm{Irr}(G|\varphi\theta)$.*

Proof. Let $I = I_G(\varphi) = I_G(\varphi\theta)$. For $\delta \in \mathrm{char}(I)$, observe that $(\delta\chi_I)^G = \delta^G\chi$ (see [Hu, V, 16.8] or [Is, Ex 5.3]). Now Theorem 6.16 of [Is] yields
that $\alpha \to \alpha\chi_I$ is a bijection from $\mathrm{Irr}(I|\varphi)$ onto $\mathrm{Irr}(I|\varphi\theta)$. Employing the
Clifford correspondence (Theorem 0.8), $\alpha \to (\alpha\chi_I)^G = \alpha^G\chi$ is a bijection
from $\mathrm{Irr}(I|\varphi)$ onto $\mathrm{Irr}(G|\varphi\theta)$. Since $\alpha \to \alpha^G$ is a bijection from $\mathrm{Irr}(I|\varphi)$
onto $\mathrm{Irr}(G|\varphi)$, the lemma follows. $\square$

To employ the above, we would like conditions sufficient to extend characters. Theorem 0.13 is quite useful, in part due to uniqueness (e.g. see
Lemma 0.18).

0.11 Proposition. *Suppose that G/N is cyclic and $\varphi \in \mathrm{IBr}_p(N)$ is G-
invariant. Then $\varphi = \beta_N$ for some $\beta \in \mathrm{IBr}_p(G)$.*

Proof. See [HB, Theorem VII, 9.9]. $\square$

0.12 Proposition. *Suppose $N \unlhd G$, $\theta \in \mathrm{Irr}(N)$ and θ extends to P whenever P/N is a Sylow subgroup of G/N. Then θ extends to G.*

Proof. See [Is, Corollary 11.31]. $\square$

Let $\mu \in \text{char}(G)$ and let $X: G \to GL(n, \mathbb{C})$ be a representation of G affording μ. For $g \in G$, let $\det(\mu)(g) = \det(X(g))$. This is independent of the choice of X and $\det(\mu)$ is a linear character of G. We let $o(\mu)$ be the order of the linear character $\det(\mu)$ of G. The following theorem can often be combined with Proposition 0.12 to extend characters. Note that $\det(\varphi + \mu) = \det(\varphi)\det(\mu)$, and $\det(\varphi\mu) = (\det\varphi)^{\mu(1)}\det(\mu)^{\varphi(1)}$.

0.13 Theorem. *Suppose that $N \trianglelefteq G$, $\theta \in \text{Irr}(N)$ is G-invariant and $(o(\theta)\theta(1), |G/N|) = 1$. There exists a unique extension $\chi \in \text{Irr}(G)$ of θ satisfying $(o(\chi), |G:N|) = 1$. Also, $o(\chi) = o(\theta)$.*

Proof. See [Is, Corollary 8.16]. □

In Theorem 0.13, we call χ the *canonical* extension of θ to G. The uniqueness in Theorem 0.13 is quite useful, often for inductive purposes. For example, we use it to prove the Fong–Swan Theorems. It is also used in the proof of Lemma 0.18, which guarantees the existence of characters of p'-degree and is helpful in Fong reduction. First however, we look at " Glauberman's Lemma", Glauberman correspondence, and some consequences thereof.

0.14 Lemma (Glauberman). *Suppose that A acts on G by automorphisms and $(|A|, |G|) = 1$. Assume that both A and G act on a set Ω and that G acts transitively on Ω. In addition, suppose that $(\omega g)a = (\omega a)g^a$ for all $a \in A$, $g \in G$, and $\omega \in \Omega$. Then*

(a) *A has fixed points in Ω; and*

(b) *$\mathbf{C}_G(A)$ acts transitively on the set of fixed points of A in Ω.*

Proof. See [Is, Lemmas 13.8 and 13.9]. Note that the hypothesis $(\omega g)a = (\omega a)g^a$ is equivalent to the condition that the semi-direct product GA acts on Ω (consistently with the actions of G and A). □

If a group A acts on G via automorphisms, we let $\text{Irr}_A(G) = \{\chi \in \text{Irr}(G) \mid \chi^a = \chi \text{ for all } a \in A\}$.

0.15 Theorem. *Whenever A acts on G by automorphisms with $(|A|, |G|) = 1$ and A solvable, there is a uniquely defined bijection $\rho(G, A) : \mathrm{Irr}_A(G) \to \mathrm{Irr}(C)$ where $C = \mathbf{C}_G(A)$ such that*

    (i) *If A is a p-group and $\chi \in \mathrm{Irr}_A(G)$, then $\chi\rho(G, A)$ is the unique $\beta \in \mathrm{Irr}(C)$ satisfying $[\chi_C, \beta] \not\equiv 0 \pmod p$.*

    (ii) *If $T \trianglelefteq A$, then $\rho(G, A) = \rho(G, T)\, \rho(\mathbf{C}_G(T), A/T)$.*

Proof. See [Is, Theorem 13.1]. $\qquad\qquad\qquad\qquad\qquad\qquad\square$

By "uniquely defined" above, we mean there is only one such map (indeed, else (ii) would be meaningless) and this map is independent of choices made in the algorithm implied by the theorem. This map is known as the *Glauberman correspondence*. If A acts on G with $(|A|, |G|) = 1$, but A not solvable, then $|G|$ is odd. Isaacs [Is 2] has exhibited a "uniquely defined" correspondence whenever A acts on G, $(|A|, |G|) = 1$, and $|G|$ is odd. Moreover, this agrees with the Glauberman correspondence when both are defined [Wo 2]. The combined map is thus referred as the *Glauberman–Isaacs correspondence*. The following appears in [Is, Theorem 13.29] and has a couple of uses in this section alone.

0.16 Lemma. *Suppose A acts on G with A solvable and $(|A|, |G|) = 1$. Suppose $N \trianglelefteq G$ is A-invariant, $\chi \in \mathrm{Irr}_A(G)$ and $\theta \in \mathrm{Irr}_A(N)$. Then $[\chi_N, \theta] \neq 0$ if and only if $[\chi\rho(G, A)_{N \cap C}, \theta\rho(N, A)] \neq 0$.*

Much of the following lemma is a consequence of Glauberman's lemma above. In fact, no more is required for G solvable or for parts (a), (b) and (c) in the general case. For G non-solvable, parts (d), (e) and (f) also employ the Glauberman correspondence and more. All parts appear somewhere (possibly as exercises) in Chapter 13 of [Is]. Due to its importance here (particularly when G is solvable), we give a sketch.

0.17 Lemma. *Suppose that $N \leq G \trianglelefteq \Gamma$ with $(|\Gamma : G|, |G : N|) = 1$ and $N \trianglelefteq \Gamma$. Let $H/N \leq \Gamma/N$ with $G \cap H = N$ and $\Gamma = GH$. Suppose that*

$\theta \in \mathrm{Irr}\,(G)$ is Γ-invariant and $\varphi \in \mathrm{Irr}\,(N)$ is H-invariant. Then

(a) θ_N has an H-invariant irreducible constituent α;

(b) If $\mathbf{C}_{G/N}(H/N) = 1$, then α is unique;

(c) If $\mathbf{C}_{G/N}(H/N) = G/N$, then every irreducible constituent of θ_N is H-invariant;

(d) φ^G has a Γ-invariant irreducible constituent η;

(e) If $\mathbf{C}_{G/N}(H/N) = 1$, then η is unique; and

(f) If $\mathbf{C}_{G/N}(H/N) = G/N$, then every irreducible constituent of φ^G is Γ-invariant.

Proof. (a, b, c). By Clifford's Theorem, G/N transitively permutes the set X of irreducible constituents of θ_N. Also H/N permutes X and acts on G/N. Since $(|H/N|, |G/N|) = 1$, Glauberman's Lemma 0.14 applies: some element of X is H/N invariant and $\mathbf{C}_{G/N}(H/N)$ transitively permutes the H-invariant elements of X. Parts (a), (b) and (c) follow.

(d, e, f). Arguing by induction on $|G/N| = |\Gamma : H|$ we may assume without loss of generality that G/N is a chief factor of Γ (note, for part (e), part (a) is employed along with the inductive hypothesis). Let $I = I_\Gamma(\varphi)$, so that $H \leq I \leq \Gamma$ and $I = (I \cap G)H$. Now $I \cap G = I_G(\varphi)$ is H-invariant. Also $\xi \to \xi^G$ gives a bijection from $\mathrm{Irr}\,(I \cap G|\varphi)$ onto $\mathrm{Irr}\,(G|\varphi)$ and furthermore ξ is H-invariant if and only if ξ^G is H-invariant. The result follows by induction should $I \cap G < G$. Thus $G \leq I$ and φ is Γ-invariant.

First suppose that G/N is abelian. The group $\mathrm{Irr}\,(G/N)$ of linear characters acts on $\mathrm{Irr}\,(G|\varphi)$ by multiplication and this action is transitive (see [Is, Exercise 6.2]). Now $\Gamma/G \cong H/N$ acts on both $\mathrm{Irr}\,(G/N)$ and $\mathrm{Irr}\,(G|\varphi)$. Here Glauberman's Lemma 0.14 yields (d), (e) and (f).

Finally we may assume that G/N is non-abelian and thus non-solvable. Since $I_\Gamma(\varphi) = \Gamma$, we may in fact assume that $N \leq \mathbf{Z}(\Gamma)$ (see [Is, Theorem 11.28]). Then there exists $Z \leq N$ and G_1, $N_1 \trianglelefteq \Gamma$ such that $G = G_1 \times Z$, $N = N_1 \times Z$ and $(|H/N|, |G_1|) = 1$. We may write $\varphi = \varphi_1 \times \lambda$ (uniquely) with $\varphi_1 \in \mathrm{Irr}\,(N_1)$ and $\lambda \in \mathrm{Irr}\,(Z)$. Also $\beta \to \beta\lambda$ defines a bijection from $\mathrm{Irr}\,(G_1|\varphi_1)$ onto $\mathrm{Irr}\,(G|\varphi)$. Since $Z_1 \le Z(G)$, this map commutes with H. Without loss of generality $\lambda = 1_Z$ and $Z = 1$.

We now have that $(|\Gamma : G|, |G|) = 1$. Choose $S \le H$ with $\Gamma = GS$ and $1 = G \cap S$. Of course $\eta \in \mathrm{Irr}\,(G)$ is H-invariant if and only if it is S-invariant and $\mathbf{C}_{G/N}(H/N) = \mathbf{C}_{G/N}(S)$. Since G/N is not solvable, the Odd-order Theorem implies that S is solvable. Setting $C = \mathbf{C}_G(S)$, the Glauberman correspondences apply: both $\rho_G : \mathrm{Irr}_S(G) \to \mathrm{Irr}\,(C)$ and $\rho_N : \mathrm{Irr}_S(N) \to \mathrm{Irr}\,(C \cap N)$ are bijections. Let $\mu = \varphi\rho_N$, and $\chi \in \mathrm{Irr}_S(G)$. By Lemma 0.13, $[\chi_N, \varphi] \ne 0 \iff [(\chi\rho_G)_{N \cap C}, \mu] \ne 0$. Hence $|\mathrm{Irr}_S(G|\varphi)| = |\mathrm{Irr}\,(C|\mu)| \ne 0$. This proves (d). If $\mathbf{C}_{G/N}(S) = 1$, then $C \le N$ and $N \cap C = N$. Then $|\mathrm{Irr}_S(G|\varphi)| = |\mathrm{Irr}\,(C|\mu)| = 1$. This proves (f). Finally, we may assume for (e) that S centralizes G/N and hence G. Part (e) is then trivial. $\qquad\square$

We combine the extendability Theorem 0.13 with Glauberman correspondence for the next lemma which has numerous uses. For example, it can be used to show that when G is p-solvable and $\varphi \in \mathrm{Irr}\,(\mathbf{O}_{p'}(G))$ is G-invariant, then the unique p-block covering $\{\varphi\}$ is indeed a block of maximal defect (see Theorem 0.28 below).

0.18 Lemma. *Suppose that G/N is p-solvable and $\varphi \in \mathrm{Irr}\,(N)$. Assume $p \nmid o(\varphi)\, \varphi(1)$ and $p \nmid |G : I_G(\varphi)|$. Then there exists $\chi \in \mathrm{Irr}\,(G|\varphi)$ such that $p \nmid \chi(1)$.*

Proof. We argue by induction on $|G : N|$. For $\psi \in \mathrm{Irr}\,(I_G(\varphi))$, we have $p \nmid \psi(1)$ if and only if $p \nmid \psi^G(1)$ because $p \nmid |G : I_G(\varphi)|$. Employing Clifford's Theorem and the inductive hypothesis, we can assume that φ is G-invariant.

Let M/N be a chief factor of G. If M/N is a p-group, Theorem 0.13 shows there exists a unique extension $\theta \in \mathrm{Irr}\,(M)$ of φ satisfying $p \nmid o(\theta)$. Since φ is G-invariant, θ is also G-invariant by uniqueness. Certainly $p \nmid \theta(1)$ and the inductive hypothesis ensures the existence of $\alpha \in \mathrm{Irr}\,(G|\theta)$ such that $p \nmid \alpha(1)$. Since $\alpha \in \mathrm{Irr}\,(G|\varphi)$, the result follows in this case. So we may assume that M/N is a p'-group.

Let $P \in \mathrm{Syl}_p(G)$. By Lemma 0.17 (d), there exists $\mu \in \mathrm{Irr}\,(M|\varphi)$ such that μ is P-invariant. In particular, $p \nmid |G : I_G(\mu)|$. Also $p \nmid \mu(1)$ because $p \nmid |M : N|\,\varphi(1)$. Now μ_N is a sum $\varphi_1 + \cdots + \varphi_m$ of (not necessarily distinct) $\varphi_i \in \mathrm{Irr}\,(N)$ that are M-conjugate to φ. Hence $o(\varphi_i) = o(\varphi)$ is a p'-number for all i. Now

$$(\det \mu)_N = \det(\mu_N) = \prod_{i=1}^{m} \det(\varphi_i).$$

Thus $(\det \mu)_N$ has p'-order as a linear character of N. Since $p \nmid |M : N|$, indeed $p \nmid o(\mu)$. Now we can apply the inductive hypothesis to conclude the existence of $\chi \in \mathrm{Irr}\,(G|\mu)$ such that $p \nmid \chi(1)$. The proof is complete as $\chi \in \mathrm{Irr}\,(G|\varphi)$. $\square$

The following well-known lemma appeared as "Lemma 1.2.3" of the important Hall–Higman paper [HH]. It appears in [HB] as Lemma IX, 1.3. Recall G is π-separable (for a set of primes π) if it has a normal series, where each factor group is a π-group or π'-group. For completeness, we mention that G is π-solvable if G has a normal series with each factor group either a π'-group or solvable π-group. Of course, p-separability is equivalent to p-solvability. By the Odd-order Theorem, G π-separable implies that G is π-solvable or π'-solvable. When G is π-separable, the analogues of the Sylow theorems (existence, containment, and conjugacy) hold for Hall π-subgroups.

0.19 Lemma. *If G is π-separable, then* $\mathbf{C}_G(\mathbf{O}_{\pi,\pi'}(G)/\mathbf{O}_\pi(G)) \subseteq \mathbf{O}_{\pi,\pi'}(G)$.

The following well-known result will be used repeatedly, often without reference. A proof is given in [Hu, Satz V, 5.17]. Alternatively, use Clifford's Theorem 0.1, Proposition 0.4, and Lemma 0.5 to reduce to the case where $|G| = p$, $\mathcal{F}$ is algebraically closed, and $\dim_{\mathcal{F}}(V) = 1$. Then the corresponding representation $X : G \to \mathcal{F}$ must be trivial because 1 is the only p^{th} root of unity in $\mathcal{F}$.

0.20 Proposition. *If V is a completely reducible and faithful $\mathcal{F}[G]$-module and $\operatorname{char}(\mathcal{F}) = p$, then $\mathbf{O}_p(G) = 1$.*

Recall $\operatorname{cf}(G)$ denotes the set of class functions of G, i.e. the set of complex-valued functions on G that are constant on conjugacy classes of G. Fix a prime p and let $\operatorname{cf}^0(G)$ be the set of complex-valued functions defined on p-regular elements of G that are constant on G-conjugacy classes. For $\alpha \in \operatorname{cf}(G)$, we denote the restriction of α to p-regular elements by $\alpha^0 \in \operatorname{cf}^0(G)$. Now $\operatorname{Irr}(G)$ and $\operatorname{IBr}_p(G)$ are bases for the vector spaces $\operatorname{cf}(G)$ and $\operatorname{cf}^0(G)$, respectively (see [Is, Theorems 2.8 and 15.10] or [NT, Theorems 3.6.2 and 3.6.5]). For $\chi \in \operatorname{Irr}(G)$, we have that χ^0 is a positive $\mathbb{Z}$-linear combination of Brauer characters (see [NT, p. 233] or [Is, Theorem 15.8]), and it easily follows that each $\varphi \in \operatorname{IBr}_p(G)$ is a constituent of ψ^0 for some $\psi \in \operatorname{Irr}(G)$. (Actually, more can be said: φ is a $\mathbb{Z}$-linear combination of $\{\chi^0 \mid \chi \in \operatorname{Irr}(G)\}$.)

The set $\operatorname{IBr}_p(G) \cup \operatorname{Irr}(G)$ is a disjoint union of *p-blocks* of G. This is often done by decomposing the group algebra $\mathcal{F}[G]$ (for a sufficiently large field $\mathcal{F}$ of characteristic p) into a direct sum $\bigoplus I_j$ of indecomposable two-sided ideals. Each ideal is then called a block. Each indecomposable $\mathcal{F}[G]$-module V is associated with a unique I_j (determined by $V I_j \neq 0$). In such a way the characters of G are partitioned into p-blocks (see Section 3.6 of [NT] for more details). The following theorem characterizes blocks sufficiently for most of our purposes.

For a p-block B of G, we let $\operatorname{Irr}(B)$ denote $B \cap \operatorname{Irr}(G)$ and $\operatorname{IBr}_p(B) = B \cap \operatorname{IBr}_p(G)$. We also let $k(B) = |\operatorname{Irr}(B)|$, $k(G) = |\operatorname{Irr}(G)|$, $l(B) = |\operatorname{IBr}_p(B)|$ and $l(G) = |\operatorname{IBr}_p(G)|$. Also $\operatorname{bl}(G)$ denotes the set of p-blocks of G.

0.21 Theorem. *Let χ, $\beta \in \operatorname{Irr}(G)$ and $\mu \in \operatorname{IBr}_p(G)$. Then*

(a) *χ and β lie in the same p-block of G if and only if there exist $\psi_1, \ldots, \psi_t \in \operatorname{Irr}(G)$ with $\chi = \psi_1$ and $\beta = \psi_t$ such that ψ_i^0 and ψ_{i+1}^0 have a common irreducible Brauer character as a constituent $(1 \leq i \leq t - 1)$.*

(b) *If μ is a constituent of χ^0, then χ and μ are in the same block of G.*

Proof. See [NT, Theorem 3.6.19] or [Is, Theorem 15.27]. □

Theorem 0.21 completely determines the p-blocks of G. This "linking process" also has an analogue of part (a) for Brauer characters: given φ, $\mu \in \mathrm{IBr}_p(G)$, then φ and μ lie in the same p-block of G if and only if there exist $\gamma_1, \ldots, \gamma_t \in \mathrm{IBr}_p(G)$ with $\varphi = \gamma_1$, $\mu = \gamma_t$ such that γ_i and γ_{i+1} are both constituents of some β_i^0, $\beta_i \in \mathrm{Irr}(G)$, $1 \le i \le t - 1$. This analogue is an easy consequence of Theorem 0.21 (a, b).

When $p \nmid |G|$, $\mathrm{Irr}(G) = \mathrm{IBr}_p(G)$ and the p-blocks of G are singletons.

0.22 Proposition. *Suppose that N is a normal p'-subgroup of G and $\varphi \in \mathrm{IBr}_p(N)$. Then $\mathrm{Irr}(G|\varphi) \cup \mathrm{IBr}_p(G|\varphi)$ is a (disjoint) union of p-blocks of G.*

Proof. Suppose that χ, $\psi \in \mathrm{Irr}(G)$ and that χ^0 and ψ^0 have a common irreducible constituent $\eta \in \mathrm{IBr}_p(G)$. The irreducible constituents of η_N are common to both $\chi_N^0 = \chi_N$ and $\psi_N^0 = \psi_N$, i.e. χ_N and ψ_N have a common irreducible constituent μ. By Clifford's Theorem, the irreducible constituents of χ_N and ψ_N coincide (up to multiplicities). If B is the p-block containing χ, then $\mathrm{Irr}(B) \subseteq \mathrm{Irr}(G|\mu)$. Now also Theorem 0.21 (b) shows that $\mathrm{IBr}_p(B) \subseteq \mathrm{IBr}_p(G|\mu)$, as also $\mu \in \mathrm{IBr}_p(N)$. This proposition now follows. □

Let B be a p-block of a p-solvable group G. The first steps of *Fong reduction* are to observe that B covers $\{\varphi\}$ for some $\varphi \in \mathrm{Irr}(\mathbf{O}_{p'}(G))$ and that the Clifford correspondence gives a bijection between a block of $I_G(\varphi)$ and B (see 0.22 and 0.25). Should $I_G(\varphi) = G$, the next steps are to show $B = \mathrm{Irr}(G|\varphi) \cup \mathrm{IBr}_p(G|\varphi)$ and the defect groups of B are Sylow subgroups of G (Theorem 0.28). Finally one must give the Brauer correspondence in the two cases (i) $I_G(\varphi) < G$ and (ii) $I_G(\varphi) = G$ (see Theorem 0.29 and 0.30). When $I_G(\varphi) = G$, we use Lemma 0.18 to show the defect groups are

Sylow p-subgroups of G. This appears to be different and a little easier than what appears in the literature. Also, when $I_G(\varphi) = G$, instead of stating $B = \mathrm{Irr}\,(G|\varphi) \cup \mathrm{IBr}_p(G)$, many texts construct a block $\hat{B}$ of a group $\hat{G}$ with a bijection between $\hat{B}$ and B. This gets to be a little awkward when one also needs to keep track of the Brauer correspondence. Not that our proof is much different, rather we find the statement more convenient to use. It should be evident from 0.29 and 0.30 that there is a strong connection between the Brauer and Glauberman correspondences. Our proofs of Fong reduction use just standard results of block theory. Some results of Fong reduction can be found in [NT], but not the interconnections with the Brauer correspondence. The latter we need to do McKay's conjecture in Section 15. Finally, we do the Fong–Swan Theorem with a proof somewhat different than what appears in the literature.

0.23 Proposition. *Suppose that $\psi \in \mathrm{Irr}\,(H)$ and $\psi \in b \in \mathrm{bl}(H)$. If $H \leq G$ and $\psi^G \in \mathrm{Irr}\,(G)$, then $b^G \in \mathrm{bl}\,(G)$ (in the sense of Brauer induction), and $\psi^G \in b^G$.*

Proof. For $\chi \in \mathrm{Irr}\,(G)$ and $\mathcal{C}$ a conjugacy class of G, we let $\hat{\mathcal{C}}$ be the class sum in $\mathbb{C}[G]$ and define $\omega_\chi(\hat{\mathcal{C}}) = \frac{\chi(g)|\mathcal{C}|}{\chi(1)} \in \mathbb{C}$ where g is any element of $\mathcal{C}$. By definition of induced blocks (see [Is, p. 282] or [NT, p. 320]), it suffices to show that whenever $\mathcal{C} \cap H = \mathcal{C}_1 \cup \cdots \cup \mathcal{C}_t$ for classes $\mathcal{C}_i$ of H, then

$$\omega_{\psi^G}(\mathcal{C}) = \sum_{i=1}^{t} \omega_\psi(\mathcal{C}_i). \tag{0.1}$$

(Actually, it is sufficient to show that these two agree modulo a maximal ideal of a large enough subring of the ring of algebraic integers. Of course, (0.1) is stronger.) If $\mathcal{C} \cap H = \emptyset$, then both sides are zero. Without loss of generality, $\mathcal{C} \cap H \neq \emptyset$ and we let $x_1 \in \mathcal{C}_1, \ldots, x_t \in \mathcal{C}_t$. Set $x - x_1$. Observe that $\psi^G(x) = |\mathbf{C}_G(x)| \sum_{i=1}^{t} \frac{\psi(x_i)}{|C_H(x_i)|}$ (see [Is, p. 64]). Now

$$\omega_{\psi^G}(\mathcal{C}) = \frac{\psi^G(x)|\mathcal{C}|}{\psi^G(1)} = \frac{|\mathcal{C}|}{\psi^G(1)}|C_G(x)| \sum_{i=1}^{t} \frac{\psi(x_i)}{|C_H(x_i)|}$$

$$= \frac{|G|}{|G : H|\psi(1)} \sum_{i=1}^{t} \frac{\psi(x_i)|\mathcal{C}_i|}{|H|}$$

$$= \sum_{i=1}^{t} \frac{\psi(x_i)|\mathcal{C}_i|}{\psi(1)} = \sum_{i=1}^{t} \omega_\psi(\mathcal{C}_i),$$

as desired.   □

Associated with a p-block B of G is a G-conjugacy class of p-subgroups D of G. Then D is called a *defect group* of B, and we say that B has p-*defect* d if $|D| = p^d$. Let $p^m \| |G|$ (that is p^m is an exact divisor of $|G|$). For $\chi \in B$, there is a unique integer h, called the *height* of χ, such that $p^{m-d+h}\|\chi(1)$. We see that h is non-negative.

0.24 Lemma. *Let B be a block of G with p-defect d. Let $p^m \| |G|$. Then*

(i) $p^{m-d} \mid \chi(1)$ *for all* $\chi \in \mathrm{Irr}\,(B) \cup \mathrm{IBr}_p(B)$,

(ii) $p^{m-d}\| \beta(1)$ *for some* $\beta \in \mathrm{Irr}(B)$.

Proof. See [NT, Theorem 5.1.11 (iii) and p. 245] or [Is, Theorem 15.41].

□

Choose $\chi \in \mathrm{Irr}(B)$ with height zero and write $\chi^0 = \sum a_\alpha \alpha$ ($\alpha \in \mathrm{IBr}_p(B)$). Evaluating both sides at 1, there exists $\mu \in \mathrm{IBr}_p(B)$ of height zero (with $p \nmid a_\mu$).

0.25 Lemma. *Let $\theta \in \mathrm{Irr}\,(K)$ where $K \trianglelefteq G$ and $K \leq \mathbf{O}_{p'}(G)$. Set $I = I_G(\theta)$. Let $\alpha, \beta \in \mathrm{Irr}\,(I|\theta) \cup \mathrm{IBr}_p(I|\theta)$. Then*

(i) α *and* β *lie in the same p-block of I if and only α^G and β^G lie in the same p-block of G;*

(ii) If $b \in \mathrm{bl}(I|\theta)$, then $B := \{\varphi^G \mid \varphi \in b\}$ is a p-block in $\mathrm{bl}(G|\theta)$. Also $B = b^G$ in the sense of Brauer induction;

(iii) If $b \in \mathrm{bl}(I|\theta)$, then b and b^G have a common defect group; and

(iv) α and α^G have the same height.

Proof. (i) Assume that α^G and β^G lie in the same p-block $B \in \mathrm{bl}(G|\theta)$. To show that α and β lie in the same p-block of I, we may assume that α and β are ordinary characters and also that $(\alpha^G)^0$ and $(\beta^G)^0$ have a common irreducible constituent $\mu \in \mathrm{IBr}_p(G|\theta)$. Writing $\alpha^0 = \sum_i a_i \sigma_i$ where $a_i > 0$ and $\sigma_i \in \mathrm{IBr}_p(I|\theta)$, we have that

$$(\alpha^G)^0 = (\alpha^0)^G = (\sum_i a_i \sigma_i)^G = \sum_i a_i \sigma_i^G$$

and each $\sigma_i^G \in \mathrm{IBr}_p(G|\theta)$. Thus $\mu = \sigma_i^G$ for some i. Likewise $\mu = \gamma^G$ for some irreducible constituent $\gamma \in \mathrm{IBr}_p(I|\theta)$ of β^0. By the uniqueness in the Clifford correspondence (Theorem 0.8), $\sigma_i = \gamma$ is a common constituent of α^0 and β^0. Hence α and β lie in the same p-block of I. The proof of the converse direction is essentially identical.

(ii) That B is a block of G is immediate from part (i). That $B = b^G$ follows from Proposition 0.23.

(iii) Since $B = b^G$, a defect group of b is contained in a defect group of B ([NT, Lemma 5, 3.3] or [Is, Lemma 15.43]). It suffices to show that B and b have the same defect. Let d be the defect of b, let $p^i = |I|_p$ and $p^m = |G|_p$. Then p^{i-d} is the largest power of p dividing $\psi(1)$ for all $\psi \in \mathrm{Irr}(b)$ by Lemma 0.24. By (i), p^{m-d} is the largest power of p dividing $\chi(1)$ for all $\chi \in \mathrm{Irr}(B)$. Thus B has defect d (again Lemma 0.24). This proves (iii).

(iv) Say $\alpha \in b \in \mathrm{bl}(I|\theta)$ and b has defect d. Let h be the height of α. Then $p^{i-d+h} \| \alpha(1)$ and so $p^{m-d+h} \| \alpha^G(1)$. But $\alpha^G \in b^G$ by (ii) and b^G has defect d by (iii). Thus h is the height of α^G. $\qquad \square$

If B is a p-block of G and $\chi \in \mathrm{Irr}(B)$, we can write $\chi^0 = \sum_{\varphi \in \mathrm{IBr}_p(G)} d_{\chi\varphi}\varphi$

for non-negative integers $d_{\chi\varphi}$, called *decomposition numbers*. The $k(B) \times l(B)$ matrix D_B is the *decomposition matrix for* B. Similarly, we have a $k(G) \times l(G)$ *decomposition matrix for* G, namely $D = \begin{pmatrix} D_{B_1} & & \\ & \ddots & \\ & & D_{B_t} \end{pmatrix}$ where $B_1, \ldots, B_t$ are the p-blocks of G. The matrices $C_B := D_B^T D_B$ and $C := D^T D$ are *Cartan matrices* for B and G respectively. Of course, $C = \begin{pmatrix} C_{B_1} & & \\ & \ddots & \\ & & C_{B_t} \end{pmatrix}$. In Lemma 0.25, character induction is not only a height-preserving bijection from b to B, but indeed decomposition numbers are preserved (the proof is trivial). In particular, b and B have the same decomposition matrices and Cartan matrices. (Of course, D_b is unique only up to row permutations and column permutations.) Generally though, Brauer induction does not even preserve block size.

Definition. If $N \trianglelefteq G$, $b \in \mathrm{bl}(N)$ and $B \in \mathrm{bl}(G)$, then B *covers* b if there exists $\varphi \in \mathrm{Irr}(b)$ and $\chi \in \mathrm{Irr}(B)$ with $[\chi_N, \varphi] \neq 0$.

0.26 Proposition. *Suppose that* $N \trianglelefteq G$, $b \in \mathrm{bl}(N)$ *and* $B \in \mathrm{bl}(G)$ *covering* b.

   (i) *If* $\varphi \in \mathrm{Irr}(b)$, *then there exists* $\psi \in \mathrm{Irr}(B)$ *with* $[\psi_N, \varphi] \neq 0$.

   (ii) *If* $\gamma \in \mathrm{IBr}_p(b)$, *then there exists* $\sigma \in \mathrm{IBr}_p(B)$ *such that* γ *is a constituent of* σ_N.

Proof. Choose $\theta \in \mathrm{Irr}(b)$ and $\chi \in \mathrm{Irr}(B)$ with $[\chi_N, \theta] \neq 0$. Without loss of generality, we may assume that θ^0 and φ^0 have a common irreducible constituent $\mu \in \mathrm{IBr}(b)$. Since $[\chi_N, \theta] \neq 0$, there exists $\zeta \in \mathrm{IBr}_p(B)$ such that ζ is a constituent of χ^0 and μ is a constituent of ζ_N. Now ζ is a constituent of μ^G by Nakayama reciprocity (see comment preceding Proposition 0.7). Now $\varphi^0 = \mu + \sum a_i \nu_i$ with $a_i > 0$ and $\nu_i \in \mathrm{IBr}_p(N)$, it follows that ζ is an irreducible constituent of $(\varphi^0)^G = (\varphi^G)^0$ and hence also an irreducible constituent of ψ^0 for some $\psi \in \mathrm{Irr}(G|\varphi)$. Note that $\psi \in B$ because $\zeta \in B$. This proves (i).

To prove (ii), we may choose $\varphi \in \mathrm{Irr}(b)$ such that γ is a constituent of φ^0. By (i), choose $\psi \in \mathrm{Irr}(B)$ with $[\psi_N, \varphi] \neq 0$. Then γ is an irreducible constituent of $(\psi^0)_N = (\psi_N)^0$. Hence γ is an irreducible constituent of σ_N for some irreducible constituent σ of ψ^0. Since $\psi \in B$, indeed $\sigma \in B$. $\qquad\square$

0.27 Corollary. *Suppose that* G/N *is a* p-*group,* $b \in \mathrm{bl}_p(N)$ *and* $\varphi \in \mathrm{IBr}_p(N)$. *Then*

   (i) $\mathrm{IBr}_p(G|\varphi) = \{\chi\}$ *for some* χ,

   (ii) *A unique block of* G *covers* b.

   (iii) *If* φ *is* G-*invariant, then* $\chi_N = \varphi$.

Proof. By Nakayama reciprocity ([HB, VII, 4.13]), $\mathrm{IBr}_p(G|\varphi)$ is not empty. Suppose $\alpha, \beta \in \mathrm{IBr}_p(G|\varphi)$. If $\varphi_1, \ldots, \varphi_t$ are the distinct G-conjugates of φ, then $\alpha_N = e \sum_{i=1}^{t} \varphi_i$ and $\beta_N = f \sum_{1}^{t} \varphi_i$, for positive integers e and f, by Clifford's Theorem. Because G/N is a p-group, every p-regular element of G lies in N and so $\alpha = \alpha_N = (e/f)\beta_N = (e/f)\beta$. By the linear independence of $\mathrm{IBr}_p(G)$, indeed $\alpha = \beta$, proving (i). Part (ii) follows from part (i) and Proposition 0.26 (ii).

We prove (iii) by induction on $|G/N|$. Choose $N \leq M \trianglelefteq G$ with $|G/M| = p$. By (i), $\mathrm{IBr}_p(M|\varphi) = \{\mu\}$ for some μ. Since φ is G-invariant and μ is unique, μ also is G-invariant. By Proposition 0.11, μ extends to G and so $\chi_M = \mu$. By the inductive hypothesis $\mu_N = \varphi$. Thus χ extends φ. $\qquad\square$

0.28 Theorem. *Suppose that* G *is* p-*solvable,* $K = \mathbf{O}_{p'}(G)$ *and* $\varphi \in \mathrm{Irr}(K)$ *is* G-*invariant. Then there is a unique block of* G *covering* $\{\varphi\}$, *i.e.* $\mathrm{Irr}(G|\varphi) \cup \mathrm{IBr}_p(G|\varphi)$ *is a* p-*block* B *of* G. *Furthermore the Sylow* p-*subgroups of* G *are the defect groups of* G.

Proof. As noted in Lemma 0.22, we have that $\mathrm{Irr}(G|\varphi) \cup \mathrm{IBr}(G|\varphi)$ is a union of p-blocks of G. By Lemma 0.18, there exists $\chi \in \mathrm{Irr}(G|\varphi)$ such that

$p \nmid \chi(1)$. Since $p \nmid \chi(1)$, χ is a character of height zero for a block of G whose defect groups are Sylow subgroups of G (see Lemma 0.24). Thus it suffices to prove that $\{\varphi\}$ is covered by a unique block.

We argue by induction on $|G : K|$. If G/K is a p-group, the result follows from Corollary 0.27. Hence, by p-solvability, we may choose $\mathbf{O}_{p'p}(G) \leq M \lhd G$ such that G/M is a p-group or is a p'-group. Now $K = \mathbf{O}_{p'}(M)$ and the inductive hypothesis implies there is a unique block b_0 of M covering $\{\varphi\}$ (i.e. $b_0 = \mathrm{Irr}\,(M|\varphi) \cup \mathrm{IBr}_p(M|\varphi)$). It suffices to show there is a unique p-block of G covering b_0. By Corollary 0.27, we may assume that G/M is a p'-group. By Lemma 0.18, there exists $\theta \in \mathrm{Irr}\,(b_0)$ such that $p \nmid \theta(1)$.

Let α, $\beta \in \mathrm{Irr}\,(G|\theta)$. By Proposition 0.26, it suffices to show that α and β necessarily lie in the same p-block of G. To this end, it suffices to show that the algebraic integer

$$|\mathcal{C}| \left(\frac{\alpha(g)}{\alpha(1)} - \frac{\beta(g)}{\beta(1)} \right)$$

is divisible by p whenever $\mathcal{C}$ is a conjugacy class of G and $g \in \mathcal{C}$; see [NT, Theorem 3.6.24] or [Is, Definition 15.17]. Since α, $\beta \in \mathrm{Irr}\,(G|\theta)$ with $\theta \in \mathrm{Irr}(M)$, $M \unlhd G$, indeed $\alpha(g)/\alpha(1) = \beta(g)/\beta(1)$ for all $g \in M$. Thus we assume that $g \notin M \geq \mathbf{O}_{p'p}(G)$. By Lemma 0.19, g does not centralize $\mathbf{O}_{p'p}(G)/\mathbf{O}_p(G)$. Thus $\mathbf{C}_G(g)$ does not contain a Sylow p-subgroup of G, i.e. $p \mid |\mathcal{C}|$. Because $p \nmid |G : M|\theta(1)$, we have that $p \nmid \alpha(1)\,\beta(1)$. Thus p does divide $|\mathcal{C}|(\frac{\alpha(g)}{\alpha(1)} - \frac{\beta(g)}{\beta(1)})$. Hence α and β lie in the same p-block of G. $\square$

Let D be a p-subgroup of G. Brauer's First Main Theorem states that $b \mapsto b^G$ is a bijection from $\{b \in \mathrm{bl}(\mathbf{N}_G(D)) \mid D$ is a defect group for $b\}$ onto $\{B \in \mathrm{bl}(G)|D$ is a defect group for $B\}$. (See [NT, Theorem 9.2.15] or [Is, Theorem 15.45] and note [Is, 15.44] should read $PC(P) \leq H \leq \mathbf{N}(P)$.) This is called the *Brauer correspondence*. Given $B_0 \in \mathrm{bl}(G)$, the Brauer correspondent of B_0 is uniquely defined up to conjugation: if Q is a defect group for B, then the Brauer correspondent of B is the unique $b_0 \in \mathrm{bl}(\mathbf{N}_G(Q))$ with defect group Q satisfying $b_0^G = B$. The G-conjugates of Q form the set of defect groups of B.

0.29 Theorem. *Let G be p-solvable, $K = \mathbf{O}_{p'}(G)$ and $P \in \mathrm{Syl}_p(G)$. Assume that $\varphi \in \mathrm{Irr}(K)$ is G-invariant and $B \in \mathrm{bl}(G)$ covers $\{\varphi\}$. Set $C = \mathbf{C}_K(P)$ and let $\mu = \varphi\rho(K,P) \in \mathrm{Irr}(C)$ be the Glauberman correspondent of φ. Then there is a unique block b of $\mathbf{N}_G(P)$ covering $\{\mu\}$. Furthermore $b^G = B$.*

Notes. By Theorem 0.28, B is the unique block of G covering $\{\varphi\}$ and P is a defect group for B. Since $P \in \mathrm{Syl}_p(G)$, in fact P is a defect group of every block of $\mathbf{N}_G(P)$. In particular, b is the Brauer correspondent of B.

Proof. By Theorem 0.15, $\phi_C = \epsilon\mu + p\Lambda$ for a (possibly zero) character Λ of C and a p'-integer ϵ. For $x \in \mathbf{N}_G(P)$, $\varphi_C = (\varphi^x)_C = \epsilon\mu^x + p\Lambda^x$ and $\mu^x = \mu$. Thus μ is invariant in $\mathbf{N}_G(P)$.

Now $\mathbf{O}_{p'}(\mathbf{N}_G(P))$ centralizes P and thus centralizes $\mathbf{O}_{p'p}(G)/\mathbf{O}_{p'}(G)$. By Lemma 0.19, $\mathbf{O}_{p'}(\mathbf{N}_G(P)) \subseteq \mathbf{O}_{p'p}(G)$. Thus $\mathbf{O}_{p'}(\mathbf{N}_G(P)) \leq \mathbf{O}_{p'}(G) \cap \mathbf{C}_G(P) = \mathbf{C}_K(P) = C$. Hence $C = \mathbf{O}_{p'}(\mathbf{N}_G(P))$. Since $\mu \in \mathrm{Irr}(C)$ is invariant in $\mathbf{N}_G(P)$, Theorem 0.28 shows that there is a unique block b of $\mathbf{N}_G(P)$ covering μ.

We need to show that $b^G = B$. Observe that b^G and $b^{K\mathbf{N}_G(P)}$ are defined by Brauer's First Main Theorem. Then $(b^{K\mathbf{N}_G(P)})^G = b^G$ (see [NT, Lemma 5.3.4]). If $\alpha \in \mathrm{Irr}(b^{K\mathbf{N}_G(P)})$, then some irreducible constituent of α^G lies in b^G by [NT, Lemma 5.3.4]. Since θ is G-invariant, it follows that $b^{K\mathbf{N}_G(P)}$ covers $\{\theta\}$ if and only if b^G covers $\{\theta\}$ (i.e. $b^G = B$). Now $\mathbf{O}_{p'p}(G) \leq K\mathbf{N}_G(P)$ and thus $K = \mathbf{O}_{p'}(K\mathbf{N}_G(P))$. By Theorem 0.28, there is a unique block of $K\mathbf{N}_G(P)$ covering $\{\theta\}$. We may thus assume without loss of generality that $G = K\mathbf{N}_G(P)$.

Since $\mu \in \mathrm{Irr}(C)$ is invariant in $\mathbf{N}_G(P)$, there exists $\zeta \in \mathrm{Irr}(b)$ such that $p \nmid \zeta(1)$. Write $\zeta^G = \Lambda + \sum_{\chi \in \mathrm{Irr}(G|\theta)} a_\chi\chi$ for a (possibly zero) character Λ of G with $[\Lambda_K, \theta] = 0$. Since $\mathrm{Irr}(B) = \mathrm{Irr}(G|\theta)$, we have $b^G = B$ if and only if $p \nmid \sum_{\chi \in \mathrm{Irr}\,(G|\theta)} a_\chi \chi(1)$ by [NT, Lemma 5.3.4]. But

$$\sum_{\chi \in \mathrm{Irr}\,(G|\theta)} a_\chi \chi(1) = \theta(1) \sum_{\chi \in \mathrm{Irr}\,(G|\theta)} a_\chi [\chi_K, \theta] = \theta(1)[\zeta^G{}_K, \theta].$$

Since $G = K \mathbf{N}_G(P)$, $\zeta^G{}_K = \zeta_{K \cap \mathbf{N}_G(P)}{}^K = \zeta_C{}^K = (\zeta(1)/\mu(1))\mu^K$. Thus

$$\sum_{\chi \in \mathrm{Irr}\,(G|\theta)} a_\chi \chi(1) = \frac{\theta(1)\zeta(1)}{\mu(1)}[\mu^G, \theta]$$

is a p'-number by Theorem 0.15. Hence $b^G = B$. $\qquad\square$

Let $B \in \mathrm{bl}\,(G)$, so that B covers $\{\mu\}$ for some $\mu \in \mathrm{Irr}(\mathbf{O}_{p'}(G))$. Theorem 0.29 describes the Brauer correspondent of B provided μ is G-invariant. One loose end needs to be tied up: namely how the Brauer correspondence works when μ is not G-invariant, i.e. what is the relationship between the Brauer correspondence and the correspondence in Lemma 0.25.

0.30 Corollary. *Suppose that* $K \trianglelefteq G$, $p \nmid |K|$, *and* $\theta \in \mathrm{Irr}\,(K)$. *Suppose* $B \in \mathrm{bl}\,(G)$ *covering* $\{\theta\}$ *and* $I = I_G(\theta)$. *Choose* $D \leq I$ *such that* D *is a defect group for* B *and let* $b \in \mathrm{bl}\,(\mathbf{N}_G(D))$ *be the Brauer correspondent of* B. *Set* $C = \mathbf{C}_K(D)$ *and* $\mu = \theta\rho(K, D)$. *Then*

(i) *b covers $\{\mu\}$;*

(ii) *$I \cap \mathbf{N}_G(D) = I_{\mathbf{N}_G(D)}(\mu)$;*

(iii) *If B_0 and b_0 are the unique blocks of I and $I \cap \mathbf{N}_G(D)$ (respectively) with $B_0^G = B$ and $b_0^{\mathbf{N}_G(D)} = b$ (see 0.25), then $b_0^I = B_0$.*

Proof. (ii). By Theorem 0.15, $\theta_C = \epsilon\mu + p\Lambda$ with $\Lambda \in \mathrm{char}\,(C)$ and $p \nmid \epsilon$. If $x \in \mathbf{N}_G(D)$, then θ^x is D-invariant, $C^x = C$, and $\theta^x = \epsilon\mu^x + p\Lambda^x$. So $\theta^x \rho(K, D) = \mu^x$. Since $\rho(K, D)$ is 1–1, it follows that $I_G(\theta) \cap \mathbf{N}_G(D) = I_{\mathbf{N}_G(D)}(\mu)$.

(i), (iii). By induction on $|G : K|$. If, say, $D = 1$ then $B = b$ and $\theta = \mu$, whence the result is trivial. So we may assume that $p \mid |G : K|$.

Let $b^* \in \mathrm{bl}\,(I \cap \mathbf{N}_G(D))$ be the Brauer correspondent of B_0. Then $(b^*)^I = B_0$ and $B_0^G = B$, whence $(b^*)^G = B$. Then $(b^*)^{\mathbf{N}_G(D)} \in \mathrm{bl}\,(\mathbf{N}_G(D))$ and $((b^*)^{\mathbf{N}_G(D)})^G = B$ (see [NT, Lemma 5.3.4]).

Assume that $I < G$. The inductive hypothesis implies that b^* covers $\{\mu\}$. By (ii) and Theorem 0.28, $(b^*)^{\mathbf{N}_G(D)}$ has D as a defect group and covers $\{\mu\}$. Since $((b^*)^{\mathbf{N}_G(D)})^G = B$ and D is a defect group of $(b^*)^{\mathbf{N}_G(D)}$, Brauer's First Main Theorem implies that $(b^*)^{\mathbf{N}_G(D)} = b$. Now the uniqueness of b_0 yields that $b_0 = b^*$. In this case, namely $I < G$, the result follows. Thus we assume that $I = G$. Part (iii) is now trivial. We need just show that b covers $\{\mu\}$.

Let $M = \mathbf{O}_{p'}(G)$. We may assume that $M > K$, since otherwise (i) follows from Theorem 0.29. We may choose $\alpha \in \mathrm{Irr}\,(M)$ such that B covers $\{\alpha\}$. By Lemma 0.25, $I_G(\alpha)$ contains a defect group for B. Replacing α by a G-conjugate, we may assume without loss of generality that $D \leq I_G(\alpha)$. Since θ is G-invariant and $\mathrm{Irr}\,(B) \subseteq \mathrm{Irr}\,(G|\theta)$, $\alpha \in \mathrm{Irr}(M|\theta)$. Because $M \trianglelefteq G$ and $\mu = \theta\rho(K, D)$, $\alpha\rho(M, D) \in \mathrm{Irr}\,(\mathbf{C}_M(D) \mid \mu)$ (see Lemma 0.16). By the inductive hypothesis, b covers $\{\alpha\rho(M, D)\}$. Thus b covers $\{\mu\}$. $\square$

If G is p-solvable and $\varphi \in \mathrm{IBr}_p(G)$, the Fong–Swan Theorem asserts the existence of $\chi \in \mathrm{Irr}\,(G)$ such that $\chi^0 = \varphi$. This is not true without the p-solvability hypothesis. While indeed χ may not be unique, Isaacs has shown there is a canonically defined set $Y(G) \subseteq \mathrm{Irr}\,(G)$ such that $\chi \mapsto \chi^0$ is a bijection from $Y(G)$ onto $\mathrm{IBr}_p(G)$. We give below a reasonably short proof of the Fong–Swan theorem. This proof, somewhat in the flavor of Isaacs', forgoes uniqueness for the sake of brevity. The next lemma is taken from [Is 4, Theorem 3.1].

0.31 Lemma. *Suppose that G/N is a p'-group, $\mu \in \mathrm{Irr}\,(N)$, $\mu^0 \in \mathrm{IBr}_p(N)$ and $I_G(\mu) = I_G(\mu^0)$. Then $\chi \mapsto \chi^0$ defines a bijection from $\mathrm{Irr}\,(G|\mu)$ onto $\mathrm{IBr}_p(G|\mu^0)$.*

Proof. We argue by induction on $|G/N|$ and let $I = I_G(\mu) = I_G(\mu^0)$. If

$I < G$, the inductive hypothesis implies that $\psi \mapsto \psi^0$ is a bijection from $\mathrm{Irr}\,(I|\mu)$ onto $\mathrm{IBr}_p(I|\mu^0)$. The Clifford correspondence then implies that $\psi \mapsto (\psi^0)^G$ from $\mathrm{Irr}\,(I|\mu)$ onto $\mathrm{IBr}_p(G|\mu^0)$. But $\psi \mapsto \psi^G$ is a bijection from $\mathrm{Irr}\,(I|\mu)$ onto $\mathrm{Irr}\,(G|\mu)$ and $(\psi^G)^0 = (\psi^0)^G$. The result follows in this case when $I < G$. We thus assume that μ and μ^0 are G-invariant.

Let $X = \mathrm{Irr}\,(G|\mu)$ and $Y = \mathrm{IBr}_p(G|\mu^0)$. Write $\mu^G = \sum_{\chi \in X} a_\chi \chi$, each $a_\chi > 0$, and $\mu^{0G} = \sum_{\varphi \in Y} b_\varphi \varphi$, each $b_\varphi > 0$. For $\chi \in X$, $(\chi^0)_N = (\chi_N)^0 = a_\chi \mu^0$ and thus $\chi^0 = \sum_{\varphi \in Y} d_{\chi\varphi}\varphi$ for non-negative $d_{\chi\varphi}$ and at least one positive $d_{\chi\varphi}$. Now $\sum_{\varphi \in Y} b_\varphi \varphi = (\mu^0)^G = (\mu^G)^0 = (\sum_{\chi \in X} a_\chi \chi)^0 = \sum_{\chi \in X} a_\chi (\sum_{\varphi \in Y} d_{\chi\varphi}\varphi) = \sum_{\varphi \in Y}(\sum_{\chi \in X} a_\chi d_{\chi\varphi})\varphi$. By linear independence, $b_\varphi = \sum_{\chi \in X} a_\chi d_{\chi\varphi}$ for each $\varphi \in Y$.

By Frobenius reciprocity, $(\mu^G)_N = \sum a_\chi \chi_N = \sum a_\chi^2 \mu$ and $(\mu^0)^G{}_N = (\mu^G{}_N)^0 = \sum a_x^2 \mu^0$. By Proposition 0.7, $\varphi_N = b_\varphi \mu^0$ for each $\varphi \in Y$ and so $(\mu^{0G})_N = \sum_{\varphi \in Y} b_\varphi \varphi_N = \sum_{\varphi \in Y} b_\varphi^2 \mu^0$. Hence $\sum_{\chi \in X} a_\chi^2 = \sum_{\varphi \in Y} b_\varphi^2$. By the equality of the last paragraph

$$
\begin{aligned}
\sum_{\chi \in X} a_\chi^2 = \sum_{\varphi \in Y} b_\varphi^2 &= \sum_{\varphi \in Y}\left(\sum_{\chi \in X} a_\chi d_{\chi\varphi}\right)^2 \\
&\geq \sum_{\varphi \in Y}\sum_{\chi \in X}(a_\chi d_{\chi\varphi})^2 \qquad (0.2) \\
&\geq \sum_{\chi \in X} a_\chi^2 \sum_{\varphi \in Y} d_{\chi\varphi}^2 .
\end{aligned}
$$

Each $a_\chi > 0$, each $d_{\chi\varphi}$ is a non-negative integer, and given $\chi \in X$, some $d_{\chi\varphi} \geq 1$. Consequently, for each $\beta \in X$, there is exactly one $\gamma \in Y$ for which $d_{\beta\gamma} = 1$ while all other $d_{\beta\varphi}$ are zero. In particular, $\chi^0 \in \mathrm{IBr}_p(G|\mu^0)$. Furthermore, we must have equality throughout (0.2). Hence for each $\varphi \in Y$

$$
\left(\sum_{\chi \in X} a_\chi d_{\chi\varphi}\right)^2 = \sum_{\chi \in X}(a_\chi d_{\chi\varphi})^2 .
$$

Since each $a_\chi > 0$, at most one $d_{\chi\varphi} \neq 0$. Thus $\chi \mapsto \chi^0$ defines a 1–1 map from X into Y. Now if $\chi^0 = \gamma \in Y$, then $a_\chi \mu^0 = \chi_N^0 = \gamma_N$ and

$a_\chi = b_\gamma$. Since $\sum_{\chi \in X} a_\chi^2 = \sum_{\varphi \in Y} b_\varphi^2$, in fact $|X| = |Y|$. This proves the lemma. $\qquad\qquad\qquad\qquad\qquad\qquad\qquad\qquad\qquad\qquad\qquad\qquad\quad$ $\square$

0.32 Theorem. *Suppose that G/N is p-solvable, $\theta \in \mathrm{Irr}\,(N)$, $\theta^0 = \varphi \in$ IBr$\,(N)$, and $I_G(\theta) = I_G(\varphi)$. Assume also that $p \nmid o(\theta)\,\theta(1)$. If $\beta \in$ IBr$_p(G|\varphi)$, there exists $\psi \in \mathrm{Irr}\,(G|\theta)$ such that $\psi^0 = \beta$.*

Proof. By induction on $|G : N|$. Let $\eta \in$ IBr$_p(I_G(\varphi)|\varphi)$ with $\eta^G = \beta$. If $I_G(\varphi) < G$, then the inductive hypothesis implies that there does exist $\sigma \in \mathrm{Irr}\,(I_G(\varphi) \mid \theta)$ with $\sigma^0 = \eta$. Since $I_G(\varphi) = I_G(\theta)$, indeed $\sigma^G \in \mathrm{Irr}\,(G|\theta)$ and $(\sigma^G)^0 = (\sigma^0)^G = \eta^G = \beta$. We are done in this case. So we can assume that φ and θ are G-invariant.

Let M/N be a chief factor of G. Choose $\zeta \in$ IBr$_p(M|\varphi)$ such that $\psi \in$ IBr$_p(G|\zeta)$. Assume first that M/N is a p'-group. By Lemma 0.31, there is a unique $\alpha \in \mathrm{Irr}\,(M|\theta)$ such that $\alpha^0 = \zeta$. Since $I_G(\theta) = I_G(\varphi) = G$, the uniqueness of α implies that $I_G(\alpha) = I_G(\zeta)$. Now $\alpha_N = \theta_1 + \cdots + \theta_t$ for not necessarily distinct θ_i that are M-conjugate to θ. Then $p \nmid o(\theta_i)$ and $(\det \theta)_N = (\det \theta_N) = \prod_{i=1}^t \det(\theta_i)$ has p'-order. Since M/N is a p'-group, $o(\theta)$ is a p'-number. Now the inductive hypothesis applies and there exists $\gamma \in \mathrm{Irr}\,(G|\alpha)$ with $\gamma^0 = \beta$. Since $\mathrm{Irr}\,(G|\alpha) \subseteq \mathrm{Irr}\,(G|\theta)$, we are done in the case where M/N is a p'-group. Since G/N is p-solvable, we have then that M/N is a p-group.

By Theorem 0.13, there is a unique extension $\mu \in \mathrm{Irr}\,(M)$ of θ satisfying $p \nmid o(\mu)$. By uniqueness, $I_G(\mu) = I_G(\theta) = G$. Now $(\mu^0)_M = (\mu_M)^0 = \theta^0 = \varphi$ is irreducible and thus $\mu^0 \in$ IBr$_p(M|\varphi)$. By Corollary 0.27, $\{\mu^0\} =$ IBr$_p(M|\varphi)$ and so $I_G(\mu^0) = G = I_G(\mu)$. Now $\beta \in$ IBr$_p(G|\mu^0)$ because $\{\mu^0\} =$ IBr$_p(M|\zeta)$. The inductive hypothesis implies the existence of $\psi \in \mathrm{Irr}\,(G|\mu)$ with $\psi^0 = \beta$. We are done because $\mathrm{Irr}\,(G|\mu) \subseteq \mathrm{Irr}\,(G|\theta)$. $\qquad$ $\square$

By setting $N = 1$ above, we get the usual form of the Fong–Swan Theorem.

0.33 Corollary. *If $\varphi \in \mathrm{IBr}_p(G)$ and G is p-solvable, then there exists $\psi \in \mathrm{Irr}(G)$ with $\psi^0 = \varphi$.*

We close this section with a counting argument that will be used repeatedly.

0.34 Lemma. *Let G be a Frobenius group with Frobenius kernel K and complement H. Suppose V is an $\mathcal{F}[G]$-module such that $\mathbf{C}_V(K) = 0$ and $\mathrm{char}(\mathcal{F}) \nmid |K|$. If $J \leq H$, then $\dim \mathbf{C}_V(J) = |H : J| \dim \mathbf{C}_V(H)$. In particular $\dim(V) = |H| \dim \mathbf{C}_V(H)$.*

Proof. See [Is, Theorem 15.16]. The second statement is obtained by setting $J = 1$. □

Chapter I
SOLVABLE SUBGROUPS OF LINEAR GROUPS

§1 Quasi-Primitive Linear Groups

An irreducible $\mathcal{F}[G]$-module V is called *imprimitive* if V can be written $V = V_1 \oplus \cdots \oplus V_n$ for $n > 1$ subspaces (not submodules) V_i that are permuted (transitively) by G. If $H = \mathrm{stab}_G(V_1)$, then $V \cong V_1^G$ (induced from H) (e.g. see [Is, Theorem 5.9]). We say V is *primitive*, if V is not imprimitive, or equivalently if V is not induced from a submodule of a proper subgroup of G.

An irreducible G-module V is called *quasi-primitive* if V_N is homogeneous for all $N \trianglelefteq G$. It is a consequence of Clifford's Theorem that a primitive module V is quasi-primitive. As would be expected, G is a quasi-primitive linear group if G has a faithful quasi-primitive module. In this case, every abelian normal subgroup of G is cyclic (by Lemma 0.5). This limits the structure of G and particularly that of the Fitting subgroup $\mathbf{F}(G)$, as is described by a theorem of P. Hall [Hu, III, 13.10]. This section uses Hall's Theorem to give a thorough look at solvable quasi-primitive linear groups.

At times, it is more convenient to weaken the quasi-primitive condition to V_N homogeneous for all characteristic subgroups N of G. We then call V a *pseudo-primitive* G-module. In Section 10, we will see a pseudo-primitive but not quasi-primitive module for a solvable group G. In that example, $|G| = 3^2 \cdot 2$ and $|V| = 4^3$.

The structure of a quasi-primitive solvable linear group is a little cleaner when the underlying field is algebraically closed, but we still will study the more general case because many of our applications will occur when the field is finite.

1.1 Proposition. *Let P be a dihedral, quaternion or semi-dihedral 2-group with $|P| = 2^n$ ($n \geq 3$). Then*

(a) $|P/\Phi(P)| = 4$ *and* $|\mathbf{Z}(P)| = 2$; *and*

(b) *if P is not isomorphic to the quaternion group of order 8, then P has a characteristic cyclic subgroup of index 2.*

Proof. By definition, P has a cyclic subgroup $A = \langle a \rangle$ of index 2 and order 2^{n-1} with n at least 3 (3, 4 respectively). Now $\Phi(P) \leq A$ and $A/\Phi(P)$ is elementary abelian. Since $A = \langle a \rangle$ is cyclic, $\Phi(P)$ is $\langle a \rangle$ or $\langle a^2 \rangle$. But P is not cyclic and so $|P/\Phi(P)| \neq 2$. Hence $\Phi(P) = \langle a^2 \rangle$ has index 4 in P.

Now there exists $y \in P$ such that $a^y = a^j$ where $j = -1$ ($-1, -1 + 2^{n-2}$ respectively). Since A is abelian and $|P : A| = 2$, $\mathbf{Z}(P) \leq A$. Direct computation shows that $|\mathbf{Z}(P)| = 2$. If $|P| \geq 16$, then $\langle a^2 \rangle \not\leq \mathbf{Z}(P)$ and thus $A = \mathbf{C}_P(\langle a^2 \rangle) = \mathbf{C}_P(\Phi(P))$ is a characteristic subgroup of P. If P is dihedral of order 8, then $\langle a \rangle$ is the unique cyclic subgroup of order 4. $\quad\square$

We let Q_n, D_n and SD_n denote the quaternion, dihedral and semi-dihedral groups of order $n = 2^m$, with m at least 3 (3, 4 respectively).

1.2 Theorem. *Let $P \neq 1$ be a p-group with every characteristic abelian subgroup cyclic. Let $Z \leq \mathbf{Z}(P)$ with $|Z| = p$. Then there exist $E, T \leq P$ such that*

(i) $P = ET$, $E \cap T = Z$ *and* $T = \mathbf{C}_P(E)$;

(ii) E *is extra-special or* $E = Z$;

(iii) $\exp(E) = p$ *or* $p = 2$;

(iv) T *is cyclic or* $p = 2$, $|T| \geq 16$, *and T is dihedral, quaternion or semi-dihedral;*

(v) *There exists U characteristic in P such that $U \leq T$, $|T : U| \leq 2$, $U = \mathbf{C}_T(U)$ and U is cyclic; and*

(vi) $EU = \mathbf{C}_P(U)$ *is characteristic in P.*

Proof. By P. Hall's Theorem ([Hu, III, 13.10]), there exist E, $T \leq P$ satisfying (i), (ii) and (iii). Furthermore T is cyclic, dihedral, quaternion or semi-dihedral. Now, if $|T| = 8$ and T is non-abelian, then $P = ET$ is extra-special (see [Hu, III, 13.8]) and the conclusion of the theorem is satisfied with P and Z playing the roles of E and T (respectively). Thus we assume that T is cyclic, or $16 \mid |T|$ and T is quaternion, dihedral or semi-dihedral. If T is cyclic, we complete the theorem by letting $U = T = \mathbf{Z}(P)$. Assume T is not cyclic. By Proposition 1.1, there exists U characteristic in T such that $|T : U| = 2$, $U = \langle u \rangle$ is cyclic, and $\Phi(T) = \langle u^2 \rangle$. Since E is extra-special, $\Phi(E) = Z = \mathbf{Z}(P) \leq \langle u^2 \rangle = \Phi(T)$ holds. Therefore, $[E, T] = 1$ implies that $\langle u^2 \rangle = \Phi(P)$ is characteristic in P. Since $\mathbf{Z}(T) < \langle u^2 \rangle$, $\mathbf{C}_T(\Phi(P)) = U$ and hence $\mathbf{C}_T(U) = U$. Thus $\mathbf{C}_P(U) = EU = \mathbf{C}_P(\Phi(P))$ and $U = \mathbf{Z}(EU)$ are characteristic in P. $\qquad\square$

1.3 Corollary. *Let P be a p-group with every abelian normal subgroup of P cyclic. Then P is cyclic, quaternion, dihedral or semi-dihedral. Also, $P \not\cong D_8$.*

Proof. Assume not and let $Z \leq \mathbf{Z}(P)$ with $|Z| = p$. By Theorem 1.2, there exist E, $T \trianglelefteq P$ with E extraspecial, $Z < E$, $T = \mathbf{C}_P(E)$, $T \cap E = Z$ and T cyclic, dihedral, quaternion or semi-dihedral. Now $|E| = p^{2n+1}$ for an integer n and there exists $Z \leq A \trianglelefteq E$ with A abelian of order p^{n+1} (see [Hu, III, 13.7]). Since $A \trianglelefteq P$, A is cyclic. But the exponent of E is p or 4. Thus $|A| = 4$ and $|E| = 8$. We may assume that $P > E$ and thus $T > Z$. Then there exists $Z \leq S \trianglelefteq T$ with S cyclic of order 4. Since $S \trianglelefteq P$ and $S \leq \mathbf{C}_P(A)$, AS is a normal abelian subgroup, which is not cyclic. This contradiction completes the proof. $\qquad\square$

Under the hypotheses of the above corollary, P has a cyclic subgroup of index p, as is proven in [Hu, III, 7.6], but this is a weaker conclusion. Satz I.14.9 of [Hu] lists all p-groups with a cyclic maximal subgroup, but observe that the groups in (a) and (b3) of that list have normal subgroups that are abelian but not cyclic.

1.4 Corollary. *Assume that every characteristic abelian subgroup of G is cyclic. Let $p_1, \ldots, p_l$ be the distinct prime divisors of $|F|$ for $F = \mathbf{F}(G)$ and let $Z \leq \mathbf{Z}(F)$ with $|Z| = p_1 \cdots p_l$. Then there exist $E, T \leq G$ such that*

(i) *$F = ET$, $Z = E \cap T$ and $T = \mathbf{C}_F(E)$.*

(ii) *The Sylow subgroups of E are extra-special or cyclic of prime order.*

(iii) *$\exp(E) \mid 2p_1 \cdots p_l$.*

(iv) *If a Sylow p_i-subgroup T_i of T is not cyclic, then $p_i = 2$, $|T_i| \geq 16$ and T_i is quaternion, dihedral or semi-dihedral.*

(v) *There exists $U \operatorname{char} G$ with U cyclic, $U \leq T$, $U = \mathbf{C}_T(U)$ and $|T : U| \leq 2$.*

(vi) *$EU = \mathbf{C}_F(U)$ is characteristic in G.*

(vii) *If every characteristic abelian subgroup of G is in $\mathbf{Z}(F)$, then $U = T = \mathbf{Z}(F)$.*

Proof. Let $P_1, \ldots, P_l$ be the Sylow subgroups of F. For each i, write $P_i = E_i T_i$ as in Theorem 1.2. Set $E = \prod_i E_i$ and $T = \prod_i T_i$. Parts (i)–(vi) now immediately follow. Furthermore, U is characteristic in F and in G. Thus, for (vii), we have by hypothesis that $U \leq \mathbf{Z}(F)$. Since $|T : U| \leq 2$ and $U = \mathbf{C}_T(U)$, $T = U = \mathbf{Z}(F)$. $\qquad\square$

We assume that G is solvable and is as in the above corollary. By a theorem of Gaschütz (see Theorem 1.12 below), $F/\Phi(G) = \mathbf{F}(G/\Phi(G))$ is a completely reducible and faithful G/F-module. Suppose now that $F \leq G'$ and that every abelian characteristic subgroup A of G is cyclic. Since the automorphism group of a cyclic group is abelian, we have $A \leq \mathbf{Z}(G')$ and $A \leq \mathbf{Z}(F)$. By Corollary 1.4 (vii), $U = T = \mathbf{Z}(F)$, and $\Phi(G)Z/\Phi(G) \cong Z/(\Phi(G) \cap Z)$ is centralized by G'. Hence Gaschütz's Theorem implies that $F/\Phi(G)Z$ is a completely reducible and faithful G'/F-module.

1.5 Lemma. *Assume that $Z \leq \mathbf{Z}(E)$, Z is cyclic and E/Z abelian. Let $A \leq \operatorname{Aut}(E)$ with $[E, A] \leq Z$ and $[Z, A] = 1$. Then $|A| \mid |E/Z|$.*

Proof. For $a \in A$, define $\varphi_a \colon E/Z \to Z$ by $\varphi_a(Zx) = [x, a]$. Since $[Z, A] = 1$

and $[E, A] \leq Z \leq \mathbf{Z}(E)$, φ_a is well-defined and $\varphi_a \in \text{Hom}(E/Z, Z)$. View $\text{Hom}(E/Z, Z)$ as a group (with pointwise multiplication). Then $a \mapsto \varphi_a$ is a monomorphism of A into $\text{Hom}(E/Z, Z)$, because A acts faithfully on E. It thus suffices to show that $|\text{Hom}(E/Z, Z)| \mid |E/Z|$.

Write $E/Z = D_1 \times \cdots \times D_n$ for cyclic groups D_i. Since $\text{Hom}(E/Z, Z) \cong \prod_i \text{Hom}(D_i, Z)$, it will be sufficient to show that $|\text{Hom}(D_i, Z)| \mid |D_i|$. Since D_i and Z are both cyclic, $|\text{Hom}(D_i, Z)| = (|D_i|, |Z|)$. $\qquad \square$

1.6 Corollary. *Assume that* $Z \leq E \leq G$, $Z = \mathbf{Z}(E)$ *is cyclic and* E/Z *is abelian. Let* $A = \mathbf{C}_G(Z)$, $B = \mathbf{C}_G(E)$ *and* $C = \mathbf{C}_A(E/Z)$. *Then* $Z = B \cap E$ *and* $EB = C$.

Proof. Note that $B \cap E = Z$ and so $|EB/B| = |E/(B \cap E)| = |E/Z|$. The hypotheses imply that C/B acts trivially on both E/Z and Z, while acting faithfully on E. Lemma 1.5 implies that $|C/B| \leq |E/Z|$. But $B \leq EB \leq C$ and $|EB/B| = |E/Z|$. Thus $EB = C$. $\qquad \square$

1.7 Corollary. *Suppose that* E/Z *is abelian,* Z *is cyclic,* $Z \leq F \leq E$ *and* $Z = \mathbf{Z}(F) = \mathbf{Z}(E)$. *Then* $E/Z = F/Z \times \mathbf{C}_E(F)/Z$.

Proof. Apply Corollary 1.6 with E replacing G and F replacing E. $\qquad \square$

1.8 Corollary. *Suppose that* $E \trianglelefteq G$, $Z = \mathbf{Z}(E)$ *is cyclic and* E/Z *is abelian. Assume that whenever* $Z < D \leq E$ *and* $D \trianglelefteq G$, *then* D *is non-abelian. Then* $E/Z = E_1/Z \times \cdots \times E_n/Z$ *for chief factors* E_i/Z *of* G *with* $Z = \mathbf{Z}(E_i)$ *for each* i *and* $E_i \leq \mathbf{C}_G(E_j)$ *for* $i \neq j$.

Proof. Let $Z < D \leq E$ with $D \trianglelefteq G$. Since $Z \leq \mathbf{Z}(D) \trianglelefteq G$ and $\mathbf{Z}(D)$ is abelian, $\mathbf{Z}(D) = Z$. By Corollary 1.7, $E/Z = D/Z \times \mathbf{C}_E(D)/Z$. By choosing D so that D/Z is a chief factor of G and by arguing via induction on $|E/Z|$, the conclusion easily follows. $\qquad \square$

1.9 Theorem. *Assume that every abelian normal subgroup of G is cyclic. Let $P \neq 1$ be a normal p-subgroup of G. If $p = 2$, also assume that G is solvable. Let $Z \leq \mathbf{Z}(P)$ with $|Z| = p$. Then there exist $E, T \trianglelefteq G$ such that*

(i) $ET = P$, $E \cap T = Z$ and $T = \mathbf{C}_P(E)$;

(ii) E *is extra-special or* $E = Z$;

(iii) $\exp(E) = p$ *or* $p = 2$;

(iv) T *is cyclic, or* $p = 2$ *and* T *is dihedral, quaternion or semidihedral;*

(v) *if* T *is not cyclic, then there exists* $U \trianglelefteq G$ *with* U *cyclic,* $U \leq T$, $|T : U| = 2$ *and* $\mathbf{C}_T(U) = U$; *and*

(vi) *if* $E > Z$, *then* $E/Z = E_1/Z \times \cdots \times E_n/Z$ *for chief factors* E_i/Z *of G and with* $Z = \mathbf{Z}(E_i)$ *for each i and $E_i \leq \mathbf{C}_G(E_j)$ for $i \neq j$.*

Proof. Assume that p is odd. By Theorem 1.2, there exist $E, T \leq G$ satisfying (i) through (v). Since $T = \mathbf{Z}(P)$ and $E = \{x \in P \mid x^p = 1\}$, we have $E, T \trianglelefteq G$. If $Z < D \leq E$ and $D \trianglelefteq G$, then $\exp(D) = p$ and D is not cyclic. Thus D is not abelian. By Corollary 1.8, part (vi) follows. We are done if p is odd. We thus assume that $p = 2$ and proceed by induction on $|P|$.

By Theorem 1.2, $P = FS$ where F is extra-special, $F \cap S = Z$, S is cyclic, or $|S| \geq 16$ and S is quaternion, dihedral or semi-dihedral. Furthermore, there exists U characteristic in P with $U \leq S$, U cyclic, $|S : U| \leq 2$ and $U = \mathbf{C}_S(U)$. First assume that $U < S$. Now $\mathbf{C}_P(U) = FU$ is a characteristic subgroup of P of index 2. Also $U = \mathbf{Z}(FU)$ is cyclic of order at least 8. The inductive hypothesis applied to $FU \trianglelefteq G$ implies there exists an extra-special subgroup $E \trianglelefteq G$ (possibly $E = Z$) such that $FU = EU$, $E \cap U = Z$ and conclusion (vi) holds. Let $T = \mathbf{C}_P(E) \geq U$. Since P centralizes $F/Z = F/(F \cap U) \cong FU/U \cong E/Z$, Corollary 1.6 implies that $ET = P$ and $E \cap T = Z$. Thus U is a cyclic subgroup of index 2 in T with $|U| \geq 8$. Since $\mathbf{Z}(P) = \mathbf{Z}(FS) = Z$, T is not cyclic and Theorem 1.2 implies that T is quaternion, dihedral or semi-dihedral. Also, $T = \mathbf{C}_P(E)$ is normal in G. We are done if $U < S$. We hence assume that $U = S = \mathbf{Z}(P)$ is cyclic. We also assume that $F > Z$.

Suppose that $|S| \geq 8$. Let $V \leq S$ have index 2 and set $m = \exp(V)$. Observe that $FV = \{x \in P \mid x^m = 1\}$ is a characteristic subgroup of P of index 2. The inductive hypothesis applies to FV and we argue similarly to the last paragraph (in this case $T = S = \mathbf{C}_P(E) = \mathbf{Z}(P)$ is cyclic). Thus we may assume $|S| \leq 4$.

Suppose that $|S| = 2$ and so $P = F$ is extra-special. We may in this case assume there exists $Z < W \leq P$ with $W \trianglelefteq G$ and W abelian, since otherwise the conclusion of the theorem is reached by setting $E = F$, $T = Z$ and applying Corollary 1.8. Now W is cyclic and hence of order 4. Since $Z = \mathbf{Z}(P)$, $|P : \mathbf{C}_P(W)| = 2$ and $\mathbf{C}_P(W) \trianglelefteq G$. By the inductive hypothesis, there exists an extra-special group $E \trianglelefteq G$ (possibly $E = Z$) satisfying (vi) such that $EW = \mathbf{C}_P(W)$ and $E \cap W = Z$. Let $T = \mathbf{C}_P(E) \trianglelefteq G$. By Corollary 1.6, $ET = P$ and $E \cap T = Z$. Observe that $|T : W| = 2$ and $T \cong D_8$ or Q_8. We are done when $|S| = 2$ (with $U = W$). We hence assume that $|S| = 4 = \exp(P)$.

Let H/S be a chief factor of G with $H \leq P$. Since $\exp(H) = 4 < |H|$, H is not cyclic and thus H is non-abelian by the hypothesis. By Corollary 1.8, P/S is a completely reducible G-module. Thus we have $P/S = H_1/S \times \cdots \times H_n/S$ for minimal normal subgroups H_i/S of G/S with $S = \mathbf{Z}(H_i)$ for each i and $H_i \leq \mathbf{C}_P(H_j)$ for $i \neq j$.

Assume there exist $E_i \trianglelefteq G$ with $H_i = SE_i$ and $E_i \cap S = Z$. Then $Z = \mathbf{Z}(E_i)$ for each i, $E_i \leq \mathbf{C}_P(E_j)$ for $i \neq j$ and $E_i/Z \cong H_i/S$ is a chief factor of G. Let $E = \prod_i E_i$, so that $E/Z \cong E_1/Z \times \cdots \times E_n/Z$. Since E is a central product of the extra-special groups E_i, E is also extra-special. The conclusion of the theorem is then satisfied by setting $T = S$.

To complete the proof, we need just show there exists $L \trianglelefteq G$ such that

$LS = H$ and $L \cap S = Z$. Let $A = \mathbf{C}_G(S)$, so that $|G/A| \leq 2$. Let $C = \mathbf{C}_A(H/S) \geq P$. Since H/S is a chief factor of G and a non-cyclic 2-group, $C < A$. By Corollary 1.6, $C = H \cdot \mathbf{C}_G(H)$ and so C centralizes the abelian group H/Z. Let M/C be a chief factor of G with $M \leq A$. Since G is solvable, M/C is a q-group for some prime q. Since H/S is a chief factor of G, observe that $\mathbf{C}_{H/S}(M/C) = 1$ and $q \neq 2$. We apply Fitting's Lemma 0.6 to the coprime action of M/C on H/Z and obtain $H/Z = [H/Z, M/C] \times \mathbf{C}_{H/Z}(M/C) = [H/Z, M/C] \times S/Z$. Set $L/Z = [H/Z, M/C]$ to complete the proof. $\qquad \square$

1.10 Corollary. *Suppose $G \neq 1$ is solvable and every normal abelian subgroup of G is cyclic. Let $F = \mathbf{F}(G)$ and let Z be the socle of the cyclic group $\mathbf{Z}(F)$. Set $A = \mathbf{C}_G(Z)$. Then there exist $E, T \trianglelefteq G$ with*

(i) *$F = ET$, $Z = E \cap T$ and $T = \mathbf{C}_F(E)$.*

(ii) *A Sylow q-subgroup of E is cyclic of order q or extra-special of exponent q or 4.*

(iii) *$E/Z = E_1/Z \times \cdots \times E_n/Z$ for chief factors E_i/Z of G with $E_i \leq \mathbf{C}_G(E_j)$ for $i \neq j$.*

(iv) *For each i, $\mathbf{Z}(E_i) = Z$, $|E_i/Z| = p_i^{2n_i}$ for a prime p_i and an integer n_i, and $E_i = \mathbf{O}_{p_i'}(Z) \cdot F_i$ for an extra-special group $F_i = \mathbf{O}_{p_i}(E_i) \trianglelefteq G$ of order $p_i^{2n_i+1}$.*

(v) *There exists $U \leq T$ of index at most 2 with U cyclic, $U \trianglelefteq G$ and $\mathbf{C}_T(U) = U$.*

(vi) *G is nilpotent if and only if $G = T$.*

(vii) *$T = \mathbf{C}_G(E)$ and $F = \mathbf{C}_A(E/Z)$.*

(viii) *$E/Z \cong F/T$ is a completely reducible G/F-module and faithful A/F-module (possibly of mixed characteristic).*

(ix) *$A/\mathbf{C}_A(E_i/Z) \lesssim Sp(2n_i, p_i)$.*

(x) *If every normal abelian subgroup of G is central in F, then $T = \mathbf{Z}(F)$ is cyclic.*

Proof. Parts (i)–(v) follow from Theorem 1.9. To prove (vi), we assume that G is nilpotent and show that $G = T$. Since G is nilpotent, $G = \mathbf{F}(G) = F = ET$. Since $T \leq \mathbf{C}_G(E)$, every subgroup of E/Z is normal in G/Z. If E_i/Z is a chief factor of G with $E_i \leq E$, then $|E_i/Z|$ is a prime and E_i is abelian, contradicting (iv). Thus $E = Z$ and by (i), $T = F = G$. This proves (vi).

Let $B = \mathbf{C}_G(E) \leq A$ and $C = \mathbf{C}_A(E/Z)$. Since $\mathbf{C}_F(E) = T$, we have $BF \leq C$ and Corollary 1.6 yields $C = EB = FB$. To establish (vii) it thus suffices to show that $B = T$. Assume not and let X/T be a chief factor of G with $X \leq B$. Now $X \not\leq F$, since otherwise $Z < X \cap E \leq \mathbf{Z}(E)$, a contradiction. As X/T is an r-group for a prime r, we write $X = TR$ for $R \in \mathrm{Syl}_r(X)$. Since X is not nilpotent and T is, we may choose a prime $q \neq r$ and $Q \in \mathrm{Syl}_q(T)$ with $[Q, R] \neq 1$. If, on the one hand, $q = 2$, then (v) ensures the existence of a cyclic subgroup $U \leq Q$ with $|Q: U| = 2$ and $U \trianglelefteq G$. Therefore, R has to centralize Q. If $q > 2$, then R centralizes the socle of the cyclic group Q, because $R \leq B \leq \mathbf{C}_G(Z)$. Since $r \neq q$, $R \leq \mathbf{C}_G(Q)$. Part (vii) follows from this contradiction.

Theorem 1.9 implies that E/Z is completely reducible as a G-module and hence also as an A-module. Part (viii) now follows from (vii).

Now E_i/Z has a non-degenerate symplectic form $(\ ,\ )$ over $GF(p_i)$, namely the commutator map $(xZ, yZ) = [x, y]$, with $E_i' = \mathbf{O}_{p_i}(Z)$ identified as $GF(p_i)$ (see [Hu, III, 13.7 (b)]). This form is preserved by $A = \mathbf{C}_G(Z)$ and part (ix) follows.

Let U be a cyclic subgroup of T of index at most 2 with $U \trianglelefteq G$. By hypothesis (x), $U \leq \mathbf{Z}(F)$ and T is abelian. Thus $T = \mathbf{Z}(F)$ is cyclic. This completes the proof of the corollary. $\square$

Note that the additional hypothesis of (x) is satisfied provided that $F \leq G'$. Moreover, when $A = G$ in Corollary 1.10, we see below that E/Z has a complement H/Z in G/Z satisfying $\mathbf{C}_G(E) \leq H$. Since $T = \mathbf{C}_G(E)$, it then follows that F/T is complemented in G/T as well.

1.11 Lemma. *Suppose that* $Z \leq E \trianglelefteq G$, $Z = \mathbf{Z}(E)$ *is cyclic and central in* G, *and* $E/Z = E_1/Z \times \cdots \times E_m/Z$ *for chief factors* E_i/Z *of* G. *Assume that* G *is solvable and* $E_i \leq \mathbf{C}_G(E_j)$ *if and only if* $i \neq j$. *Then there exists* $H \leq G$ *with* $EH = G$, $E \cap H = Z$ *and* $\mathbf{C}_G(E) \leq H$.

Proof. We argue by induction on $|E/Z|$. The result is trivial when $E = Z$ and we thus assume that $m \geq 1$. We let $C = \mathbf{C}_G(E)$ and $B = \mathbf{C}_G(E/Z)$. Since $Z \leq \mathbf{Z}(G)$, Corollary 1.6 yields $B = EC$ and $E \cap C = Z$. In particular, $B/C \cong E/Z$ as G-modules.

First suppose that $m = 1$. Then E/Z is a faithful irreducible G/B-module. Since $Z = \mathbf{Z}(E)$, $|E/Z|$ is not prime and so $B < G$. Let M/B be a chief factor of G. Then M/B is a p-group for a prime p, and furthermore $p \nmid |E/Z| = |B/C|$. Thus if $P/C \in \mathrm{Syl}_p(M/C)$, then $BP = M$ and $B \cap P = C$. Now $\mathbf{C}_{E/Z}(P/Z) = \mathbf{C}_{E/Z}(M/B) = 1$, because $E/Z \cong B/C$ is a faithful irreducible G/B-module and $M/B \trianglelefteq G/B$. The Frattini argument implies that $G = M \cdot \mathbf{N}_G(P/C)$. Setting $J = \mathbf{N}_G(P/C)$, we have that

$$EJ = E(CPJ) = (EC)PJ = BPJ = MJ = G.$$

Since E/Z is abelian, $E \cap J \trianglelefteq EJ = G$. Thus $E \cap J$ is Z or E. In the latter case, $G = J = \mathbf{N}_G(P/C)$ and $P/C \trianglelefteq G/C$, whence P/C and M/B act trivially on $B/C \cong E/Z$, a contradiction. Thus $E \cap J = Z$ and the result follows in this case by setting $H = J$. We may assume that $m > 1$.

By the last paragraph, there exists $J \leq G$ satisfying $E_1 \cdot J = G$, $E_1 \cap J = Z$ and $\mathbf{C}_G(E_1) \leq J$. In particular, $E_2 \cdots E_m \leq J$. Observe that each E_i/Z $(i > 1)$ is even a chief factor in J, because $E_1 \cdot J = G$ and $[E_1, E_i] = 1$. Setting $F = E_2 \cdots E_m$, the inductive hypothesis applied to J implies that there exists $H \leq J$ such that $FH = J$, $H \cap F = Z$ and $\mathbf{C}_J(F) \leq H$. Now

$$EH = E_1FH = E_1J = G,$$

$$E \cap H = E \cap J \cap H = (E_1F \cap J) \cap H = (E_1 \cap J) \cdot F \cap H = F \cap H = Z, \text{ and}$$

$$\mathbf{C}_G(E) = \mathbf{C}_G(E_1) \cap \mathbf{C}_G(F) \leq J \cap \mathbf{C}_G(F) = \mathbf{C}_J(F) \leq H. \qquad \square$$

Related to Lemma 1.11 is a theorem due to Gaschütz, which has frequent use later in the text. Recall that the Frattini subgroup $\Phi(G)$ is a normal nilpotent subgroup of G.

1.12 Theorem (Gaschütz). *Let G be solvable. Then*

$$\mathbf{F}(G/\Phi(G)) = \mathbf{F}(G)/\Phi(G)$$

is a completely reducible and faithful $G/\mathbf{F}(G)$-module (possibly of mixed characteristic). Furthermore, $G/\Phi(G)$ splits over $\mathbf{F}(G)/\Phi(G)$.

Proof. See [Hu, III, 4.2, 4.4 and 4.5].   □

§2   Semi-Linear and Small Linear Groups

We begin this section with semi-linear and affine semi-linear groups. These groups play an important role in the study of solvable linear groups and solvable permutation groups (e.g. see Theorem 2.1 and the paragraph following it). We conclude the section by characterizing solvable irreducible subgroups of $GL(n, q)$ for small values of q^n. In between, some standard arguments in representation theory are presented. In many of these arguments, we require that the underlying field has positive characteristic or is algebraically closed in order to guarantee trivial Schur indices (see Proposition 0.4).

Let V be the Galois field $GF(q^m)$ for a prime power q. Of course V is a vector space over $GF(q)$ of dimension m. Fix $a \in V \setminus \{0\} = V^\#$, $w \in V$ and $\sigma \in \mathcal{G} := \mathrm{Gal}\,(GF(q^m)/GF(q))$. We define a mapping

$$T: V \to V \quad \text{by} \quad T(x) = ax^\sigma + w.$$

Then T is a permutation on V and T is trivial if and only if $a = 1$, $\sigma = 1$ and $w = 0$. Thus we have the following subgroups of $\mathrm{Sym}\,(V)$:

(i) $A(V) = \{x \mapsto x + w \mid w \in V\}$ consisting of translations.

(ii) The *semi-linear group*

$$\Gamma(V) = \{x \mapsto ax^\sigma \mid a \in GF(q^m)^\#, \ \sigma \in \mathcal{G}\}.$$

(iii) The subgroup $\Gamma_0(V) = \{x \mapsto ax \mid a \in \mathcal{G}^\#\}$ of $\Gamma(V)$, consisting of multiplications.

(iv) The *affine semi-linear group*

$$A\Gamma(V) = \{x \mapsto ax^\sigma + w \mid a \in GF(q^m)^\#, \ \sigma \in \mathcal{G}, \ w \in V\}.$$

Clearly, $A(V)$ acts regularly on V and $A(V) \cong V$ as vector spaces over $GF(q)$. Now both $A(V)$ and V are $\Gamma(V)$-modules, where $\Gamma(V)$ acts on $A(V)$ by conjugation and on V by semi-linear mappings. Hence, as is easily checked, $A(V) \cong V$ as $GF(q)[\Gamma(V)]$-modules. Observe that $\Gamma(V)$ and even $\Gamma_0(V)$ act transitively on the non-zero elements of $A(V)$ and V. In fact, $A\Gamma(V)$ is the semi-direct product of $A(V)$ and $\Gamma(V)$ (and is isomorphic to the semi-direct product of V and $\Gamma(V)$). Also $\Gamma(V)$ is a point-stabilizer (for zero) in the doubly transitive permutation group $A\Gamma(V)$. Note that $\Gamma_0(V)$ is cyclic of order $q^m - 1$ and $\Gamma(V)/\Gamma_0(V) \cong \mathcal{G}$ is cyclic of order m. If $\sigma \in \mathcal{G}$ has order n, then $|\mathbf{C}_V(\sigma)| = |\mathbf{C}_{A(V)}(\sigma)| = q^{m/n}$ and $|\mathbf{C}_{\Gamma_0(V)}(\sigma)| = q^{m/n} - 1$.

We will also write $\Gamma(q^m)$ for $\Gamma(V)$, etc.. Our notation is different from that in [Hu; II, 1.18(d)], where Γ is used to denote what we call $A\Gamma$. Observe that e.g. $\Gamma(8^2)$ and $\Gamma(4^3)$ are distinct proper subgroups of $\Gamma(2^6)$. For the most part, we will assume that the base field $GF(q)$ is the prime field.

Theorem 2.1 will turn out to be critical for many topics in this book.

2.1 Theorem. *Suppose that G acts faithfully on a $GF(q)$-vector space V of order q^m, q a prime power. Assume that G has a normal abelian subgroup A for which V_A is irreducible. Then G may be identified as a subgroup of $\Gamma(q^m)$ (i.e. the points of V may be labelled as the elements of $GF(q^m)$ in such a way that $G \leq \Gamma(q^m)$) and $A \leq \Gamma_0(q^m)$.*

Proof. Let $D = \mathrm{End}_A(V)$. By Schur's Lemma, D is a division ring. Since V is finite, D is finite and hence a field. Now $A \leq \mathbf{C}_G(A) \leq D^\#$. Thus

$\mathbf{C}_G(A)$ is a cyclic normal subgroup of G. Without loss of generality, we may assume that $A = \mathbf{C}_G(A)$.

Since every D-invariant subspace of V is also A-invariant, V is an irreducible D-vector space, i.e. $\dim_D(V) = 1$. In particular, $D \cong GF(q^m)$. In order to label the points of V by the elements of D, we fix some $w \in V^{\#}$. We then identify $v \in V$ with the unique $d \in D$ such that $v = wd$. For $f \in D$, the vector vf corresponds to df, and so scalar multiplication on V agrees with field multiplication in D. Since $A \leq D$, $A \leq \Gamma_0(q^m)$ follows.

Let $g \in G$. We wish to show that $g \in \Gamma(q^m)$. Let $b := 1g \in D^{\#}$, i.e. b corresponds to $wg \in V^{\#}$. Then $h := gb^{-1} \in GL(m, q)$ and $1h = 1$. Since $b^{-1} \in \Gamma_0(q^m)$, it suffices to show that $h \in \Gamma(q^m)$. As $A \trianglelefteq G$ and $D^{\#} = \mathbf{C}_{GL(V)}(A)$, $D^{\#}$ is G-invariant and thus $\langle h \rangle$-invariant. Now $D^{\#} = \langle a \rangle$ for some a, because $D^{\#}$ is cyclic. Let $h^{-1}ah = a^m$ for some $m \in \mathbb{Z}$, so that $h^{-1}a^i h = a^{im}$ for all i. It suffices to show that $h \in \mathrm{Gal}\,(GF(q^m)/GF(q)) \leq \Gamma(q^m)$. Certainly, h acts $GF(q)$-linearly. Let x, $y \in D^{\#}$, say $x = a^t$ and $y = a^s$. Because $1h = 1$, we have that

$$xh = (1a^t)h = 1h^{-1}(a^t h) = 1(h^{-1}a^t h) = 1a^{tm} = a^{tm}.$$

Likewise, $yh = a^{sm}$ and $(xy)h = a^{(s+t)m} = (xh)(yh)$. This shows that h is a field automorphism of $GF(q^m)$. $\square$

Alternatively, Theorem 2.1 follows from [Hu, II, 3.11] with $s = 1$.

Suppose that H is a primitive solvable permutation group on a finite set Ω with point stabilizer H_α ($\alpha \in \Omega$). Then H has a unique minimal normal subgroup M, $H = M \cdot H_\alpha$, $M \cap H_\alpha = 1$, $\mathbf{C}_H(M) = M$ and M acts regularly on Ω. Consequently, $|\Omega| = |M| = p^m$ is a prime power. Moreover the mapping $m \mapsto \alpha m$ ($m \in M$) is an H_α permutation isomorphism between M and Ω; here H_α acts on M by conjugation. We may consider H as a subgroup of the affine linear group $AGL(m, p)$, where M is the normal subgroup consisting of all translations (see II, 2.2, II, 3.2 and II, 3.5 of [Hu]). In particular, H is doubly transitive if and only if H_α acts transitively on

$M^\#$. (More will be said about solvable doubly transitive permutation groups in §6.)

Suppose that H_α has a normal abelian subgroup A that acts irreducibly on M. As a consequence of Theorem 2.1, the points of Ω and M (respectively) may be labelled as the elements of $GF(p^m)$ in such a way that $H_\alpha \leq \Gamma(p^m)$, $A \leq \Gamma_0(p^m)$, and hence $H \leq A\Gamma(p^m)$ (cf. also [Hu, II, 3.12]).

If V is a finite faithful $GF(q)$-module for a group G such that G may be identified as a subgroup of $\Gamma(V)$ (i.e. after a labelling of the points of V), we will write $G \leq \Gamma(V)$. This may be a little sloppy, but of course this will only be done when there has been no previous labelling of the points of V. Note that $\Gamma(V)$ of course depends on a particular labelling. We will combine the last theorem and the next lemma in a convenient corollary.

2.2 Lemma. *Let V be a faithful irreducible $\mathcal{F}[G]$-module, and let A be a normal abelian self-centralizing subgroup of G such that V_A is homogeneous. If $\mathrm{char}(\mathcal{F}) \neq 0$ or $\mathcal{F}$ is algebraically closed, then V_A is irreducible.*

If in particular G is solvable, $F := \mathbf{F}(G)$ is abelian and V_F homogeneous, then V_F is irreducible.

Proof. Since V_A is homogeneous and A is abelian, A is in fact cyclic. Write $V_A = eW$ for a faithful irreducible $\mathcal{F}[A]$-module W and $e \in \mathbb{N}$. Our aim is to show that $e = 1$.

Let $\mathcal{K}$ be an algebraically closed extension of $\mathcal{F}$, with $\mathcal{K} = \mathcal{F}$ should $\mathrm{char}(\mathcal{F}) = 0$. Now $V \otimes_{\mathcal{F}} \mathcal{K} = V_1 \oplus \cdots \oplus V_t$ and $W \otimes_{\mathcal{F}} \mathcal{K} = W_1 \oplus \cdots \oplus W_s$ for distinct absolutely irreducible G-modules V_i and distinct absolutely irreducible A-modules W_j (see Proposition 0.4). In particular,

$$(V_1 \oplus \cdots \oplus V_t)_A = (V \otimes_{\mathcal{F}} \mathcal{K})_A = V_A \otimes_{\mathcal{F}} \mathcal{K} = e(W_1 \oplus \cdots \oplus W_s).$$

Since W_1 is a faithful absolutely irreducible module for the cyclic group A and since $\mathbf{C}_G(A) = A$, $\mathbf{I}_G(W_1) = A$. Hence W_1^G is an irreducible G-module by Clifford's Theorem (see Theorem 0.1). Also, W_1 has say $l =$

$|G : A|$ G-conjugates. Without loss of generality, the G-conjugates of W_1 are $W_1, \dots, W_l$. We may also assume that $W_1^G = V_1$. Therefore, $W_j^G = V_1$ for $1 \leq j \leq l$, and since $\dim_{\mathcal{K}}(V_1) = l \cdot \dim_{\mathcal{K}}(W_1)$, $(V_1)_A = W_1 \oplus \cdots \oplus W_l$. Likewise, if W_k is a constituent of $(V_i)_A$, $W_k^G = V_i$. Thus W_1 is not a constituent of $(V_i)_A$ for $i \geq 2$. Hence $e = 1$.

The supplement follows since $\mathbf{C}_G(F) \leq F$ in solvable groups G. $\qquad \square$

2.3 Corollary. *Suppose that V is a faithful irreducible $GF(q)[G]$-module for a solvable group G and a prime power q. Let $F = \mathbf{F}(G)$.*

(a) *If F is abelian and V_F is homogeneous, then $G \leq \Gamma(V)$.*

(b) *If V is quasi-primitive and $F = T$ (T as in Cor. 1.10), then $G \leq \Gamma(V)$.*

Proof. (a) By Lemma 2.2, V_F is irreducible and so Theorem 2.1 implies that $G \leq \Gamma(V)$.

(b) We adopt the notation of Corollary 1.10 which applies to G because V is quasi-primitive. Since by our hypothesis $E/F = F/T = 1$, part (viii) of Corollary 1.10 implies that $\mathbf{C}_G(Z) = A = F = T$. Now $Z \leq U$, and consequently $U = \mathbf{C}_T(U) = \mathbf{C}_G(U)$, by Corollary 1.10 (v). Since U is cyclic, Lemma 2.2 and Theorem 2.1 yield $G \leq \Gamma(V)$. $\qquad \square$

2.4 Lemma. *Let F be a group with center Z. Suppose that V is a faithful irreducible $\mathcal{F}[F]$-module where $\mathcal{F}$ is a field that has positive characteristic or is algebraically closed. Assume that $\mathrm{char}(\mathcal{F}) \nmid |F|$. Let W be an irreducible Z-submodule of V. Then $\dim_{\mathcal{F}}(V) = te \cdot \dim_{\mathcal{F}}(W)$ for integers t and e, with $e = \chi(1)$ for a faithful irreducible ordinary character χ of F.*

Proof. Since $Z = \mathbf{Z}(F)$, Clifford's Theorem implies that $V_Z \cong f \cdot W$ for an integer f. Let $\mathcal{K}$ be a Galois extension of $\mathcal{F}$ such that $\mathcal{K}$ contains an $|F|^{\mathrm{th}}$ root of unity. If $\mathcal{F}$ is algebraically closed, we choose $\mathcal{F} = \mathcal{K}$. Then all irreducible $\mathcal{K}[F]$- and $\mathcal{K}[Z]$-representations are absolutely irreducible. Now

$$V \otimes_{\mathcal{F}} \mathcal{K} = V_1 \oplus \cdots \oplus V_l$$

for absolutely irreducible $\mathcal{K}[F]$-modules V_i that are the distinct Galois conjugates of V_1, i.e. the V_i afford $\mathcal{K}$-representations X_i of F that are conjugate under the Galois group $\mathrm{Gal}(\mathcal{K}/\mathcal{F})$ and $\{X_1, \ldots, X_l\}$ form an orbit (see Proposition 0.4). Similarly,

$$W \otimes_{\mathcal{F}} \mathcal{K} = W_1 \oplus \cdots \oplus W_s$$

for absolutely irreducible $\mathcal{K}[Z]$-modules W_j, whose representations form an orbit under $\mathrm{Gal}(\mathcal{K}/\mathcal{F})$. Now

$$(V_1 \oplus \cdots \oplus V_l)_Z = (V \otimes_{\mathcal{F}} \mathcal{K})_Z = V \otimes_{\mathcal{F}} \mathcal{K} \cong f(W_1 \oplus \cdots \oplus W_s).$$

For each i, $(V_i)_Z$ has a unique W_j as a constituent, because $Z = \mathbf{Z}(F)$. If an element of $\mathrm{Gal}(\mathcal{K}/\mathcal{F})$ maps W_1 to W_j, then it must map the set $\{V_k \mid (V_k)_Z \cong W_1 \oplus \cdots \oplus W_1\}$ to the set $\{V_i \mid (V_i)_Z \cong W_j \oplus \cdots \oplus W_j\}$. Thus the number of distinct V_i for which a given W_j is a constituent in $(V_i)_Z$ is l/s (i.e. the number is independent of the particular choice of j). In particular, $s \mid l$. Since the W_j are absolutely irreducible modules for the abelian group Z, $\dim_{\mathcal{K}}(W_j) = 1$ for all j. Observe that V_1 is a faithful $\mathcal{K}[F]$-module, because V was assumed to be a faithful $\mathcal{F}[F]$-module. Since $\mathrm{char}(\mathcal{K}) \nmid |F|$, $\dim_{\mathcal{K}}(V_1) = \chi(1)$ for a faithful $\chi \in \mathrm{Irr}(F)$. Now

$$\dim_{\mathcal{F}}(V) = \dim_{\mathcal{K}}(V \otimes_{\mathcal{F}} \mathcal{K}) = l \cdot \dim_{\mathcal{K}}(V_1) = l \cdot \chi(1)$$

and

$$\dim_{\mathcal{F}}(W) = \dim_{\mathcal{K}}(W \otimes_{\mathcal{F}} \mathcal{K}) = s \cdot \dim_{\mathcal{K}}(W_1) = s.$$

Thus

$$\dim_{\mathcal{F}}(V) = (l/s) \cdot \chi(1) \cdot \dim_{\mathcal{F}}(W).$$

Set $t = l/s \in \mathbb{N}$ to complete the proof. $\square$

2.5 Corollary. *Let V be a faithful irreducible $\mathcal{F}[G]$-module for a field $\mathcal{F}$. If char $(\mathcal{F}) = 0$, assume that $\mathcal{F}$ is algebraically closed. Suppose that $P \trianglelefteq G$ and P is a non-abelian p-group. Then $p \mid \dim_{\mathcal{F}}(V)$.*

Proof. Because $\mathbf{O}_p(G) \neq 1$, $p \nmid$ char $(\mathcal{F})$. Now $V_P = V_1 \oplus \cdots \oplus V_n$ for irreducible P-modules V_i such that $P/\mathbf{C}_P(V_i) \cong P/\mathbf{C}_P(V_j)$ for all i, j. Since $\bigcap_i \mathbf{C}_P(V_i) = 1$ and P is non-abelian, each $P/\mathbf{C}_P(V_i)$ is non-abelian. By Lemma 2.4, $p \mid \dim_{\mathcal{F}}(V_i)$ for all i and so $p \mid \dim_{\mathcal{F}}(V)$. $\qquad\square$

The next result, which again is a consequence of Lemma 2.4, applies to the Fitting subgroup of quasi- and pseudo-primitive linear groups (cf. Corollaries 1.4 and 1.10).

2.6 Corollary. *Assume that $H = EU$ where $U = \mathbf{Z}(H)$ is cyclic, $U \cap E = \mathbf{Z}(E)$, E is nilpotent and the Sylow subgroups of E are extra-special or of prime order. Let V be a faithful irreducible $\mathcal{F}[H]$-module and W an irreducible submodule of V_U. If char $(\mathcal{F}) = 0$, assume that $\mathcal{F}$ is algebraically closed. Then $\dim_{\mathcal{F}}(V) = e \cdot \dim_{\mathcal{F}}(W)$ with $e^2 = |H : U|$.*

Proof. First observe that char$(\mathcal{F}) \nmid |H|$. Since E is a direct product of extra-special groups and groups of prime order, $|H/U| = |E/\mathbf{Z}(E)| = e^2$ for an integer e and each faithful $\varphi \in \mathrm{Irr}(E)$ has degree e (see [Hu, V, 16. 14]). Since $U = \mathbf{Z}(H)$ and $H = EU$, every $\varphi \in \mathrm{Irr}(E)$ is H-invariant. But H/E even is cyclic and so every $\chi \in \mathrm{Irr}(H|\varphi)$ extends φ (see Proposition 0.11). In particular, if χ is faithful, so is φ and $\chi(1) = \varphi(1) = e = |H : U|^{1/2}$. By Lemma 2.4 and its proof,

$$\dim_{\mathcal{F}}(V) = et \cdot \dim_{\mathcal{F}}(W)$$

where t is the number of irreducible non-isomorphic constituents of $V \otimes_{\mathcal{F}} \mathcal{K}$ whose restriction to $U = \mathbf{Z}(H)$ is a multiple of a fixed irreducible constituent W_1 of $W \otimes_{\mathcal{F}} \mathcal{K}$. Recall that all irreducible constituents of $V \otimes_{\mathcal{F}} \mathcal{K}$ and $W \otimes_{\mathcal{F}} \mathcal{K}$ are absolutely irreducible.

Let $\chi_U = e \cdot \lambda$ for a faithful $\lambda \in \mathrm{Irr}(U)$. By [Is, Exercise 6.3], which is restated below as Proposition 12.3, $\mathrm{Irr}(H|\lambda) = \{\chi\}$. Thus $t = 1$. $\qquad\square$

2.7 Example. Let $C \neq 1$ be a cyclic group and assume that the prime power q is coprime to $|C|$. Let l be the smallest positive integer such that $|C| \mid q^l - 1$. Then every faithful irreducible $GF(q)[C]$-module has dimension l.

Proof. Let V be a faithful irreducible $GF(q)[C]$-module. Every C-orbit on $V^\#$ has size $|C|$. Thus $|C| \mid |V| - 1$. Set $|V| = q^k$. Since $q^k \equiv 1 \pmod{|C|}$, $l \mid k$.

Let $\mathcal{K} = GF(q^l)$. Then $V \otimes_{GF(q)} \mathcal{K} = V_1 \oplus \cdots \oplus V_t$ for distinct irreducible $\mathcal{K}[C]$-modules V_i that are afforded by representations which are Galois conjugate under $\mathrm{Gal}(\mathcal{K}/GF(q))$ (see Proposition 0.4). Hence $t \leq [\mathcal{K} : GF(q)] = l$. Since $\mathcal{K}$ contains a primitive $|C|^{\mathrm{th}}$ root of unity, every irreducible $\mathcal{K}[C]$-representation is absolutely irreducible and thus $\dim_\mathcal{K}(V_i) = 1$ for each i. Hence $k = \dim_{GF(q)}(V) = t \cdot \dim_\mathcal{K}(V_1) = t \leq l \leq k$ and $\dim_{GF(q)}(V) = l$. $\qquad\square$

The next lemma gives some structure about imprimitive linear groups. What then follows is information about "small" linear groups. Often, with solvable groups, *ad hoc* arguments are needed to handle "small" cases. We collect some information for later use. There is some overlap with what appears in Suprunenko's book [Su], but our approach is slightly different.

Suppose that U is a faithful irreducible H-module for a group $H \neq 1$ and that S is a transitive subgroup of the symmetric group S_n $(n > 1)$. Then $U^n := U + \cdots + U$ is a faithful irreducible module for the wreath product $H \mathrm{wr} S$. Also, U^n is an imprimitive $(H \mathrm{wr} S)$-module.

2.8 Lemma. *Let V be a faithful irreducible $\mathcal{F}[G]$-module and suppose $V = V_1 \oplus \cdots \oplus V_n$ $(n > 1)$ is a system of imprimitivity for G. Let $\gamma \colon G \to S_n$ be the homomorphism induced by the permutation action of G on the V_i.*

Set $S = \gamma(G)$, which is a transitive subgroup of S_n. Let finally $H = \mathbf{N}_G(V_1)/\mathbf{C}_G(V_1)$. Then G is isomorphic to a subgroup of $H \operatorname{wr} S$ as linear groups.

Note. The conclusion is stronger than just $G \leq H \operatorname{wr} S$. But rather G as a linear group on V is isomorphic to a subgroup M of the linear group $H \operatorname{wr} S$ on V_1^n. It is the stronger form that we desire. Note further that then $G \leq \operatorname{Aut}(V_1) \operatorname{wr} S_n$ as a linear group.

Proof. Fix a basis B_1 for V_1, let $r = \dim(V_1)$ and $I = \mathbf{N}_G(V_1)$. For $x \in I$, let $X(x)$ be the matrix afforded by x relative to B_1. Thus

$$X : I \to GL(r, \mathcal{F})$$

is a representation of I with kernel $\mathbf{C}_G(V_1)$. Let $K = \{X(x) \mid x \in I\}$ so that $K \leq GL(r, \mathcal{F})$ and $K \cong I/\mathbf{C}_G(V_1) = H$.

Consider the following subsets of $GL(rn, \mathcal{F})$:

$$M = \left\{ \begin{pmatrix} k_1 & & & 0 \\ & k_2 & & \\ & & \ddots & \\ 0 & & & k_n \end{pmatrix} \ \middle| \ k_i \in K \leq GL(r, \mathcal{F}) \right\}$$

and N the set of those matrices obtained by applying a "block-preserving" permutation of the columns of an element of M by an element of $s \in S \leq S_n$. By choosing an appropriate basis for V_1^n, we have a representation

$$Z : H \operatorname{wr} S \to N$$

that is faithful and onto. Furthermore, $H \operatorname{wr} S$ and N are isomorphic as linear groups. Thus we intend to show that G is isomorphic as a linear group to a subgroup of N.

Since G transitively permutes $V_1, \dots, V_n$, we may choose a complete set $\{g_1 = 1, g_2, \dots, g_n\}$ of coset representatives of I in G with $V_1 g_i = V_i$. We extend B_1 to a basis B of V by $B = B_1 \cup B_1 g_2 \cup \dots \cup B_1 g_n$. We let Y be the representation afforded by G relative to B so that

$$Y : G \to GL(rn, \mathcal{F})$$

is faithful. It suffices to show that $Y(g) \in N$ for all $g \in G$.

Fix $g \in G$. For each i, there exists $h_i \in I$ such that $g_i g = h_i g_j$, where $j = i \cdot \gamma(g)$. Now the matrix

$$C = \begin{pmatrix} X(h_1) & & 0 \\ & \ddots & \\ 0 & & X(h_n) \end{pmatrix}$$

is in M and $Y(g)$ is the matrix obtained by a "block-preserving" permutation $s = \gamma(g) \in S$ to the columns of C. Thus $Y(g) \in N$. $\qquad\square$

If V is an imprimitive irreducible faithful G-module, then V is induced from an irreducible module W of a maximal subgroup H of G. Then

$$G \leq (H/\mathbf{C}_H(W)) \text{ wr } S$$

where S is a primitive permutation group on $t := |G : H|$ letters. In fact, $S = G/\mathrm{core}_G(H)$. The maximality forces H to be $\mathbf{N}_G(W)$.

2.9 Lemma. *Suppose that $H \leq G$ and V is a faithful $\mathcal{F}[G]$-module that is irreducible as an $\mathcal{F}[H]$-module. Assume that $\mathrm{char}(\mathcal{F}) \neq 0$ or $\mathcal{F}$ is algebraically closed. Then $\mathbf{C}_G(H)$ is cyclic.*

Proof. Without loss of generality, $G = H \cdot \mathbf{C}_G(H)$ and $H \trianglelefteq G$. Let $\mathcal{K}$ be an algebraically closed extension of $\mathcal{F}$, with $\mathcal{K} = \mathcal{F}$ should $\mathrm{char}(\mathcal{F}) = 0$.

It again follows that

$$V \otimes_{\mathcal{F}} \mathcal{K} = V_1 \oplus \cdots \oplus V_t$$

for non-isomorphic absolutely irreducible faithful $\mathcal{K}[G]$ modules V_i (see Proposition 0.4). Likewise,

$$V_H \otimes_{\mathcal{F}} \mathcal{K} = W_1 \oplus \cdots \oplus W_l$$

for non-isomorphic absolutely irreducible faithful $\mathcal{K}[H]$-modules W_i. Since $V_H \otimes_{\mathcal{F}} \mathcal{K} = (V \otimes_{\mathcal{F}} \mathcal{K})_H$, we have that $(V_1)_H$ is a direct sum of non-isomorphic absolutely irreducible $\mathcal{K}[H]$-modules that are G-conjugate. Because $G = H \cdot \mathbf{C}_G(H)$, each W_i is G-invariant. Thus $(V_1)_H$ is an absolutely irreducible $\mathcal{K}[H]$-module and Schur's Lemma yields

$$\mathcal{K} = \mathrm{End}_{\mathcal{K}[H]}(V_1) = \mathbf{C}_{\mathrm{End}(V_1)}(H).$$

Therefore, $\mathbf{C}_G(H)$ is isomorphic to a finite subgroup of the multiplicative group of the field $\mathcal{K}$. Hence $\mathbf{C}_G(H)$ is cyclic. $\square$

2.10 Lemma. *Suppose that V is a faithful quasi-primitive $\mathcal{F}[G]$-module for a solvable group G and a finite field $\mathcal{F}$. Corollary 1.10 applies and we let E, T, Z and $F = \mathbf{F}(G)$ be as in that Corollary. Set $e^2 = |E/Z|$. Then*

(i) *If $\dim_{\mathcal{F}}(V) = e \cdot \dim_{\mathcal{F}}(W)$ for an irreducible Z-submodule W of V, then $T = \mathbf{Z}(F) = \mathbf{C}_G(E)$ and T is cyclic.*

(ii) *Suppose that $E \neq Z$. Hence there exists $1 \neq D \trianglelefteq G$ such that $E = DZ$ and all Sylow subgroups of D are extra-special. If $\dim_{\mathcal{F}}(V) = e \cdot \dim_{\mathcal{F}}(Y)$ for an irreducible $\mathbf{Z}(D)$-submodule Y of V, then $T = \mathbf{Z}(F)$ and $D/\mathbf{Z}(D) \cong E/Z \cong F/T$ is a faithful completely reducible G/F-module.*

(iii) *If $\dim_{\mathcal{F}}(V) = e$, then $T \leq \mathbf{Z}(GL(V))$ and $F/T \cong E/Z$ is a faithful completely reducible G/F-module.*

(iv) *If $\dim_{\mathcal{F}}(V)$ is a prime, then $\dim_{\mathcal{F}}(V) = e$ or $G \leq \Gamma(V)$.*

Proof. We shall freely use the assertions of Corollary 1.10.

(i) Let V_0 be an irreducible E-submodule of V. Since V is quasi-primitive, $V_E \cong V_0 \oplus \cdots \oplus V_0$, and Corollary 2.6 yields $\dim_{\mathcal{F}}(V_0) = e \cdot \dim_{\mathcal{F}}(W)$. Thus V_E is irreducible. Applying Lemma 2.9, $\mathbf{C}_G(E)$ is a cyclic normal subgroup of G and thus $\mathbf{C}_G(E) = \mathbf{C}_F(E) = T$. Since $F = ET$, it follows that $T = \mathbf{Z}(F) = \mathbf{C}_G(E)$ is cyclic.

(ii) Observe that $|D/\mathbf{Z}(D)| = |E/Z| = e^2$. Hence the same argument as in (i) shows that V_D is irreducible. Since $F = DT$, repeating the arguments in (i) yields that $T = \mathbf{Z}(F) = \mathbf{C}_G(D)$. Let $B = \mathbf{C}_G(\mathbf{Z}(D))$. By Lemma 1.5, $|\mathbf{C}_B(D/\mathbf{Z}(D))/T|$ divides $|D/\mathbf{Z}(D)| = |F/T|$. But $F \leq \mathbf{C}_B(D/\mathbf{Z}(D))$ and so $F = \mathbf{C}_B(D/\mathbf{Z}(D))$. Set $C = \mathbf{C}_G(D/\mathbf{Z}(D))$. Then $C \cap B = F$ and C/F acts faithfully on $\mathbf{Z}(D)$. Assume that $F < C$ and choose $Q \in \mathrm{Syl}_q(C)$ for a prime divisor q of $|C/F|$. Then there exists a Sylow p-subgroup P of D such that $[\mathbf{Z}(P), Q] \neq 1$. In particular we have $p \neq q$, since otherwise $\mathbf{Z}(P) \leq \mathbf{Z}(Q)$. Note that $\mathbf{Z}(P) = \mathbf{\Phi}(P)$ and Q centralizes $P/\mathbf{\Phi}(P)$, because $Q \leq C$. But then $[P, Q] = 1$, a contradiction. Thus $F = C$ and G/F acts faithfully on $D/\mathbf{Z}(D) \cong E/Z \cong F/T$. That the action is completely reducible follows from Corollary 1.10.

(iii) Assume now that $\dim_{\mathcal{F}}(V) = e$. Recall that $\dim_{\mathcal{F}}(V_0) = e \cdot \dim_{\mathcal{F}}(W)$ for an irreducible E-submodule V_0 of V. By (i), V_E is irreducible and $T = \mathbf{Z}(F)$ is cyclic. Since $E \leq F$, also V_F is irreducible. Let X be an irreducible T-submodule of V. We again apply Corollary 2.6 and obtain

$$e = \dim_{\mathcal{F}}(V) = e \cdot \dim_{\mathcal{F}}(X).$$

Thus $\dim_{\mathcal{F}}(X) = 1$, and since $V_T \cong X \oplus \cdots \oplus X$, T acts on V by $\mathcal{F}$-scalar multiples of the identity. This implies $T \leq \mathbf{Z}(GL(V))$; in particular, $Z \leq T \leq \mathbf{Z}(G)$ and part (iii) follows from Corollary 1.10.

(iv) Assume that $\dim_{\mathcal{F}}(V)$ is a prime. Since $e \mid \dim_{\mathcal{F}}(V)$, we may assume that $e = 1$ and hence that $F = T$. Then $G \leq \Gamma(V)$ by Corollary 2.3(b). $\square$

By Corollary 2.6, the hypothesis on dimensions in Lemma 2.10 (ii) will be satisfied if V_D is irreducible.

2.11 Theorem. *Let G be a solvable irreducible subgroup of $GL(2, q)$, q a prime power. Then $\mathrm{dl}(G) \leq 4$ and one of the following occurs:*

(a) $G \leq Z_{q-1}\mathrm{wr}\, Z_2$;

(b) $G \leq \Gamma(q^2)$; or

(c) $\mathbf{F}(G) = QT$ where $Q_8 \cong Q \trianglelefteq G$, $T = \mathbf{Z}(\mathbf{F}(G)) = \mathbf{Z}(G)$ is cyclic,

$T \leq \mathbf{Z}(GL(2,q))$, $T \cap Q = \mathbf{Z}(Q)$ and $G/\mathbf{F}(G) \cong Z_3$ or S_3. Also $q \neq 2$.

If G is quasi-primitive, then (b) or (c) must occur.

Proof. Suppose that the underlying module V is not primitive. Since $\dim_{GF(q)} V = 2$, Lemma 2.8 yields that $G \leq Z_{q-1} \mathrm{wr}\, Z_2$ and $\mathrm{dl}(G) \leq 2$. Thus, by Clifford's Theorem, we may assume that G is quasi-primitive. Let $A \trianglelefteq G$ with A abelian. If V_A is irreducible, then $G \leq \Gamma(q^2)$, by Theorem 2.1, and so $\mathrm{dl}(G) \leq 2$. We may thus assume that $V_A \cong V_0 \oplus V_0$ for a 1-dimensional A-module V_0. Hence every abelian normal subgroup A of G is cyclic of order dividing $q - 1$, and central even in $GL(2,q)$. For the rest of the proof we will as well assume that $G \not\leq \Gamma(q^2)$.

We now apply Corollary 1.10 and let E, T, Z and $F = \mathbf{F}(G)$ be as in that Corollary; in particular, E, $T \trianglelefteq G$, $F = ET$ and $E \cap T = Z$. By Lemma 2.10 (iv), $2 = \dim_{GF(q)}(V) = |E : Z|^{1/2}$, and hence Lemma 2.10 (iii) implies that $T \leq \mathbf{Z}(GL(2,q))$ and that E/Z is a faithful completely reducible G/F-module. Then clearly $T = \mathbf{Z}(F) = \mathbf{Z}(G)$ is cyclic. Since the order of each irreducible constituent of E/Z is a square and since $|E/Z| = 4$, E/Z is irreducible and thus $G/F \cong Z_3$ or S_3. Now $E = Q \times Z_{2'}$ for an extra-special group $Q \trianglelefteq G$ of order 8 and $T \cap Q = \mathbf{Z}(Q)$. Since Q admits an automorphism of order 3, in fact $Q \cong Q_8$ (see Proposition 1.1 (b)) and the theorem follows, because $\mathrm{dl}(G) \leq 4$ and $\mathbf{O}_q(G) = 1$. $\qquad\qquad \square$

If we replace 2 above by an odd prime, the same arguments apply. Furthermore, we can use the above theorem to describe $G/\mathbf{F}(G)$. Much of the reason to include 2.11, 2.12 and the rest of this section is to avoid repetitious *ad hoc* arguments later.

2.12 Theorem. *Let G be a solvable irreducible subgroup of $GL(p,q)$ for a prime p and a prime power q. Then $\mathrm{dl}(G) \leq 6$ and one of the following occurs:*

    (a) $G \leq Z_{q-1} \mathrm{wr}\, S$ where $Z_p \leq S \leq Z_p \cdot Z_{p-1} \leq S_p$ and $q \neq 2$;

(b) $G \leq \Gamma(q^p)$; or

(c) $\mathbf{F}(G) = DT$ for an extra-special group $D \trianglelefteq G$ with $|D| = p^3$, T is cyclic with $T \leq \mathbf{Z}(GL(p,q))$, $T \cap D = \mathbf{Z}(D)$, and $D/\mathbf{Z}(D)$ is a faithful irreducible $G/\mathbf{F}(G)$-module of order p^2. Also $q \neq 2$ and $G/\mathbf{F}(G) \leq Sp(2,p) = SL(2,p)$. (Note that Theorem 2.11 applies to the action of $G/\mathbf{F}(G)$ on $D/\mathbf{Z}(D)$.)

If G is quasi-primitive, conclusion (b) or (c) must occur.

Proof. If the underlying module V is not quasi-primitive, then we may choose $C \trianglelefteq G$ maximal such that V_C is not homogeneous. Thus G/C faithfully and primitively permutes the homogeneous components $V_1, \ldots, V_m$ of V_C, $m > 1$ (see Proposition 0.2). Since $\dim_{GF(q)}(V) = p$ is a prime, $m = p$ and $\dim_{GF(q)}(V_i) = 1$. Note that $C \neq 1$ and so $q = |V_i| \neq 2$. We apply Lemma 2.8 and obtain that $G \leq Z_{q-1}\mathrm{wr}\,(G/C)$. Since G/C is a solvable primitive permutation group on p letters, $Z_p \leq G/C \leq Z_p \cdot Z_{p-1} \leq S_p$ (see [Hu, II, 3.6]). Conclusion (a) and $\mathrm{dl}(G) \leq 3$ hold now and we thus assume that V is quasi-primitive.

Let $A \trianglelefteq G$ with A abelian. By Theorem 2.1, we may assume that V_A is not irreducible, because otherwise $G \leq \Gamma(q^p)$ and $\mathrm{dl}(G) \leq 2$. Since $\dim_{GF(q)}(V) = p$ is a prime, we have that $V_A \cong p \cdot U$ for an irreducible A-module U and $\dim_{GF(q)}(U) = 1$. Therefore every normal abelian subgroup A of G is cyclic of order dividing $q - 1$, and is central even in $GL(p,q)$. In particular, $q \neq 2$.

To finish the proof we may assume that $G \not\leq \Gamma(q^p)$ and we proceed as in Theorem 2.11. Thus we apply Corollary 1.10 and Lemma 2.10, and there exist E, T, Z and $F = \mathbf{F}(G)$ such that $E, T \trianglelefteq G$, $F = ET$, $E \cap T = Z$, $p = \dim_{GF(q)}(V) = |E/Z|^{1/2}$, $T \leq \mathbf{Z}(GL(p,q))$ and E/Z is a faithful completely reducible G/F-module. Furthermore, there exists an extra-special group $D \trianglelefteq G$ of order p^3 such that $E = D \times Z_{p'}$ and $D \cap T = \mathbf{Z}(D)$. Also the order of each irreducible constituent of $E/Z \cong D/\mathbf{Z}(D)$ is a square, and since $|D/\mathbf{Z}(D)| = p^2$, G/F acts faithfully, irreducibly and symplectically on $D/\mathbf{Z}(D)$. As Theorem 2.11 applies to this action, we conclude that

$\mathrm{dl}(G) \leq \mathrm{dl}(F) + \mathrm{dl}(G/F) \leq 2 + 4 = 6$ and the proof is complete. $\qquad\square$

2.13 Corollary. Let G be a solvable irreducible subgroup of $GL(p, q)$ for primes p and q.

(a) If $q = 2$, then $G \leq \Gamma(2^p)$.

(b) If $q = p$, then $G \leq \Gamma(p^p)$ or $G \leq Z_{p-1}\mathrm{wr}\, S$ where $Z_p \leq S \leq Z_p \cdot Z_{p-1} \leq S_p$.

Proof. Assertion (a) follows directly from Theorem 2.12. To prove assertion (b), note that $\mathbf{O}_p(G) = 1$ and hence case (c) of 2.12 cannot hold. $\qquad\square$

2.14 Theorem. Let G be a solvable irreducible subgroup of $GL(pr, 2)$ where p and r are primes not necessarily distinct. After possibly interchanging p and r, one of the following occurs:

(a) $G \leq \Gamma(2^p)\mathrm{wr}\, S$ where $Z_r \leq S = Z_r \cdot Z_{r-1} \leq S_r$;

(b) $G \leq \Gamma(2^{pr})$; or

(c) $\mathbf{F}(G) = DT$ with D, $T \trianglelefteq G$, $T = \mathbf{Z}(\mathbf{F}(G))$ is cyclic, D is extraspecial of order p^3, $\mathbf{F}(G)/T \cong D/\mathbf{Z}(D)$ is a faithful irreducible $G/\mathbf{F}(G)$-module of order p^2. Furthermore, $|T| \mid 2^r - 1$ and $p \neq 2$.

In all cases, $\mathrm{dl}(G) \leq 6$. If G is quasi-primitive, then (b) or (c) must occur.

Proof. Let V be the corresponding module of order 2^{pr}. If V is not quasi-primitive, choose $C \trianglelefteq G$ maximal such that V_C is not homogeneous and write $V_C = V_1 \oplus \cdots \oplus V_l$ for $l > 1$ homogeneous components V_i of V_C. Since $C \neq 1$, $|V_i| \neq 2$ and $1 < \dim_{GF(2)}(V_i)$. Since p and r are primes and $l \cdot \dim_{GF(2)}(V_i) = pr$, we may assume without loss of generality that $\dim_{GF(2)}(V_i) = p$ and $l = r$. Now G/C faithfully and primitively permutes the V_i and hence G/C is isomorphic to a transitive subgroup of $Z_r \cdot Z_{r-1} \leq S_r$ (see [Hu, II, 3.6]). If $I = \mathbf{N}_G(V_1)$, then $V \cong V_1^G$ and V_1 is an irreducible I-module, by Clifford's Theorem. Thus Corollary 2.13 applies, and $I/\mathbf{C}_G(V_1) \le \Gamma(2^p)$. By Lemma 2.8, conclusion (a) holds and $\mathrm{dl}(G) \le 4$. We thus assume that G is quasi-primitive and also that $G \not\le \Gamma(2^{pr})$.

By Corollary 2.3, $F = \mathbf{F}(G)$ is non-abelian. Since $\mathrm{char}(V) = 2$, F has odd order. Also, every normal abelian subgroup of G is cyclic and so $T := \mathbf{Z}(F)$ is a proper cyclic subgroup of F. By Corollaries 1.10 and 2.6, $|F/T| = e^2$ for an integer $e > 1$ dividing $\dim_{GF(2)}(V)/\dim_{GF(2)}(W)$, where W is an irreducible T-submodule of V. Since $T \ne 1$ and $\mathrm{char}(V) = 2$, $\dim_{GF(2)}(W) > 1$. But $\dim_{GF(2)}(V) = pr$ and we may assume without loss of generality that $\dim_{GF(2)}(W) = r$ and $e = p$. In particular, $|T| \mid 2^r - 1$ and $|F/T| = p^2$. Now Corollary 1.10 implies that there exists an extra-special group $D \trianglelefteq G$ with $|D| = p^3$, $F = DT$ and $D/\mathbf{Z}(D) \cong F/T$ an irreducible G-module. By Corollary 2.6, $p = |D/\mathbf{Z}(D)|^{1/2} \mid \dim_{GF(2)}(V)/\dim_{GF(2)}(Y)$ for an irreducible $\mathbf{Z}(D)$-module Y of V. Since $\mathbf{Z}(D) \ne 1$, $|Y| \ne 2$ and $\dim_{GF(2)}(Y) > 1$. Thus $\dim_{GF(2)}(Y) = r$ and

$$p = e = \dim_{GF(2)}(V)/\dim_{GF(2)}(Y).$$

Apply Lemma 2.10 (ii) to obtain the faithful action of G/F on F/T. Finally observe that $\mathrm{dl}(G) \le 6$ follows from Theorem 2.11. □

2.15 Corollary. *Let G be a solvable irreducible subgroup of $GL(2n, 2)$ with a prime number n. Then one of the following occurs:*

(a) $G \le \Gamma(2^n)\mathrm{wr}\, Z_2$, *or* $G \le S_3\mathrm{wr}\, S$ *where* $Z_n \le S \le Z_n \cdot Z_{n-1} \le S_n$;

(b) $G \le \Gamma(2^{2n})$; *or*

(c) $n = 3$, $\mathbf{F}(G)$ *is extra-special of order* 3^3, $\mathbf{F}(G)/\mathbf{Z}(\mathbf{F}(G))$ *is a faithful irreducible* $G/\mathbf{F}(G)$-*module and* $|G/\mathbf{F}(G)|$ *is even, dividing* 48.

If G is quasi-primitive, conclusion (b) or (c) must occur.

Proof. Theorem 2.14 applies with $pr = 2n$. Conclusions (a) and (b) above are exactly those in that theorem. We may assume that conclusion (c) of Theorem 2.14 holds. Because $\mathbf{O}_2(G) = 1$, we have that $p = n$ and $r = 2$. Then p and $|T|$ divide 3, whence $\mathbf{F}(G)$ is extra-special of order 3^3.

Since $\mathbf{F}(G)/\mathbf{Z}(\mathbf{F}(G))$ is a faithful irreducible $G/\mathbf{F}(G)$-module of order 3^2 and $|GL(2,3)| = 48$, $2 \mid |G/\mathbf{F}(G)|$ and $|G/\mathbf{F}(G)| \mid 48$. $\square$

The final result we need about small solvable linear groups is rather technical.

2.16 Lemma. *Let G be a solvable irreducible subgroup of $GL(2n, q)$ such that G/G' is a p-group for distinct primes p and q, and such that $G' \neq 1$ is a p'-group. Then*

(i) $q^n \neq 2$, 2^2, 2^4 or 3.

(ii) *If $q^n = 2^3$, then $|G| = 3^2 \cdot 7$ and $p = 3$.*

(iii) *If $q^n = 2^5$, then $p = 5 = |G : G'|$ and $|G'| \leq 2^5 \cdot 3^5$.*

(iv) *If $q^n = 3^2$, then $p = 2$ and $|G'| = 5$, or $p = 5 = |G : G'|$ and G' is extra-special of order 2^5.*

(v) *If $q^n = 3^3$, then $p = 2$ and $|G'|$ divides $3^2 \cdot 13^2$ or $7 \cdot 13$.*

(vi) *If $n = 1$, then $p = 2$, or $p = 3$ and $G' \cong Q_8$.*

Proof. Of course $\mathbf{O}_q(G) = 1$ and so the hypotheses imply that $\mathbf{O}_r(G) \neq 1$ for some prime $r \notin \{p, q\}$. Using Theorem 2.11, conclusion (vi) easily follows. Since $|G|$ is divisible by at least two primes distinct from q, $q^n \neq 2$ or 3. For the proof of (i)–(v), we thus assume that $q = 2$ and $2 \leq n \leq 5$, or $q = 3$ and $2 \leq n \leq 3$.

Suppose now that the corresponding G-module V is not quasi-primitive. Choose $C \trianglelefteq G$ maximal such that V_C is not homogeneous and write $V_C = U_1 \oplus \cdots \oplus U_m$ for homogeneous components U_i of V_C. By Proposition 0.2, G/C faithfully and primitively permutes the U_i, and so G/C is a solvable primitive permutation group on m letters. Note that $m \mid 2n$ and thus $2 \leq m \leq 10$. For the structure of a solvable primitive permutation group, cf. the comments following Theorem 2.1; in particular, m is a prime power. There are limited possibilities for G/C. In each case, p is determined by the fact that $(G/C)/(G/C)'$ is a p-group. Since $p \neq q$ and $m \mid 2n$, we have one of the following cases:

m	G/C	p	q^n	$|U_i|$
2	Z_2	2	3^2 or 3^3	3^2 or 3^3
3	Z_3	3	2^3	2^2
3	S_3	2	3^3	3^2
4	A_4	3	2^2 or 2^4	2 or 2^2
4	S_4	2	3^2	3
5	Z_5	5	2^5	2^2
5	D_{10} or F_{20}	2	no possibility	–
8	?	?	2^4	2

We immediately rule out the cases where $|U_i| = 2$, since then $C = 1$, a contradiction. Next suppose $\{p, q\} = \{2, 3\}$. In all such cases, G/C is a $\{2, 3\}$-group. Thus $|C|$ must be divisible by r $(r \geq 5)$, and so there exists a solvable irreducible subgroup $H \leq GL(U_i)$ with $r \mid |H|$ and $G \leq H \operatorname{wr} G/C$. This only occurs in the case $|U_i| = 3^3$, $q^n = 3^3$ and $p = 2 = m$. Then $H \leq \Gamma(3^3)$ (cf. Corollary 2.13), $G \leq \Gamma(3^3) \operatorname{wr} Z_2$ and conclusion (v) holds. The remaining case is when $m = 5 = p = |G/C|$ and $|U_i| = 2^2$. Then $G \leq S_3 \operatorname{wr} Z_5$ and conclusion (iii) holds. We can thus assume that V is quasi-primitive.

For now, assume that $G \not\leq \Gamma(q^{2n})$. If $q^n = 2^3$ or 2^5, then Corollary 2.15 (c) implies that G is a $\{2, 3\}$-group, contradicting $\mathbf{O}_r(G) \neq 1$. Since V is quasi-primitive, Corollary 1.10 applies and we adopt the notation (F, T, U, A, Z) there. Set $e^2 = |F/T|$. We may assume that $e > 1$, since otherwise $G \leq \Gamma(V)$ by Corollary 2.3 (b). By Corollary 2.6, $e \mid \dim(V) = 2n$. Since $e \mid |F|$, q does not divide e. Hence $q^n \neq 2^2$, 2^4. The only remaining values now are $q^n = 3^2$ and 3^3. If W is an irreducible U-submodule of V, then $e \mid \dim(V)/\dim(W)$ and $|U| \mid |W| - 1$. Thus we have

q^n	e	$|W|$
3^3	2	3 or 3^3
3^2	2	3 or 3^2
3^2	4	3.

When $e = 4$, then $|U| = 2$, and so $T = U = Z \leq \mathbf{Z}(G)$. Also F is extra-special of order 2^5. Since $p \neq q = 3$ and $2 \mid |\mathbf{F}(G')|$, $p \geq 5$. As

$A = G$, the Sylow p-subgroup of G/F must act symplectically on F/Z. Now $|Sp(4,2)| = 2^4 \cdot 3^2 \cdot 5$, and therefore $p = 5 = |P|$, where $P \in \mathrm{Syl}_5(G)$. Since P and G/F act irreducibly on F/Z, Corollary 2.15 yields that $G/F \leq \Gamma(2^4)$. The hypotheses easily imply that $|G/F| = 5$, and conclusion (iv) holds.

Next assume that $e = 2$ and recall that $q = 3$. Now $|Z| \mid |U|$, and $|U|$ divides 26 or 8. Since $G/A \leq \mathrm{Aut}(Z)$, since $p \neq 3$ and since G/G' is a p-group, we must have that G/A is a 2-group. Also A/F acts faithfully on F/T of order 2^2. Consequently, $A/F \cong S_3$ or Z_3, and it follows that $p = 2$. By our hypotheses, G has an abelian Sylow 2-subgroup. F however has a non-abelian Sylow 2-subgroup, a contradiction. This rules out the case $e = 2$.

Hence we finally assume that $G \leq \Gamma(q^{2n})$. Since $G' \neq 1$, we have that $1 \neq G/(G \cap \Gamma_0(q^{2n})) \leq \Gamma(q^{2n})/\Gamma_0(q^{2n})$ and p must divide $2n$. This rules out $q^n = 2^2$ and 2^4. When $q^n = 2^5$, then $p = 5 = |G : G'|$ and $|G'| \leq |\Gamma_0(2^{10})| \leq 2^5 \cdot 3^5$, as desired. If $q^n = 3^2$ or 3^3, then certainly $p = 2$. Conclusions (iv) and (v) follow, since G' is a $2'$-group.

What remains is that $q^n = 2^3$ and $p = 3$. Now $\Gamma(2^6)$ has a non-abelian Sylow 3-subgroup of order 27. Since a Sylow 3-subgroup of G is abelian, $3^3 \nmid |G|$. To reach conclusion (ii) we may assume that G is non-abelian of order $3 \cdot 7$. In characteristic 2, G has two absolutely irreducible faithful representations, both of degree 3. But $G \cong \Gamma(2^3)$ has a faithful representation of degree 3 over $GF(2)$, which must be absolutely irreducible. Thus G has two faithful irreducible representations over $GF(2)$ of degree 3, none of degree 6. Thus G cannot act irreducibly on V. This completes the proof. $\square$

§3   Bounds for the Order and the Derived Length of Linear Groups

Let q be a prime. While the orders of $GL(n, q)$ and its Sylow q-subgroups are well-known (namely exponential functions of q^n), the order of a solvable irreducible subgroup is considerably smaller. The Sylow q-subgroup e.g. cannot act completely reducibly. Because a chief factor of a solvable group G gives rise to a representation of G over a finite field, bounding the order of a completely reducible subgroup of $GL(n, q)$ proves to be a useful tool on several occasions in this book. In fact, we give a cubic bound q^{3n} for solvable groups G. But we start first with nilpotent linear groups. We let $\mathfrak{F}$ and $\mathfrak{M}$ denote the sets of *Fermat* and *Mersenne primes* (respectively). The notation $(2, \mathfrak{F})$ will denote the set of ordered pairs $(2, q)$, $q \in \mathfrak{F}$.

We close this section by giving logarithmic bounds for the derived length of solvable subgroups of S_n and solvable completely reducible subgroups of $GL(n, \mathcal{F})$ (for arbitrary fields $\mathcal{F}$).

3.1 Proposition. *Suppose* $q^n - 1 = p^m$ *for primes* p *and* q *and positive integers* m *and* n. *Then*

(i) $n = 1$, $q \in \mathfrak{F}$ *and* $p = 2$;

(ii) $m = 1$, $p \in \mathfrak{M}$ *and* $q = 2$; *or*

(iii) $n = 2$, $m = q = 3$ *and* $p = 2$.

Proof. This is well-known and not difficult. For a proof, see [HB, IX, 2.7]. □

3.2 Proposition. *Suppose that* $q^m - 1 = 2^n \cdot 3$ *for a prime* q *and positive integers* m *and* n. *Then*

(i) $m = 1$; *or*

(ii) $m = 2$ *and* $q \in \{5, 7\}$.

Proof. Assume that $m > 1$ and observe that q is odd. Let $t = 1 + q + \cdots + q^{m-1}$ so that $t \mid 2^n \cdot 3$. If m is odd, then t is odd and so $t = 3$ and $q = 2$. This is a contradiction and hence we write $m = 2k$ for an integer k. Then $2^n \cdot 3 = (q^k - 1)(q^k + 1)$. Since $4 \nmid (q^k - 1, q^k + 1)$, since $2 \neq q^k \pm 1$ and q is odd, it follows that $6 = q^k \pm 1$. Hence $m = 2k = 2$ and q is 5 or 7. □

We now turn to nilpotent linear groups over finite fields. We set $\beta = \log(32)/\log(9)$ and note that $3/2 < 1.57 < \beta < 1.58 < 8/5$.

3.3 Theorem. Let $V \neq 0$ be a faithful, completely reducible and finite G-module for a nilpotent group G. Let $\text{char}(V) = q > 0$. Then

(a) $|G| \leq |V|^{\beta}/2$;

(b) $|G| \leq |V|/2$ provided that G is a p-group and $(p, q) \notin (\mathfrak{M}, 2) \cup (2, \mathfrak{F}) \cup \{2, 7\}$.

Proof. We work by induction on $|G|\,|V|$. If $V = V_1 \oplus \cdots \oplus V_m$ for non-zero G-modules V_i and $m \geq 2$, the inductive hypothesis implies that if $C_i = \mathbf{C}_G(V_i)$, then $|G/C_i| \leq |V_i|^{\beta}/2$, and in part (b) that $|G/C_i| \leq |V_i|/2$ for $i = 1, \ldots, m$. Since $\bigcap_i C_i = 1$, G is isomorphic to a subgroup of $G/C_1 \times \cdots \times G/C_m$. Then

$$|G| \leq \prod_i |G/C_i| \leq \prod_i (|V_i|^{\beta}/2) \leq |V|^{\beta}/2^m \leq |V|^{\beta}/2.$$

Similarly for part (b), we have $|G| \leq |V|/2^m \leq |V|/2$. Thus we may assume that V is an irreducible G-module.

If G is not quasi-primitive, it follows from Corollary 0.3 that there exists $C \trianglelefteq G$ of prime index p with $V_C = V_1 \oplus \cdots \oplus V_p$ for irreducible C-modules V_i. The argument in the last paragraph applied to C shows that $|C| \leq |V|^{\beta}/2^p$. Thus $|G| \leq p|V|^{\beta}/2^p$. Since $2^{x-1} \geq x$ for all $x \geq 2$, $|G| \leq |V|^{\beta}/2$. Similarly for part (b), we have $|G| \leq p|V|/2^p \leq |V|/2$. Thus we may assume that V is quasi-primitive.

Every normal abelian subgroup of G is cyclic. Since G is nilpotent, Corollary 1.3 implies that $G = S \times T$ where T is cyclic of odd order and S is a 2-group that is cyclic, dihedral, quaternion or semi-dihedral. In particular, G has a cyclic normal subgroup U of index at most 2. If U has k orbits on $V^{\#}$, then $|V| - 1 = k \cdot |U|$ because V_U is homogeneous. Since $x^{3/2} - 2x + 2 \geq 0$ for all $x \geq 2$, we have that $|U| \leq |V| - 1 \leq |V|^{3/2}/2 \leq |V|^{\beta}/2$.

To prove (a), we assume that $|G\colon U| = 2$. Since $x^{3/2} - 4x + 4 \geq 0$ for all $x \geq 16$, it follows that either $|V| < 16$, or

$$|G| \leq 2|U| \leq 2(|V| - 1) \leq |V|^{3/2}/2 \leq |V|^{\beta}/2.$$

It remains to consider that $|V| < 16$, G is not cyclic and $|V|$ is odd. But then $|V| = 3^2$ and a nilpotent subgroup of $GL(2,3)$ has order at most $16 = |V|^{\beta}/2$. This proves (a).

In case (b) we assume that G is a p-group, $|U| = p^n$ and $|V| = q^m$. Now $|V| - 1 = k|U|$. If $k = 1$ or if $k = p = 2$, it follows from Proposition 3.1 that $(p,q) \in (2,\mathfrak{F}) \cup (\mathfrak{M}, 2)$. Thus $k \geq 2$. If $U = G$, then $|G| = |U| \leq |V|/2$. If $k \geq 4$, then $|G| \leq 2|U| \leq |V|/2$. The only possibility then is $k = 3$ and $|G\colon U| = 2 = p$. Thus $q^m - 1 = |V| - 1 = 3|U| = 3 \cdot 2^n$. By Proposition 3.2, $q \in \{5,7\}$ or $m = 1$. The hypotheses of (b) imply that $m = 1$. But then G is cyclic and $G = U$, a contradiction. $\square$

3.4 Corollary. *Assume that G is a group of order $p^a q^b$ for primes p and q and $a, b \in \mathbb{N}$.*

(a) *If $p^a > q^{b\beta}/2$, then $\mathbf{O}_p(G) \neq 1$.*

(b) *If $p^a > q^b/2$ and $(p,q) \notin (\mathfrak{M}, 2) \cup (2,\mathfrak{F}) \cup (2,7)$, then $\mathbf{O}_p(G) \neq 1$.*

Proof. Assume that $\mathbf{O}_p(G) = 1$. By Burnside's "well-known" $p^a q^b$-Theorem ([Hu, V, 7.3]), G is solvable. Hence $\mathbf{O}_q(G) \neq 1$ and $\mathbf{F}(G) = \mathbf{O}_q(G) =: Q$. Then $P \in \mathrm{Syl}_p(G)$ acts faithfully on Q and thus on $Q/\Phi(Q)$, because $p \neq q$. Since $Q/\Phi(Q)$ is a completely reducible and faithful P-module, Theorem 3.3(a) yields $|P| \leq |Q/\Phi(Q)|^{\beta}/2 \leq q^{b\beta}/2$. This proves part (a). Part (b) follows analogously from Theorem 3.3(b). $\square$

Part (b) is often referred to as Burnside's "other" $p^a q^b$-Theorem, although Burnside omitted $(p,q) \neq (2,7)$, which is necessary because $2^{23} \mid |GL(8,7)|$. Observe that both (a) and (b) are equivalent to number theoretical statements about prime power divisors of $(q^n - 1) \ldots (q - 1)$. Coates, Dwan and

Rose [CDR] corrected Burnside's proof by giving a number theoretical argument for (b). A group theoretical proof of both (a) and (b) appeared in [Wo 4], as did Theorem 3.5 (a), (b) below. The group theoretical approach is much shorter.

We let $\alpha = (3\cdot\log(48)+\log(24))/(3\cdot\log(9))$, i.e. $9^\alpha = 48\cdot(24)^{1/3}$. Observe that $11/5 < 2.24 < \alpha < 2.25 = 9/4$. We also let $\lambda = (24)^{1/3} = 2\cdot 3^{1/3} < 3$.

3.5 Theorem. Let $V \neq 0$ be a faithful, completely reducible and finite G-module for a solvable group G. Set $\mathrm{char}\,(V) = q > 0$.

(a) Then $|G| \le |V|^\alpha/\lambda$.

(b) If $2 \nmid |G|$ or if $3 \nmid |G|$, then $|G| \le |V|^2/\lambda$.

(c) If $2 \nmid |G|$ and $q \neq 2$, then $|G| \le |V|^{3/2}/\lambda$.

Proof. We proceed by induction on $|G|\,|V|$.

Step 1. We may assume that V is irreducible.

Proof. If not, write $V = V_1 \oplus \cdots \oplus V_m$ for irreducible G-modules V_i and set $C_i = \mathbf{C}_G(V_i)$. Then $\bigcap_i C_i = \mathbf{C}_G(V) = 1$ and G is isomorphic to a subgroup of $\mathbf{X}_i G/C_i$, whence $|G| \le \prod_i |G/C_i|$. Then the inductive hypothesis for (a) implies that $|G/C_i| \le |V_i|^\alpha/\lambda$ and hence $|G| \le |V|^\alpha/\lambda^m \le |V|^\alpha/\lambda$. Parts (b) and (c) follow similarly in this case.

Step 2. We may assume that V is quasi-primitive.

Proof. If not, we choose $N \trianglelefteq G$ maximal such that V_N is not homogeneous and write $V_N = U_1 \oplus \cdots \oplus U_m$ for the homogeneous components U_i of V_N. Then G/N faithfully and primitively permutes $U_1, \ldots, U_m$ (see Proposition 0.2). Let M/N be a chief factor of G. Then $|M/N| = m$ and M/N is a faithful irreducible G/M-module (cf. the comments about primitive permutation groups following Theorem 2.1). Using the inductive hypothesis and the argument in Step 1, we have

$$|N| \leq \begin{cases} |V|^{\alpha}/\lambda^m \\ |V|^2/\lambda^m, & \text{if } 2 \nmid |G| \quad \text{or } 3 \nmid |G| \\ |V|^{3/2}/\lambda^m, & \text{if } 2 \nmid |G| \text{ and } q \neq 2 \end{cases} \tag{3.1}$$

and that

$$|G/M| \leq \begin{cases} m^{\alpha}/\lambda \\ m^2/\lambda, & \text{if } 2 \nmid |G| \text{ or } 3 \nmid |G|. \end{cases}$$

Thus

$$|G| \leq \begin{cases} (m^{\alpha+1}/\lambda^m) \cdot |V|^{\alpha}/\lambda \\ (m^3/\lambda^m) \cdot |V|^2/\lambda, & \text{if } 2 \nmid |G| \quad \text{or } 3 \nmid |G| \\ (m^3/\lambda^m) \cdot |V|^{3/2}/\lambda, & \text{if } 2 \nmid |G| \text{ and } q \neq 2. \end{cases}$$

Since $3 < \alpha + 1 < 10/3$, we may assume that $m^{10/3} > \lambda^m$, i.e. $m^{10} > (24)^m$. Thus $2 \leq m \leq 5$. On the other hand, it suffices via inequality (3.1) to show that $|G/N| \leq \lambda^{m-1}$. If $m = 2, 3$ or 5, then G/N is a solvable primitive permutation group of prime degree and [Hu, II, 3.6] yields $G/N \leq Z_m \cdot Z_{m-1} \leq S_m$. Consequently, $|G/N| \leq m(m-1) \leq \lambda^{m-1}$. If $m = 4$, then $G/N \leq S_4$ and $|G/N| \leq 24 = \lambda^3$. This step is complete.

Step 3. Set $|V| = q^n$. We may then assume that $n \geq 2$ and $q^n \geq 16$.

Proof. First assume that $|V| = q$. Then

$$|GL(V)| = q - 1 \leq q^2/3 \leq |V|^2/\lambda,$$

yielding (a) and (b). To prove (c) in this case, we assume that $q > 2$ and let $S \in \text{Hall}_{2'}(GL(V))$. Now $|S| \leq q/2 \leq q^{3/2}/3 \leq |V|^{3/2}/\lambda$, and we are done if $|V| = q$.

If $|V| = 4$, then $|GL(V)| = 6 \leq 4^{11/5}/3 \leq |V|^{\alpha}/\lambda$, and each Sylow subgroup of $GL(V)$ has order at most $3 \leq |V|^2/\lambda$. If $|V| = 8$, then Corollary 2.13(a) implies that $|G| \leq 21 \leq |V|^2/\lambda$. We may thus assume that $|V| = 9$. Then $|GL(V)| = 48 = 9^{\alpha}/\lambda = |V|^{\alpha}/\lambda$. Furthermore, a Hall $2'$-subgroup of $GL(V)$ has order $3 \leq |V|^{3/2}/\lambda$, and a Hall $3'$-subgroup has order $2^4 \leq 9^2/3 \leq |V|^2/\lambda$. This step is proven.

Step 4. We may assume that

(i) $G \not\leq \Gamma(q^n)$;

(ii) $n > 3$;

(iii) if $q = 2$, then $n \geq 8$.

Proof. (i) Suppose that $G \leq \Gamma(q^n)$. Then $|G| < nq^n$. To prove (a) and (b), we can assume that $nq^n > q^{2n}/\lambda$ and thus $3n > q^n \geq 2^n$. This can only happen when $q = 2$ and $n \leq 3$. But we handled these cases in Step 3. To prove (c) here, we assume that $nq^n > q^{3n/2}/\lambda$ and thus $3n > q^{n/2}$. By Step 3, $n > 1$ and so it follows that $q^n = 3^2, 5^2, 3^3$ or 3^4. A Hall $2'$-subgroup S of $\Gamma(q^n)$ has order 1, 3, 39 or 5 (respectively) and $|S| \leq q^{3n/2}/3$. This yields (i).

(ii) Recall that $n \geq 2$. Suppose that $n = 2$ or 3. Since V is quasi-primitive and $G \not\leq \Gamma(q^n)$, it follows from Theorem 2.12 that G has normal subgroups $F = \mathbf{F}(G)$ and $T = \mathbf{Z}(F)$ such that $|T| \mid q-1$, F/T is elementary abelian of order n^2, and F/T is a faithful irreducible symplectic G/F-module. When $n = 3$, note that $|G/F|$ must be even. Thus in both cases $n = 2$ and $n = 3$, $|G|$ is even and part (c) vacuously holds. When $n = 2$, $q \geq 5$ by Step 3, and

$$|G| = |T|\,|F/T|\,|G/F| \leq (q-1) \cdot 4 \cdot 6 \leq q^4/3 \leq |V|^2/\lambda.$$

When $n = 3$, then

$$|G| = |T|\,|F/T|\,|G/F| \leq (q-1) \cdot 9 \cdot 24 \leq 216 \cdot q.$$

We have that $q \neq 2$ by Step 3. Furthermore $q \neq 3$, because $\mathbf{O}_3(G) \neq 1$. Thus $216 \leq 5^5/3 \leq q^5/3$ and $|G| \leq |V|^2/\lambda$.

(iii) We now assume that $q = 2$, $4 \leq n \leq 7$ and $G \not\leq \Gamma(q^n)$. By Corollary 2.13, n is not a prime. Since V is quasi-primitive, Corollary 2.15 implies that $n = 6$ and $|G| \leq 3^3 \cdot 48 \leq 2^{12}/3 \leq |V|^2/\lambda$. This completes Step 4.

Step 5. Conclusion.

Proof. Since V is quasi-primitive, we may apply Corollary 1.10 to conclude there exist normal subgroups $F = \mathbf{F}(G)$, Z, U, T, E and A with $Z = \text{socle}(U) = \mathbf{Z}(E)$, U cyclic, $|T\colon U| \leq 2$, $U = \mathbf{C}_T(U)$, $F = ET$ and $E \cap T = Z$. Furthermore, $F \leq A = \mathbf{C}_G(Z)$ and $E/Z \cong F/T$ is a completely reducible and faithful A/F-module of order e^2 for an integer e. Since V is quasi-primitive, V_{EU} is a direct sum of $t \geq 1$ isomorphic faithful irreducible EU-modules. By Corollary 2.6, it follows that $V_U \cong te \cdot W$ where W is a faithful irreducible U-module. Note that $|U| \mid |W| - 1$. By Step 4(i) and Corollary 2.3, $e \geq 2$.

Since $A = \mathbf{C}_G(Z)$ and Z is cyclic, $|G/A| \leq |Z| \leq |U|$. If $T > U$, then $|T\colon U| = 2$, $|Z|$ is even and $|G/A| \leq |Z|/2 \leq |U|/2$. In all cases,

$$|G/A|\,|T| \leq |U|^2.$$

If $e = 2$, then $|G|$ is even and

$$|G| = |G/A|\,|T|\,|A/F|\,|F/T| \leq |U|^2 \cdot 6 \cdot 4 \leq 24 \cdot |V| \leq |V|^2/3,$$

since $|V| \geq 81$ by Step 4 (ii), (iii). Should $e = 3$, then $3 \mid |F|$ and $q \neq 3$. Also by Step 4, $|V| \geq 256$. Since $|F/T| = 3^2$, Corollary 1.10 implies that F/T is an irreducible G/F-module. Thus $2 \mid |G/F|$ and $6 \mid |G|$. Now $|V| = |W|^{3t} \geq |U|^3$ and

$$|G| = |G/A|\,|T|\,|A/F|\,|F/T| \leq |U|^2 \cdot 48 \cdot 9 \leq 432 \cdot |V|^{2/3} \leq |V|^2/3,$$

since $|V| \geq 256$. We may assume in the following that $e \geq 4$.

Because F/T is a faithful completely reducible A/F-module of order $e^2 > 1$, the inductive hypothesis implies that

$$|A/F| \leq \begin{cases} e^{2\alpha}/\lambda \\ e^4/\lambda & \text{if } 3 \nmid |G| \\ e^3/\lambda & \text{if } 2 \nmid |G|. \end{cases}$$

Since $|U| < |W|$ and $|G/A||T| \leq |U|^2$, it follows that

$$|G| \leq \begin{cases} e^{2\alpha+2} \cdot |W|^2/\lambda \\ e^6 \cdot |W|^2/\lambda & \text{if } 3 \nmid |G| \\ e^5 \cdot |W|^2/\lambda & \text{if } 2 \nmid |G|. \end{cases}$$

Recall that $|V| = |W|^{te}$.

To prove (c), we assume that $|G||V|$ is odd and $e^5|W|^2 > |V|^{3/2} \geq |W|^{3e/2}$. Then $e^{10} > |W|^{3e-4}$. Because $|U| \mid |W| - 1$ and $|U||W|$ is odd, it follows that $|W| \geq 7$ and $e^{10} > 7^{3e-4}$. This implies $e < 5$, a contradiction, because e is odd and $e \geq 4$. Part (c) follows.

We now prove (a) and (b). If (b) is false, then $e^6|W|^2 > |W|^{2te}$ and

$$e^3 > |W|^{te-1}. \tag{3.2}$$

If (a) is false, then $e^{2\alpha+2}|W|^2 > |W|^{te\alpha}$, and so $e^{2+(2/\alpha)} > |W|^{te-(2/\alpha)}$. Since $\alpha > 2$, inequality (3.2) also holds in this case. Since $|U| \mid |W| - 1$, we have that $|W| \geq 3$ and $e^3 > 3^{e-1}$. Thus $e \leq 5$. If $e = 5$, then $5 \mid |U|$ and $|W| \geq 11$. Now (3.2) gives a contradiction and hence $e = 4$. Then $2 \mid |U|$ and inequality (3.2) implies that $|W| = 3$, $t = 1$ and $|V| = 3^4$. Now $T = U = Z$ has order 2 and F is extra-special of order 2^5. Since $A = C_G(Z) = G$, F/Z is a faithful completely reducible G/F-module. By Corollary 1.10, F/Z is irreducible or the direct sum of two irreducible G/F-modules of order 2^2. Thus $|G/F|$ divides 60 or 72 (see Corollary 2.15). If $|G/F| \leq 60$, then $|G| \leq 60 \cdot 2^5 \leq 3^7 = |V|^2/3 \leq |V|^2/\lambda$. Thus $|G/F| = 72$, $6 \mid |G|$, and $|G| = 72 \cdot 2^5 \leq (3^4)^{11/5}/3 \leq |V|^\alpha/\lambda$. $\square$

3.6 Corollary. *Let G be a solvable primitive permutation group on the finite set Ω. Then $|G| \leq (|\Omega|^{\alpha+1})/\lambda \leq (|\Omega|^{13/4})/2$.*

Proof. Let M be a minimal normal subgroup of G. Then $|M| = |\Omega|$ and G/M acts faithfully on M (cf. the comments following Theorem 2.1). By Theorem 3.5,

$$|G| = |M||G/M| \leq |\Omega|(|\Omega|^\alpha/\lambda),$$

and the assertion follows, since $\alpha < 9/4$ and $\lambda > 2$. $\square$

Repeating some of Step 5 above, we have:

3.7 Corollary. *Assume that every normal abelian subgroup of G is cyclic. With the notation of Corollary 1.10, we have $|G| \leq e^{13/2}|U|^2/2$.*

Proof. By Corollary 1.10, G has a normal series

$$G \geq A \geq F \geq T \geq U \geq Z,$$

where U is cyclic, $Z = \text{socle}\,(U)$, $|T/U| \leq 2$ and $A = \mathbf{C}_G(Z)$. Then $|G/A| \leq |Z| \leq |U|$. If $|T/U| = 2$, then $|Z|$ is even and $|G/A| \leq |Z|/2$. In all cases, $|G/A|\,|T| \leq |U|^2$. Since F/T has order e^2 and is a completely reducible and faithful A/F-module, Theorem 3.5 implies that $|A/F| \leq (e^2)^{9/4}/2$. Consequently, $|G| = |G/A|\,|T|\,|A/F|\,|F/T| \leq |U|^2 e^{13/2}/2$. $\square$

3.8 Example. For each integer $n \geq 0$, there exists a vector space V_n over $GF\,(3)$ and a solvable group G_n such that V_n is a faithful irreducible G_n-module, $\dim(V_n) = 2 \cdot 4^n$, and $|G_n| = |V_n|^\alpha/\lambda$.

Proof. If W is a faithful irreducible H-module over a field $\mathcal{F}$, we define $W^* = W \oplus W \oplus W \oplus W$ and $H^* = H\,\text{wr}\,S_4$. Then W^* is a faithful irreducible H^*-module, $\dim_\mathcal{F}(W^*) = 4 \cdot \dim_\mathcal{F}(W)$ and $|H^*| = 24|H|^4$. Observe that if $|H| = |W|^\alpha/\lambda$, then $|H^*| = 24 \cdot (|W|^\alpha/\lambda)^4 = (24/\lambda^3)|W|^{4\alpha}/\lambda = |W^*|^\alpha/\lambda$. Of course, H^* is solvable if and only if H is.

Let V_0 have dimension 2 over $GF\,(3)$ and set $G_0 = GL(V)$. Then $|G_0| = 48 = 9^\alpha/\lambda = |V_0|/\lambda$. For $n > 0$, define iteratively V_n to be V_{n-1}^* and G_n to be G_{n-1}^*. The assertion follows from the first paragraph. $\square$

Up to this point, the results of this section appeared in [Wo 4]. While this example shows that the results of 3.5 and 3.6 are in some sense best possible, Palfy [Pl 1] has shown that the exponent in Theorem 3.5 can be improved for characteristic other than 3 and gives specific exponents for specific characteristics. For maximal solvable subgroups of $GL(n, q)$ which are not necessarily completely reducible, see A. Mann [Ma 1].

Huppert [Hu 1] was the first to give a logarithmic bound for $\mathrm{dl}(G)$ and Dixon "improved" it, i.e. Dixon gave a stronger bound, which has an error and is too strong for linear groups, but is the correct order of magnitude. Indeed it differs from below by a constant. After the proof, we will give some examples and discuss the error in [Di 1].

3.9 Theorem. *Let G be solvable.*

(a) *If G is a subgroup of the symmetric group S_n, then*
$$\mathrm{dl}(G) \leq \tfrac{5}{2} \log_3(n).$$

(b) *Let $V \neq 0$ be a faithful and completely reducible $\mathcal{F}[G]$-module over an arbitrary field $\mathcal{F}$. Set $n = \dim_{\mathcal{F}}(V)$. Then*
$$\mathrm{dl}(G) \leq 8 + \tfrac{5}{2} \log_3(n/8).$$

Proof. The proof is by induction on $|G|n$, i.e. among all counterexamples to the theorem choose one with $|G|n$ minimal. For $x > 0$, we let

$$\alpha(x) = \tfrac{5}{2}\log_3(x) \text{ and } \beta(x) - 8 + \tfrac{5}{2}\log_3(x/8) = \alpha(x) + 8 - \tfrac{5}{2}\log_3(8).$$

Note that $\alpha(x) + \alpha(y) = \alpha(xy)$ and $\alpha(x) + \beta(y) = \beta(xy)$.

First suppose that G is a subgroup of S_n and let $\Omega = \{1, \ldots, n\}$. If Ω is the disjoint union of non-empty G-invariant subsets Δ_1 and Δ_2, then let $C_i = \{g \in G \mid \omega^g = \omega \text{ for all } \omega \in \Delta_i\} \trianglelefteq G$. By the inductive hypothesis, $\mathrm{dl}(G/C_i) \leq \alpha(|\Delta_i|)$ for $i = 1, 2$. Since $C_1 \cap C_2 = 1$, it follows that

$$\mathrm{dl}(G) \leq \max\{\alpha(|\Delta_i|) \mid i = 1, 2\} \leq \alpha(n).$$

Hence we may assume that G acts transitively on Ω.

If G acts imprimitively on Ω, we may then write Ω as a disjoint union $\Omega = \Omega_1 \mathbin{\dot\cup} \ldots \mathbin{\dot\cup} \Omega_m$ with $1 < m < n$ for subsets Ω_i permuted by G. Let

$N = \{g \in G \mid \Omega_i^g = \Omega_i \text{ for all } i\} \trianglelefteq G$. Then G/N transitively and faithfully permutes the Ω_i and so the inductive hypothesis implies that $\mathrm{dl}(G/N) \leq \alpha(m)$. Now each Ω_i is N-invariant and $|\Omega_i| = n/m$ for all i, because G transitively permutes the Ω_i. As in last paragraph, it follows that $\mathrm{dl}(N) \leq \alpha(n/m)$ and altogether

$$\mathrm{dl}(G) \leq \mathrm{dl}(G/N) + \mathrm{dl}(N) \leq \alpha(m) + \alpha(n/m) = \alpha(n).$$

We can now assume that G acts primitively on Ω.

Since G is a solvable primitive permutation group, G has a minimal normal subgroup M that is a faithful G/M-module and $|\Omega| = |M| = p^l$ for a prime p and integer l. The inductive hypothesis implies that $\mathrm{dl}(G/M) \leq \beta(l)$ and thus

$$\mathrm{dl}(G) \leq 1 + \beta(l) = 9 + \alpha(l) - \alpha(8) = \alpha(3^{18/5}l/8).$$

Since $\alpha(x)$ is increasing, we have that $\mathrm{dl}(G) \leq \alpha(p^l) = \alpha(n)$ unless $3^{18/5}l/8 \geq p^l \geq 2^l$. This can only happen for those values of p^l in the next table. In each case, we may use Corollaries 2.13 and 2.15 or $\mathrm{dl}(GL(l,p))$ to give an upper bound for $\mathrm{dl}(G/M)$ and $\mathrm{dl}(G) \leq 1 + \mathrm{dl}(G/M)$. We use $[x]$ to denote greatest integer in x.

p^l	$\max \mathrm{dl}(G/M)$	$\max \mathrm{dl}(G)$	$[\alpha(p^l)]$
2	0	1	1
3 or 5	1	2	2 or 3
2^2	2	3	3
3^2	4	5	5
2^3	2 (Cor. 2.13)	3	4
2^4	4 (Cor. 2.15)	5	6
2^5	2 (Cor. 2.13)	3	7

In all cases, $\mathrm{dl}(G) \leq \alpha(p^l) = \alpha(n)$. The result follows when $G \leq S_n$.

We now assume that V is a faithful completely reducible $\mathcal{F}[G]$-module of dimension n. If $\mathcal{K}$ is any extension field of $\mathcal{F}$, then $V \otimes_{\mathcal{F}} \mathcal{K}$ is a completely reducible and faithful $\mathcal{K}[G]$-module of dimension n. Hence it involves no loss of generality to assume that $\mathcal{F}$ is algebraically closed.

If $V = V_1 \oplus V_2$ for G-modules V_i of dimension $n_i > 0$, then each V_i is completely reducible and $\mathbf{C}_G(V_1) \cap \mathbf{C}_G(V_2) = 1$. By induction, we conclude

$$\mathrm{dl}\,(G) \leq \max\{\mathrm{dl}(G/\mathbf{C}_G(V_i))|i=1,2\} \leq \max\{\beta(n_i)|i=1,2\} \leq \beta(n).$$

We thus assume that V is an irreducible G-module.

If V is not quasi-primitive, we choose $C \trianglelefteq G$ maximal such that V_C is not homogeneous and write $V_C = V_1 \oplus \cdots \oplus V_t$ for homogeneous components V_i of V_C, $t > 1$. Now G/C transitively and faithfully permutes the V_i. Thus $\dim_{\mathcal{F}}(V_i) = n/t$ for each i, and applying the inductive hypothesis to the action of C on $V_1 \oplus \cdots \oplus V_t$, we see as in the last paragraph that $\mathrm{dl}(C) \leq \beta(n/t)$. We also apply induction to the permutation action of G/C on $\{V_1, \ldots, V_t\}$ to conclude that $\mathrm{dl}(G/C) \leq \alpha(t)$. It follows altogether that

$$\mathrm{dl}(G) \leq \mathrm{dl}(G/C) + \mathrm{dl}(C) \leq \alpha(t) + \beta(n/t) = \beta(n).$$

Thus we assume that V is quasi-primitive.

Since $\mathcal{F}$ is algebraically closed and V is quasi-primitive, every normal abelian subgroup of G is cyclic and central in G. Let $F = \mathbf{F}(G)$ and $T = \mathbf{Z}(F) = \mathbf{Z}(G)$. By Corollary 1.10, $\mathrm{dl}(F) \leq 2$ and F/T is a completely reducible and faithful G/F-module. If $F = T$, then $G = F$ is abelian and $\mathrm{dl}(G) = 1 \leq \beta(n)$. We may thus assume that $e > 1$, and write $F/T = E_1/T \times \cdots \times E_m/T$ for $m \geq 1$ chief factors E_i/T of G, with each $|E_i/T| = p_i^{2k_i}$ for primes p_i and integers k_i. Let $D_i = \mathbf{C}_G(E_i/T)$ so that $\bigcap_i D_i = F$ and $\mathrm{dl}(G/F) = \max\{\mathrm{dl}(G/D_i) \mid i = 1, \ldots, m\}$. Set $e = \prod_i p_i^{k_i}$ and observe that $e \mid n$, by Corollary 2.6.

If $p_i^{k_i} = 2$, then $G/D_i \leq S_3$ and $\mathrm{dl}(G/D_i) \leq 2$. If each $p_i^{k_i}$ is 2, then $\mathrm{dl}(G) \leq 4 \leq \beta(2) \leq \beta(e) \leq \beta(n)$. Thus some $p_j^{k_j}$ is at least 3. If $p_i^{k_i} = 3$, then $G/D_i \leq SL(2,3)$ and $\mathrm{dl}(G/D_i) \leq 3$, because G acts symplectically on

E_i/T. Thus if each $p_i^{k_i}$ is at most 3, then $\mathrm{dl}(G) \leq 5 \leq \beta(3) \leq \beta(e) \leq \beta(n)$. Consequently, some $p_j^{k_j}$ is at least 4.

We claim that, for some j, $p_j^{k_j} \geq 8$ and $k_j \geq 2$. Otherwise, for each i, $k_i = 1$ or $p_i^{k_i} = 2^2$. Then Theorem 2.11 and Corollary 2.15 yield that $\mathrm{dl}(G/D_i) \leq 4$ for all i. So $\mathrm{dl}(G) \leq 6 \leq \beta(4) \leq \beta(e)$. The claim follows and $e \geq p_j^{k_j} \geq 8$.

Each k_i satisfies $k_i \leq \log_2(e) \leq e/2$, because $e \geq 8$. Thus $\dim(E_i/T) = 2k_i \leq e \leq n$ the inductive hypothesis yields that $\mathrm{dl}(G/D_i) \leq \beta(2k_i)$ for all i. Set $k = \max\{k_i \mid i = 1, \ldots, m\}$. Then

$$\mathrm{dl}(G) \leq \mathrm{dl}(F) + \mathrm{dl}(G/F) \leq 2 + \beta(2k).$$

For $k \leq 2$, $\mathrm{dl}(G) \leq 2 + [\beta(4)] = 8 = \beta(8) \leq \beta(e)$. So $k \geq 3$ and

$$\mathrm{dl}(G) \leq 2 + \beta(2k) = \alpha(3^{4/5}) + \beta(2k) = \beta(2 \cdot 3^{4/5} \cdot k).$$

Since β is an increasing function of x and $k \leq \log_2(e)$, we may assume that $2 \cdot 3^{4/5} k > e$ and $2 \cdot 3^{4/5} \log_2(e) > e$. The latter inequality implies that $e < 24$ and the first rules out the case $k = 3$ and $e = 16$. Since $k \geq 3$, the only possibilities are $e = p_1^{k_1} = 2^3$ or $e = p_1^{k_1} = 2^4$. In particular, F/T is an irreducible G/F-module of order 2^6 or 2^8 (respectively). If $e = 8$, then Corollary 2.15 implies that the derived length of G/F is at most 6, and therefore G has derived length at most $8 = \beta(e) \leq \beta(n)$.

We now have that $16 = e \leq n$ and that F/T is a faithful irreducible G/F-module of order 2^8. If F/T is an imprimitive G/F-module, then G/F is isomorphic to a subgroup of S_3 wr S_4 or H wr Z_2 for a solvable irreducible subgroup H of $GL(4,2)$. In either case, $\mathrm{dl}(G/F) \leq 5$ by Corollary 2.15. By Proposition 0.20 and Corollary 2.5, $\mathbf{F}(G/F)$ is abelian of odd order. If F/T is primitive, then $G/F \lesssim \Gamma(2^8)$ by Corollary 2.3. In all cases, $\mathrm{dl}(G/F) \leq 5$. Hence $\mathrm{dl}(G) \leq 7 < \beta(e) \leq \beta(n)$. $\qquad\square$

Suppose that G is a primitive linear group of degree n and $|\mathbf{F}(G) : \mathbf{Z}(G)| = n^2$. It is incorrectly argued in [Di 1] (when $n = 3$ and more critically when $n = 8$, see p. 156) that $G/\mathbf{F}(G)$ must have trivial center, whence the $G/\mathbf{F}(G)$-module $\mathbf{F}(G)/Z$ cannot be irreducible and primitive. Indeed, the bound given in [Di 1] for a linear group of degree 8 is $7 = [\frac{5}{2}\log_3(8) + 5/2]$, not 8 as above. However, Glasby and Howlett [GH] have constructed a solvable irreducible $G \le GL(8,3)$ with $\mathrm{dl}(G) = 8$ (even more $G \le Sp(8,3)$). In this group, $\mathbf{F}(G)$ is extra-special of order 2^7 and $\mathbf{F}_2(G)/\mathbf{F}(G) = \mathbf{F}(G/\mathbf{F}(G))$ is extra-special of order 3^3. Also $G/\mathbf{F}_2(G) \cong GL(2,3)$ acts faithfully and irreducibly on $\mathbf{F}_2(G)/L$, where $L/\mathbf{F}(G) = \mathbf{Z}(\mathbf{F}_2/\mathbf{F}(G)) \not\le \mathbf{Z}(G/\mathbf{F}(G))$.

3.10 Proposition. *Let S be a solvable permutation group on Ω (not necessarily transitive). Assume one of the following*

(i) *S is a primitive permutation group;*

(ii) *$\mathrm{dl}(S) \le 2$;*

(iii) *$\mathrm{dl}(S) = l$ and $S^{(l-1)}$ has odd order.*

If H is solvable, then $\mathrm{dl}(H\,\mathrm{wr}\,S) = \mathrm{dl}(H) + \mathrm{dl}(S)$.

Proof. Say $S \le S_n$ and let $G = H\,\mathrm{wr}\,S$. Then G has a normal subgroup $K = H \times \cdots \times H$, a direct product of n copies of H that are permuted by S. Also $G = KS$ and $1 = K \cap S$. Clearly, $\mathrm{dl}(G) \le \mathrm{dl}(K) + \mathrm{dl}(G/K) = \mathrm{dl}(H) + \mathrm{dl}(S)$. If $l = \mathrm{dl}(S)$, it suffices to show $\mathrm{dl}(G^{(l)}) \ge \mathrm{dl}(H)$. We may assume that $l \ne 0$, i.e. $S \ne 1$.

Let $M = S^{(l-1)}$ so that $M \ne 1$ is an abelian normal subgroup of S. Let $1 \ne \sigma \in M$ and assume w.l.o.g. that $\sigma(1) = 2$. Let $x = (h, 1, \dots) \in K$. Then $[\sigma, x] = \sigma^{-1}x^{-1}\sigma x = (h, h^{-1}, 1, \dots, 1)$ and $[\sigma, x] \in [K, M] \subseteq [K, G^{(l-1)}] \subseteq G^{(l-1)} \cap K$. If we let $\Pi_1 \colon K \to H$ be the projection map from K to H relative to the first component, then $\Pi_1([K, M]) = H$. Hence $\mathrm{dl}(G') \ge \mathrm{dl}([K, M]) \ge \mathrm{dl}(H)$. Thus we may assume S is non-abelian. Since M acts non-trivially on the S-orbit $\{1^S\}$ and since $M \le S'$, indeed $|1^S| \ge 3$. We thus can assume $1^S = \{1, 2, \dots, m\}$ for an integer m, $3 \le m \le n$.

First suppose $\tau \in M$ and $\tau(2) = 3$. If $w = [\sigma, x] = (h, h^{-1}, 1, \dots, 1)$, then $w \in G^{(l-1)} \cap K$ (as in the last paragraph). Since $\tau \in G^{(l-1)}$, $[\tau, w] \in G^{(l)} \cap K$.

But $[\tau, w] = (*, *, h, \dots)$. Hence $\Pi_3(G^{(l)} \cap K) = H$ and $\mathrm{dl}(G^{(l)}) \geq \mathrm{dl}(H)$, as desired. Thus $\tau(2) \in \{1, 2\}$ for all τ. In particular, $\{1, 2\}$ is an M-orbit in $\{1, 2, \dots, m\}$. Since $\sigma \in M - \{1\}$ was chosen arbitrarily and since M cannot have fixed points in the S-orbit $\{1, \dots, m\}$, we have that the M-orbits of 1^S are $\{1, 2\}, \{3, 4\}, \dots, \{m - 1, m\}$, after a possible relabelling. Thus $M/\mathbf{C}_M(1^S)$ is a non-trivial elementary abelian 2-group.

If S is a primitive permutation group on Ω, then M would be the unique minimal normal subgroup of S and act transitively on Ω (see discussion following Theorem 2.1). Since $m \geq 3$, S is not primitive by the last paragraph. To complete the proof, we may assume that $\mathrm{dl}(S) = 2$. Choose $\alpha \in S$ with $\alpha(1) = 3$. Then $[\alpha, x] = (h, 1, h^{-1}, 1, \dots)$ and $[\sigma, [\alpha, x]] = (h, h^{-1}, *, \dots)$. Since $[\sigma, [\alpha, x]] \in G'' \cap K$ it follows that $\mathrm{dl}(G'' \cap K) \geq \mathrm{dl}(H)$ and $\mathrm{dl}(G) = 2 + \mathrm{dl}(H)$. $\qquad\square$

3.11 Examples. Let H be the semi-direct product of an elementary abelian group E of order 9 and $\mathrm{Aut}(E) \cong GL(2, 3)$. Then H is a permutation group of degree 9 and derived length 5.

(a) Let $H_1 = H$ and iteratively define $H_j = H_{j-1}\mathrm{wr}H$. Then H_j is a transitive permutation group of degree 9^j. By Proposition 3.10, $\mathrm{dl}(H_j) = 5j = \frac{5}{2}\log_3(9^j)$. Hence the bound in Theorem 3.9 (a) is best possible for infinitely many n.

(b) Let V_0 be a faithful irreducible G_0-module with G_0 solvable of derived length d. Let $m = \dim(V_0)$. Iteratively define V_i and G_i by letting V_i be the direct sum of 9 copies of V_{i-1} and letting $G_i = G_{i-1}\mathrm{wr}H$. Then V_i is a faithful irreducible G_i-module with $\dim(V_i) = 9^i m$ and $\mathrm{dl}(G_i) = d + 5i$. Thus $\mathrm{dl}(G_i) = \frac{5}{2}\log_3(\dim(V_i)) + d - \frac{5}{2}\log_3(m)$. This shows that the bound $\frac{5}{2}\log_3(n) + 8 - \frac{5}{2}\log_3(8)$ is a best bound or nearly so in each characteristic, i.e. in a given characteristic, it may be possible to lower the additive constant $(8 - \frac{5}{2}\log_3(8) \cong 3.268)$, but the coefficient of $\log(n)$ cannot be reduced and the additive constant still must be non-negative. Also, to see the bound is obtained for infinitely many i in some characteristic, it suffices to find a linear group of degree 8 and derived length 8. We refer the reader to the discussion preceding Proposition 3.10.

The following is a little weaker than Theorem 3.9, but sometimes more convenient.

3.12 Corollary. *Let G be solvable.*

(a) *If $G \leq S_n$, then $\mathrm{dl}(G) \leq 2\log_2(n)$.*

(b) *If $V \neq 0$ is a faithful and completely reducible G-module over an arbitrary field $\mathcal{F}$, then $\mathrm{dl}(G) \leq 2\log_2(2n)$.*

Proof. Since $\frac{5}{2}\log_3(x) \leq 2\log_2(x)$ for all $x \geq 1$, part (a) is immediate. Observe that $8 + \frac{5}{2}\log_3(x/8) \leq 2\log_2(2x)$ for all $x \geq 8$. For $2 \leq n \leq 7$, the greatest integer in $8 + \frac{5}{2}\log_3(n/8)$ is the same as that in $2\log_2(2n)$. So it remains to verify (b) when $n = 1$ and this is trivial.   □

3.13 Remark. We have given polynomial bounds for the order of a Sylow p-subgroup of $GL(n,q)$ for $p \neq q$ (Theorem 3.3) and also for solvable completely reducible subgroups of $GL(n,q)$ (in Theorem 3.5). Since the order of a Sylow q-subgroup Q of $GL(n,q)$ is $q^{n(n-1)/2}$, its order is not a polynomial function of q^n. Of course, Q does not act completely reducibly. These bounds also show Q cannot be a subgroup of a completely reducible solvable subgroup of $GL(n,q)$. That can also be deduced from Theorem 3.9. Actually, if q^m is the order of a Sylow q-subgroup of a completely reducible q-solvable subgroup of $GL(n,q)$, then $m < n$. In [Wo 5], it is shown that

$$m \leq \sum_{i=0}^{\infty} \left[\frac{n}{(p-1)p^i} \right] \leq n \cdot \frac{p}{(p-1)^2} - \frac{1}{p-1}$$

and

$$m \leq \sum_{i=1}^{\infty} \left[\frac{n}{p^i} \right] \leq \frac{n-1}{p-1} \quad \text{if } p \text{ is not a Fermat prime.}$$

(Here [ ] denotes the greatest integer function). These bounds are in some sense best possible.

There are some interesting consequences of the bounds just mentioned. Suppose that G is p-solvable with p-length l and p-rank r. (The p-rank of G is the largest integer r such that p^r is the order of a chief factor of G.) Also, let p^b be the order of a Sylow p-subgroup. Then

   i) l is bounded above by a logarithmic function of b.

   ii) l is bounded above by a logarithmic function of r.

Proofs are given in Theorems 2.2 and 2.3 of [Wo 5]. The bounds, which are in some sense best possible, are slightly weaker for Fermat primes. The bounds for l in terms of r are stated below in Remark 14.12 (a).

Chapter II
SOLVABLE PERMUTATION GROUPS

§4 Orbit Sizes of p-Groups and the Existence of Regular Orbits

Let G be a permutation group on a finite set Ω. The orbit $\{\omega^g \mid g \in G\}$ is called *regular*, if $\mathbf{C}_G(\omega) = 1$ holds.

In this section we consider a finite p-group P which acts faithfully and irreducibly on a finite vector space V of characteristic $q \neq p$. For several questions in representation theory, it turns out to be helpful if one knows that P has a long orbit (preferably a regular orbit) in its permutation action on V. For applications, see §14.

We start with an easy, but useful, lemma.

4.1 Lemma. *Let G act on a vector space V over $GF(q)$, and let $\Delta \neq 0$ be an orbit of G on V. If $\lambda \in GF(q)^{\#}$, then $\lambda\Delta \neq \Delta$ or $o(\lambda) \mid \exp(G)$.*

Proof. Let $\lambda\Delta = \Delta$. Then, for $v \in \Delta$, there exists $g \in G$ such that $\lambda v = vg$. If $n = o(g)$, then $v = vg^n = v\lambda^n$ and $o(\lambda) \mid n$. $\qquad\square$

We recall that $\mathfrak{F}$ and $\mathfrak{M}$ denote the set of Fermat and Mersenne primes, respectively.

4.2 Lemma. *Let P be a non-trivial p-group and V a faithful, irreducible and primitive P-module over $GF(q)$ for a prime $q \neq p$. Set $|P| = p^n$ and $|V| = q^m$.*

    (a) *There always is a regular orbit of P on V, except the case where*

$(p, q) \in (2, \mathfrak{M})$ and P is dihedral or semi-dihedral. In this exceptional case, clearly P has $D_8 \cong Z_2 \mathrm{wr} Z_2$ as a subgroup.

(b) If $(p, q) \notin (2, \mathfrak{M}) \cup (2, \mathfrak{F}) \cup (\mathfrak{M}, 2)$, there even exist two regular orbits.

(c) In any case there are $v_1, v_2 \in V$ such that $\mathbf{C}_P(v_1) \cap \mathbf{C}_P(v_2) = 1$.

Proof. Since P acts primitively on V, every normal abelian subgroup of P is cyclic. Hence Corollary 1.3 implies that P is cyclic, quaternion, dihedral or semi-dihedral. In particular, P has a cyclic normal subgroup Z of index 1 or 2. Every subgroup of Z is normal in P and so $\mathbf{C}_Z(v) = 1$ for all $0 \neq v \in V$, i.e. every Z-orbit on $V - \{0\}$ is regular. Also $|\mathbf{C}_P(v)| \leq 2$ for all $v \in V - \{0\}$ and part (c) easily follows.

To show that (a) implies (b), we may assume that P has exactly one regular orbit Δ in V. If λ is a generator for $GF(q)^{\#}$, then $\lambda \Delta$ is also a regular orbit and so $\lambda \Delta = \Delta$. By Lemma 4.1, $q - 1 = o(\lambda) = p^j$ for some j. By Proposition 3.1 and our assumption on (p, q), the only possibility is $q = 2$, p odd. But then $P = Z$ has only one orbit on $V^{\#}$ and $q^m - 1 = |P|$, contradicting Proposition 3.1. Hence (a) implies (b).

It remains to prove (a). We thus assume that P has no regular orbits, $P > Z$, and $p = 2 = |\mathbf{C}_P(v)|$ for all $v \in V^{\#}$. Thus $P = Z\mathbf{C}_P(v)$ for all $v \in V$ and each Z-invariant subspace of V is indeed P-invariant, i.e. V_Z is irreducible. Also, P cannot have a unique involution, whence P is dihedral or semi-dihedral. An easy counting argument shows P has 2^{n-1} or 2^{n-2} involutions outside Z (respectively). (Definitions of dihedral and semi-dihedral appear in the proof of Proposition 1.1). We need to show $q \in \mathfrak{M}$.

Since now V_Z is irreducible, P acts semi-linearly on $V = GF(q^{2m'})$, where $m = 2m'$ by Theorem 2.1. More precisely, we have with $Z = \langle z \rangle$ and $g \in P \setminus Z$ that

$$vz = av \quad \text{and} \quad vg = bv^{q^{m'}} \quad (a, b \in GF(q^{2m'})^{\#}).$$

In particular, $g^2 = 1$ if and only if $bb^{q^{m'}} = 1$. This enables us to count the number of fixed-points of g on $V^{\#}$ (and hence of all involutions outside Z). Namely $v_0 \neq 0$ is a fixed-point of g if and only if $v_0 = bv_0^{q^{m'}}$, i.e. $v_0^{q^{m'}-1} = b^{-1}$. Since $b^{1+q^{m'}} = 1$, this equation has $q^{m'} - 1$ solutions v_0 in $GF(q^{2m'})^{\#}$.

Since $|\mathbf{C}_P(v)| = 2$ holds for all $0 \neq v \in V$, and since Z acts fixed-point-freely on V, counting of the set

$$\{(v,i) \mid v \in V^{\#},\ i \text{ involution in } P \setminus Z,\ i \in \mathbf{C}_P(v)\}$$

yields $s \cdot (q^{m'} - 1) = q^{2m'} - 1$, i.e. $q^{m'} + 1 = s$ is a power of 2. Thus $q \in \mathfrak{M}$ (Proposition 3.1). $\qquad\square$

In view of §7, the following lemma is stated in a more general version than needed in this section.

4.3 Lemma. *Let P be a p-group and V a faithful P-module over $GF(q)$ for a prime q (possibly $q = p$). Suppose that $V = V_1 \oplus \cdots \oplus V_m$ for subspaces $V_i \neq 0$ of V that are permuted by P (not necessarily transitively). Assume that $\mathbf{N}_P(V_i)/\mathbf{C}_P(V_i)$ has at least k regular orbits for an integer $k \in \mathbb{N}$ $(i - 1, \ldots, m)$. Then P has at least k regular orbits on V, unless*

(i) $k = 1$, and $q = 2$ or $(p,q) \in (2, \mathfrak{F})$; or

(ii) $k = 2$, and $(p,q) \in (2, \mathfrak{F} \cup \{2\})$.

Proof. We proceed by induction on $\dim_{GF(q)}(V)$ and assume without loss of generality that $m > 1$. If P has more than one orbit on $\{V_1, \ldots, V_m\}$, we may write $V = U_1 \oplus U_2$ for P-invariant subspaces U_i, each of which is a sum of some V_j's. We apply the inductive hypothesis to the action of $P/\mathbf{C}_P(U_i)$ on U_i $(i = 1, 2)$. If the exceptional case occurs for at least one $i \in \{1, 2\}$, we are clearly done. Hence there exist $x_1, \ldots, x_k \in U_1$ belonging to k distinct regular orbits of $P/\mathbf{C}_P(U_1)$ and $y_1, \ldots, y_k \in U_2$ belonging to k distinct regular orbits of $P/\mathbf{C}_P(U_2)$. Now

$$\mathbf{C}_P((x_i, y_j)) = \mathbf{C}_P(U_1) \cap \mathbf{C}_P(U_2) = 1$$

and $(x_i, y_j) \in V$ generates a regular P-orbit for all i, j. Note that (x_i, y_j) is conjugate to (x_k, y_l) if and only if $i = k$ and $j = l$. Thus P has at least $k^2 \geq k$ distinct regular orbits on V.

We thus assume that P transitively permutes the V_i. Let $N \trianglelefteq P$ such that $N \geq \mathbf{N}_P(V_1)$ and $|P/N| = p$. Then $V_N = W_1 \oplus \cdots \oplus W_p$ for N-invariant subspaces W_i that are transitively permuted by P/N. Observe that $\mathbf{N}_P(V_i) \leq N$ and $\mathbf{N}_P(V_i) = \mathbf{N}_N(V_i)$ for all i. As in the intransitive case we may assume by induction that $N/\mathbf{C}_N(W_i)$ has at least k regular orbits on W_i ($i = 1, \ldots, p$). Let $\mathfrak{S}_i$ contain one element of W_i from each regular orbit. Then $l := |\mathfrak{S}_i| \geq k$ and $\mathfrak{S} := \mathfrak{S}_1 \times \cdots \times \mathfrak{S}_p$ contains $l^p \geq k^p$ elements. Pick $y \in \mathfrak{S}$, say $y = (w_1, \ldots, w_p)$. Assume that $l \geq 2$ and choose $y \neq z \in \mathfrak{S}$ with $z = (u_1, w_2, \ldots, w_p)$. We claim that y or z is in a regular P-orbit. Since both y and z belong to regular N-orbits, we can assume that $\mathbf{C}_P(y)$ and $\mathbf{C}_P(z)$ have order p and complement N. Choose $a \in \mathbf{C}_P(y)$ and $b \in \mathbf{C}_P(z)$ with $o(a) = o(b) = p$ and $Na = Nb$. Then a and b induce the same non-trivial permutation on $\{W_1, \ldots, W_p\}$ and hence there exists $j > 1$ such that $w_1^a = w_j$ and $u_1^b = w_j$. Then $w_1^{ab^{-1}} = u_1$. Since $ab^{-1} \in N$, w_1 and u_1 are N-conjugate, contradicting our choice of y, $z \in \mathfrak{S}$. Thus y or z generates a regular P-orbit. Consequently, at most one element of the form $(v, w_2, \ldots, w_p)$, $v \in \mathfrak{S}_1$, is not in a regular P-conjugacy class. Hence at least $l^p - l^{p-1} = (l-1)l^{p-1}$ elements of $\mathfrak{S}$ lie in regular P-orbits. No two elements of $\mathfrak{S}$ are N-conjugate, but they may be P-conjugate. Hence there exist at least $(l-1)l^{p-1}/p$ distinct regular P-orbits in V. Should $(k-1)k^{p-1}/p \geq k$, the conclusion follows, because $l \geq k$. We thus assume that $(k-1)k^{p-2} < p$, i.e. either $k = 1$, or $k = 2$ and $p \leq 3$.

Assume that $k = 2$ and $p = 3$ and let $\{u_i, w_i\} \subseteq \mathfrak{S}_i$ ($i = 1, 2, 3$). We can find an element $(w_1, w_2, w_3) \in \mathfrak{S}$ which is not in a regular P-orbit, since otherwise P has at least $2^3/3 \geq 2$ regular orbits. The argument in the last paragraph now shows that (u_1, w_2, w_3), (w_1, u_2, w_3) and (w_1, w_2, u_3) all lie in regular P-orbits. If there are four elements of $\mathfrak{S}$ in regular P-orbits, then there are at least $4/3$ and in fact two regular orbits. Hence (u_1, u_2, u_3) and

(u_1, u_2, w_3) are both not in regular P-orbits, contradicting the argument of the last paragraph. Therefore $k = 1$, or $k = p = 2$.

Let λ generate the multiplicative group $GF(q)^{\#}$. If $k = p = 2$, the argument in the next to last paragraph shows that (u_1, u_2) or (w_1, u_2) lies in a regular P-orbit. Say (u_1, u_2) is in a regular P-orbit. Then so is $(\lambda u_1, \lambda u_2)$ and hence we may assume that $(\lambda u_1, \lambda u_2)$ is P-conjugate to (u_1, u_2). By Lemma 4.1, $q - 1 = o(\lambda) = p^j = 2^j$ for some j. Hence $q \in \mathcal{F} \cup \{2\}$. We may now assume that $k = 1$ and $N/\mathbf{C}_N(W_1)$ has exactly one regular orbit on W. If $w_1 \in \mathfrak{S}_1$, then λw_1 is N-conjugate to w_1 and so $q - 1 = o(\lambda) = p^j$ for some j. Thus $q = 2$ or $(p, q) \in (2, \mathcal{F})$. $\square$

4.4 Theorem. *Suppose that the p-group P acts irreducibly and faithfully on the $GF(q)$-vector space V of coprime characteristic q. If $(p, q) \notin (2, \mathfrak{M}) \cup (2, \mathfrak{F}) \cup (\mathfrak{M}, 2)$, then P induces at least two regular orbits on V.*

Proof. If P acts primitively on V, apply Lemma 4.2 (b). If not, we may choose $N \trianglelefteq P$ of index p such that $V_N = V_1 \oplus \cdots \oplus V_p$ for irreducible N-modules V_i which are transitively permuted by P (see Corollary 0.3). By induction, $\mathbf{N}_P(V_i)/\mathbf{C}_P(V_i) = N/\mathbf{C}_N(V_i)$ has at least two regular orbits on V_i ($i = 1, \ldots, p$). Thus the result follows by induction from Lemma 4.3. $\square$

The following examples show that in the exceptional cases of Theorem 4.4 regular orbits need not exist.

4.5 Examples. (a) Let $(p, q) \in (\mathfrak{M}, 2)$ and set $p = 2^f - 1 \geq 3$. We consider $P = Z_p \mathrm{wr}\, Z_p$ and denote by $U = GF(2^f)$ the f-dimensional module over $GF(2)$ on which $Z_p \cong GF(2^f)^{\#}$ acts fixed-point-freely. We view U as a module for the base-group $Z_p \times \cdots \times Z_p \trianglelefteq P$, where the first component acts as above and where the others centralize U. Then $V := U^P$ is a faithful irreducible P-module. But since

$$|P| = p^{p+1} > (p+1)^p - 1 = 2^{fp} - 1 = |V| - 1,$$

P has no regular orbit on V.

(b) We now consider $(p,q) \in (2,\mathfrak{F})$ where $q = 2^f + 1 \geq 3$. Let $P = Z_{2^f} \operatorname{wr} Z_2$ and denote by $U = GF(q)$ the 1-dimensional module over $GF(q)$ on which $Z_{2^f} \cong GF(q)^\#$ acts fixed-point-freely. As above we view U as a module for $Z_{2^f} \times Z_{2^f} \trianglelefteq P$. Then $V := U^P$ is a faithful and irreducible P-module. If $q > 3$, then P has no regular orbit, because

$$|P| = 2^{2f+1} = 2(q-1)^2 > q^2 - 1 = |V| - 1.$$

If $q = 3$, then $|V| = 9$ and $|P| = 8$. In this case, a regular orbit must be the only non-trivial orbit. But $(x,0)$ is not in a regular orbit ($x \neq 0$), and so P has no regular orbit.

(c) Let finally $(p,q) \in (2,\mathfrak{M})$, $q = 2^f - 1 \geq 3$. Set $V = GF(q^2)$, take $\alpha \in V$ of order $o(\alpha) = q + 1 = 2^f$ and consider $P = \langle a, b \rangle \leq \Gamma(q^2)$ such that P acts on V via $a\colon v \mapsto \alpha v$, $b\colon v \mapsto v^q$. But now for each $v_0 \in V^\#$, there is some j such that $v_0^{q-1} = \alpha^{-j}$. This however means that $v_0 b a^j = \alpha^j v_0^q = v_0$, and P has no regular orbit on V.

In the exceptional cases of Theorem 4.4 we show that there at least exists an orbit of size greater or equal to $\sqrt{|P|}$.

4.6 Lemma. *Let P be a p-group which acts faithfully on a vector space V over $GF(q)$ for a prime q (not necessarily different from p). Suppose that $V = V_1 \oplus \cdots \oplus V_m$ for subspaces $V_i \neq 0$ that are permuted by P (possibly intransitively). Assume that for each i, there exist u_i, $v_i \in V_i$ such that $\mathbf{C}_{\mathbf{N}(V_i)}(u_i) \cap \mathbf{C}_{\mathbf{N}(V_i)}(v_i) = \mathbf{C}_{\mathbf{N}(V_i)}(V_i)$. If $p > 2$ assume in addition that u_i and v_i are not conjugate in $\mathbf{N}_P(V_i)$. Then there exist u, $v \in V$ such that $\mathbf{C}_P(u) \cap \mathbf{C}_P(v) = 1$. If $p > 2$, then u and v may be chosen so as not to be P-conjugate.*

Proof. We proceed by induction on $\dim_{GF(q)}(V)$ and assume $m > 1$. If P has more than one orbit on $\{V_1, \ldots, V_m\}$, the argument is similar to the one in the proof of Lemma 4.3 and we omit details.

Thus there exists $\mathbf{N}_P(V_1) \leq N \trianglelefteq P$ with $|P/N| = p$ such that $V = W_1 \oplus \cdots \oplus W_p$ for N-invariant subspaces W_i that are transitively permuted by P/N. Also there exist $x_i, y_i \in W_i$ $(i = 1, \ldots, p)$ such that $\mathbf{C}_N(x_i) \cap \mathbf{C}_N(y_i) = \mathbf{C}_N(W_i)$. For $p > 2$, we may assume that x_i and y_i are not conjugate in N. For $g \in P$, we have that $x_i^g, y_i^g \in W_i^g$ and hence that $\mathbf{C}_N(x_i^g) \cap \mathbf{C}_N(y_i^g) = \mathbf{C}_N(W_i^g)$. Since P transitively permutes the W_i, we may assume that $x_1, x_2, \ldots, x_p$ are all P-conjugate and $y_1, y_2, \ldots, y_p$ are all P-conjugate.

Consider first the case $p > 2$. If x_i is P-conjugate to some y_j, then also $x_i^g = y_i$ for some $g \in P$. This however implies $g \in \mathbf{N}_P(W_i) = N$, a contradiction to the choice of x_i and y_i. Hence x_i is never P-conjugate to any y_j. Let $x = (x_1, y_2, \ldots, y_p)$, $y = (y_1, x_2, \ldots, x_p) \in V$. Then $\mathbf{C}_P(x) \leq N$ and $\mathbf{C}_P(y) \leq N$. Hence

$$\mathbf{C}_P(x) \cap \mathbf{C}_P(y) = \mathbf{C}_N(x) \cap \mathbf{C}_N(y)$$

$$= \bigcap_{i=1}^{p} (\mathbf{C}_N(x_i) \cap \mathbf{C}_N(y_i)) = \bigcap_{i=1}^{p} \mathbf{C}_N(W_i) = 1.$$

If x and y are P-conjugate, say $x^h = y$, then $y_2^h, y_3^h \in \{y_1, x_2, \ldots, x_p\}$ and $y_2^h = y_1 = y_3^h$, a contradiction. We are done if $p > 2$.

Let $p = 2$. Now

$$\mathbf{C}_N((x_1, x_2)) \cap \mathbf{C}_N((y_1, y_2)) = 1 = \mathbf{C}_N((x_1, y_2)) \cap \mathbf{C}_N((y_1, x_2)).$$

We may thus assume that

$$\mathbf{C}_P((x_1, x_2)) \cap \mathbf{C}_P((y_1, y_2)) = \langle a \rangle \quad \text{and} \quad \mathbf{C}_P((x_1, y_2)) \cap \mathbf{C}_P((y_1, x_2)) = \langle b \rangle$$

for involutions $a, b \in P \setminus N$. Now $x_1^a = x_2$, $y_1^a = y_2$, $x_1^b = y_2$ and $y_1^b = x_2$. Set $s = ab \in N$. Then s^2 fixes all x_i and y_j and $s^2 = 1$. Since $x_1^s = y_1$, x_1 and y_1 are conjugate by an involution $s \in N$. If x_1 and y_1 are linearly dependent, say $y_1 \in \langle x_1 \rangle$, then also $y_2 \in \langle x_2 \rangle$ by conjugation. In this case, x_i is in a regular orbit of $N/\mathbf{C}_N(W_i)$, $i = 1, 2$. Thus $\mathbf{C}_N((x_1, x_2)) = 1$, and without loss of generality $\mathbf{C}_P((x_1, x_2)) = \langle c \rangle$ for an involution $c \in P$. Choose $v \in V$

not centralized by c and note that $\mathbf{C}_P(v) \cap \mathbf{C}_P((x_1, x_2)) = 1$. Thus the assertion holds and we may assume that x_1 and y_1 are linearly independent. Now $\{x_1, y_1\}$ and $\{x_1 - y_1, y_1\}$ generate the same 2-dimensional subspace of W_1. Hence

$$\mathbf{C}_N(x_1 - y_1) \cap \mathbf{C}_N(y_1) = \mathbf{C}_N(x_1) \cap \mathbf{C}_N(y_1) = \mathbf{C}_N(W_1).$$

Replacing x_1 by $x_1 - y_1$ in the above argument, there exists an involution $t \in N$ conjugating $x_1 - y_1$ and y_1. Now $x_1^{st} = y_1^t = x_1 - y_1$ and $y_1^{st} = x_1^t = ((x_1 - y_1) + y_1)^t = y_1 + (x_1 - y_1) = x_1$. A matrix for st restricted to the subspace generated by $\{x_1, y_1\}$ thus is $\begin{pmatrix} 1 & 1 \\ -1 & 0 \end{pmatrix}$, which has order 6. This is a contradiction, because P is a 2-group. $\qquad\qquad\square$

4.7 Theorem. *Suppose that the p-group P acts irreducibly and faithfully on the $GF(q)$-vector space V of coprime characteristic q. Then there always are two vectors v_1, $v_2 \in V$ such that $\mathbf{C}_P(v_1) \cap \mathbf{C}_P(v_2) = 1$. In particular, $|P : \mathbf{C}_P(v_i)| \geq \sqrt{|P|}$ for $i = 1$ or 2.*

Proof. Assume at first that $p > 2$. We show that we can even find such vectors v_1, v_2 in different orbits. If P is primitive, then P is cyclic and thus has a regular orbit $\{v_1^g \mid g \in P\}$. Take v_1 and $v_2 = 0$. If P is imprimitive, apply induction and Lemma 4.6.

When $p = 2$, the primitive case follows from Lemma 4.2(c) and the imprimitive case follows via induction and Lemma 4.6. $\qquad\qquad\square$

We give another criterion for the existence of regular orbits. It should be clear that the hypothesis of the following theorem is rather difficult to check explicitly. But since $Z_p \mathrm{wr}\, Z_p$ has class p, it is certainly satisfied if P has class less than p.

4.8 Theorem. *Suppose that the p-group P acts irreducibly and faithfully on a $GF(q)$-vector space V of coprime characteristic q. If P does not involve* a section isomorphic to $Z_p \mathrm{wr}\, Z_p$, then P has a regular orbit in its action on V.

Proof. By Lemma 4.2 (a), we may assume that P is imprimitive. By Corollary 0.3, there exists $N \trianglelefteq P$ with $|P \colon N| = p$ such that $V_N = V_1 \oplus \cdots \oplus V_p$ for irreducible N-modules V_i which are transitively permuted by P. Since $N/\mathbf{C}_N(V_1)$ involves no section isomorphic to $Z_p \mathrm{wr}\, Z_p$, the inductive hypothesis yields the existence of a vector $u \in V_1^{\#}$ such that $\mathbf{C}_N(u) = \mathbf{C}_N(V_1)$. Let $g \in P \setminus N$, hence $g^p \in N$. Without loss of generality, we may assume that g cyclically permutes the spaces $V_1, \ldots, V_n$. We now consider the vector $v = (u, u^g, \ldots, u^{g^{p-1}}) \in V$. Since

$$\mathbf{C}_N(v) = \bigcap_{j=0}^{p-1} \mathbf{C}_N(u)^{g^j} = \bigcap_{j=0}^{p-1} \mathbf{C}_N(V_j) = \mathbf{C}_N(V) = 1,$$

it remains to assume that $|\mathbf{C}_P(v)| = p$. Replacing g by a generator of $\mathbf{C}_P(v)$, we may also assume that $g^p = 1$. Suppose now that even $\bigcap_{j=1}^{p-1} \mathbf{C}_N(V_j) = 1$ and let $w = (0, u^g, \ldots, u^{g^{p-1}}) \in V$. Then

$$\mathbf{C}_P(w) = \mathbf{C}_N(w) = \bigcap_{j=1}^{p-1} \mathbf{C}_N(u)^{g^j} = \bigcap_{j=1}^{p-1} \mathbf{C}_N(V_j) = 1.$$

Thus w generates a regular P-orbit and we may take $1 \neq y \in \bigcap_{j=1}^{p-1} \mathbf{C}_N(V_j)$ with $y^p = 1$. Let A be a matrix representation of the action of y on V_1, and let E denote the identity matrix of rank $\dim(V_1)$. Then y induces the map $(A, E, \ldots, E)$ on $V_N = V_1 \oplus \cdots \oplus V_p$. Consequently, $Z_p \mathrm{wr}\, Z_p \cong \langle y, g \rangle \leq P$ and the proof is complete. $\qquad\square$

It is not hard to extend Theorems 4.4 and 4.8 to nilpotent operator groups. For this and also other related results we refer the reader to T. Berger [Be 1], P. Fleischmann [Fl 1], R. Gow [Go 1], D. Passman [Pa 1], and B. Huppert & O. Manz [HM 2]. The step towards supersolvable groups however is much more delicate.

4.9 Remark. Let $\Gamma(p^m)$ be a semi-linear group as defined in §2 . For $q \mid m$, let S be the unique subgroup of $\mathrm{Gal}(Gf(p^m))$ of order q. In the group $\Gamma_0(p^m)$ of multiplications, we define the q-*norm*-1 *subgroup* $N(p^m, q)$ by

$$N(p^m, q) = \{x \in \Gamma_0(p^m) \mid \prod_{\sigma \in S} x^\sigma = 1\}.$$

Set $G(p^m, q) = N(p^m, q) \cdot S$. Then the following theorem holds (A. Turull [Tu 1]). Let G be a supersolvable group which acts faithfully and completely reducibly on a $GF(p)$-vector space W. Suppose that G involves no section isomorphic to

(1) $Z_r \mathrm{wr}\, Z_s$ for primes r, s, or

(2) $N(p^{q^e}, q)$ for a positive integer e and a prime q.

Then G has a regular orbit on W.

The main difficulties in proving regular orbit theorems arise when the action is imprimitive. Quasi-primitive solvable groups however are easier to handle, as we shall see now.

4.10 Proposition. *Suppose that the solvable group G acts faithfully and quasi-primitively on a finite vector space V. Therefore, every normal abelian subgroup of G is cyclic and we adopt the notations of Corollary 1.10. If $e > 118$, then G has at least two regular orbits on V.*

Proof. (a) We first show that $|\mathbf{C}_V(g)| \leq |V|^{3/4}$ for all $g \in G^\#$, and freely use the assertions of Corollary 1.10.

(1) If $g \in U^\#$, then $\mathbf{C}_V(g) = 1$, since the cyclic group U acts fixed-point-freely on V.

(2) If $g \in T \setminus U$, then $[g, u] \in U^\#$ for some $u \in U$. Observe that $|\mathbf{C}_V(g)| = |\mathbf{C}_V(g^{-1})| = |\mathbf{C}_V(g^u)|$ and $\mathbf{C}_V(g^{-1}) \cap \mathbf{C}_V(g^u) \leq \mathbf{C}_V([g, u])$. Now $\mathbf{C}_V([g, u]) = 1$ (by (1)) and therefore $|V| \geq |\mathbf{C}_V(g^{-1}) \cdot \mathbf{C}_V(g^u)| = |\mathbf{C}_V(g)|^2$, which yields the claim.

(3) If $g \in F \setminus T$, then there exists $x \in E$ such that $[g, x] \in Z^\#$. By (1), $\mathbf{C}_V([g, x]) = 1$ and the same argument as in (2) yields $|\mathbf{C}_V(g)| \leq |V|^{1/2}$.

(4) Let $g \in A \setminus F$. Then there exists $x \in E$ such that $[g, x] \in E \setminus Z$. By (3) we know that $|\mathbf{C}_V([g, x])| \leq |V|^{1/2}$. Consequently

$$|\mathbf{C}_V(g)|^2 = |\mathbf{C}_V(g^{-1})| \cdot |\mathbf{C}_V(g^x)|$$

$$\leq |\mathbf{C}_V(g^{-1}) \cdot \mathbf{C}_V(g^x)| \cdot |\mathbf{C}_V([g, x])| \leq |V| \cdot |V|^{1/2} = |V|^{3/2}.$$

(5) Finally if $g \in G \setminus A$, there exists $z \in Z$ such that $[g, z] \in Z^{\#}$. By (1), $\mathbf{C}_V([g, z]) = 1$ and $|\mathbf{C}_V(g)| \leq |V|^{1/2}$.

(b) Any $v \in V$ not contained in $\bigcup_{g \in G^\#} \mathbf{C}_V(g)$ must necessarily lie in a regular G-orbit. If G does not have two regular orbits, then (a) implies

$$(|G| - 1) \cdot |V|^{3/4} \geq \sum_{g \in G^\#} |\mathbf{C}_V(g)| \geq |V| - |G|.$$

Let W be an irreducible submodule of V_U. By Corollary 2.6, $|V| \geq |W|^e$ where e is as in Corollary 1.10. We thus obtain

$$|G| \geq (|V| + |V|^{3/4})/(|V|^{3/4} + 1) \geq |V|^{1/4} \geq |W|^{e/4}.$$

But by Corollary 3.7, $|G| \leq e^{13/2}|U|^2/2 < e^{13/2}|W|^2/2$. It follows that $e^{13/2} > 2 \cdot |W|^{e/4}/|W|^2 = 2 \cdot |W|^{e/4-2}$, and then $e^{26} > 2^4 \cdot |W|^{e-8} \geq 2^4 \cdot 3^{e-8}$. Therefore, $e \leq 118$. □

With some more care, the existence of regular orbits can certainly be established also for smaller values of e. A. Espuelas [Es 3] has shown that whenever $|G|$ is odd and $e > 1$, then there exist two regular orbits on the quasi-primitive module V of odd characteristic.

4.11 Remark. Suppose that G acts faithfully and coprimely on a solvable group H. Since $\mathbf{F}(HG) = \mathbf{F}(H)$, G also acts faithfully on $\mathbf{F}(H)$ and then even on $\mathbf{F}(H)/\Phi(\mathbf{F}(H)) =: V$. We decompose $V = V_1 \oplus \cdots \oplus V_n$ into irreducible G-modules (possibly of different characteristic). Suppose $G/\mathbf{C}_G(V_i)$ all have regular orbits on V_i, say generated by $v_i \in V_i$. Then $v = v_1 + \cdots + v_n$ generates a regular orbit of G on V. Thus there is $h \in H$ such that $\mathbf{C}_G(h) = 1$, i.e. h generates a regular G-orbit on H.

§5   Solvable Permutation Groups and the Existence of
Regular Orbits on the Power Set

When investigating a permutation group G on a finite set Ω, one can as well consider the induced action of G on the power set $\mathfrak{P}(\Omega)$ of Ω. The question we are concerned with in this section is whether G has a regular orbit on $\mathfrak{P}(\Omega)$, i.e. whether there exists a subset $\Delta \subseteq \Omega$ such that the setwise stabilizer $\mathrm{stab}_G(\Delta)$ of Δ is trivial. As the examples of the symmetric and alternating groups show, one cannot expect this to be true in general. For primitive solvable permutation groups, however D. Gluck [Gl 1] has given a complete answer.

In the following we use the structure of primitive solvable permutation groups stated in comments following Theorem 2.1. Let S denote a point stabilizer and V the unique minimal normal subgroup of G. Thus $|\Omega| = |V| = p^m$ for a prime p and $m \in \mathbb{N}$. To obtain certain consequences, in fact a slightly stronger question will be considered, namely whether there is $\Delta \subseteq \Omega$, $|\Delta| \neq |\Omega|/2$, such that $\mathrm{stab}_G(\Delta) = 1$. We call such an orbit a *strongly regular orbit*. Clearly for $p > 2$, each regular orbit is strongly regular. We denote by $n(g)$ the number of cycles of $g \in G$ on Ω, and by $s(g)$ the number of fixed points.

For 5.1 to 5.5, we assume that G is a solvable primitive permutation group on Ω.

5.1 Lemma. If $g \in G^{\#}$, then $n(g) \leq (|\Omega| + s(g))/2 \leq (p+1)|\Omega|/(2p) \leq 3|\Omega|/4$.

Proof. If $s(g) = 0$, we clearly have $n(g) \leq |\Omega|/2$. We thus may assume that g has fixed points, and without loss of generality $g \in S$. Since the actions of S on V and Ω are permutation isomorphic, it follows that $s(g) = |\mathbf{C}_V(g)|$, and since S acts faithfully on V, $s(g) \mid |V|/p = |\Omega|/p$. Therefore

$$n(g) \leq s(g) + (|\Omega| - s(g))/2 = (|\Omega| + s(g))/2 \leq (p+1)|\Omega|/(2p) \leq 3|\Omega|/4. \qquad \square$$

For a subset $X \subseteq G$, it is worthwhile to consider in the following the set

$$\mathfrak{T}(X) = \{(g, \Delta) \mid g \in X, \ \Delta \subseteq \Omega, \ g \in \text{stab}_G(\Delta)\}.$$

By an easy counting argument, which in turn relies on bounds for the order of linear solvable groups, we are left with only finitely many cases.

5.2 Proposition. If $|\Omega| \geq 81$, then G has a strongly regular orbit on the power set $\mathfrak{P}(\Omega)$.

Proof. Note that $g \in G$ stabilizes exactly $2^{n(g)}$ subsets of Ω. Consequently, Corollary 3.6 and Lemma 5.1 imply

$$|\mathfrak{T}(G^{\#})| \leq (1/2) \cdot |\Omega|^{13/4} \cdot 2^{3|\Omega|/4}.$$

Since $|\mathfrak{P}(\Omega)| = 2^{|\Omega|}$, we certainly find a regular orbit of G on $\mathfrak{P}(\Omega)$, provided that

$$(1/2) \cdot |\Omega|^{13/4} \cdot 2^{3|\Omega|/4} < 2^{|\Omega|},$$

or equivalently

$$13 \cdot \log_2(|\Omega|) < |\Omega| + 4.$$

One easily checks that this holds for $|\Omega| \geq 81$.

In order to prove the existence of a strongly regular orbit, we may therefore assume that $|\Omega|$ is a 2-power greater than or equal to $128 = 2^7$. First observe that $\binom{m}{m/2} \leq \binom{m}{m/2+1} + \binom{m}{m/2-1}$ for an even number $m \in \mathbb{N}$. Thus the number of subsets of Ω of cardinality different from $|\Omega|/2$, which equals $2^{|\Omega|} - \binom{|\Omega|}{|\Omega|/2}$, is greater than or equal to $2^{|\Omega|-1}$. Hence we get as our condition for the existence of a strongly regular orbit

$$(1/2) \cdot |\Omega|^{13/4} \cdot 2^{3|\Omega|/4} < 2^{|\Omega|-1},$$

or equivalently $13 \cdot \log_2(|\Omega|) < |\Omega|$. This holds for $|\Omega| \geq 128$. $\square$

Rather easy to handle is also the case where $|\Omega| = p$ is a prime.

5.3 Lemma. *Suppose that* $|\Omega| = p$ *is a prime number. Then* G *has no regular orbit on* $\mathfrak{P}(\Omega)$ *if and only if* $p = 3,\ 5,\ 5$ *or* 7 *and* G *is isomorphic to the Frobenius group* $F_6,\ F_{10},\ F_{20}$ *or* F_{42} *(respectively).*

Proof. If $G = V$ is cyclic of order p, then any one-point subset of Ω generates a regular orbit of G on $\mathfrak{P}(\Omega)$. Therefore we may assume that G is a Frobenius group with kernel V and complement $S \neq 1$; in particular $|S| \mid p - 1$ and $p \geq 3$. Recall the permutation isomorphism between Ω and V (where S acts on V by conjugation).

Let T be a subgroup of G of prime order q which has a fixed point on $\mathfrak{P}(\Omega)$. Then T is contained in some conjugate of S, and T fixes exactly $2^{1+(p-1)/q}$ subsets $\Delta \subseteq \Omega$. Since for a fixed prime q, G contains exactly p such subgroups T, we obtain

$$|\{\Delta \subseteq \Omega|\ \mathrm{stab}_G(\Delta) \neq 1\}| \leq |\{(T, \Delta)|\Delta \subseteq \Omega, |T|\ \text{a prime},\ T \leq \mathrm{stab}_G(\Delta)\}|$$

$$\leq p \sum_q 2^{1+(p-1)/q} =: f(p),$$

where q runs through all prime divisors of $p - 1$. Now observe that the number of prime divisors of $p - 1$ is bounded by $\log_2 p$. We thus obtain that

$$f(p) \leq p \cdot \log_2 p \cdot 2^{(p+1)/2} < 2^p,$$

provided that $p \geq 13$. Also $f(11) = 11(64 + 8) < 2^{11}$. This counting argument tells us that regular orbits on $\mathfrak{P}(\Omega)$ can only fail to exist in the case $p \leq 7$.

We start by considering the case $p = 7$. If $|G| = 42$, then G cannot have a regular orbit on $\mathfrak{P}(\Omega)$, because $\binom{7}{i} \leq \binom{7}{3} = 35 < |G|$ for all $i = 0, \ldots, 7$. If however $|G| = 21$, then every $\Delta \subseteq \Omega$ such that $|\Delta| = 2$ indeed satisfies $\mathrm{stab}_G(\Delta) = 1$. Let finally $|G| = 14$. Then each of the seven involutions of G stabilizes exactly three subsets $\Delta \subseteq \Omega$ of cardinality $|\Delta| = 3$. Thus the remaining $\binom{7}{3} - 21 = 14$ such subsets form a regular orbit under the action of G. We next consider $p = 5$. If $|G| = 10$, then each of the five involutions of G stabilizes exactly two subsets $\Delta \subseteq \Omega$ with $|\Delta| = 2$. This immediately implies that every subset $\Delta \subseteq \Omega$ has a non-trivial stabilizer in G. Consquently, also the Frobenius group of order 20 has no regular orbit on $\mathfrak{P}(\Omega)$. Since $p = 3$ forces $G \cong S_3$, the proof of the lemma is complete. $\square$

Unfortunately, the proper prime powers less than 81 require a very detailed step-by-step analysis.

5.4 Lemma. If $|\Omega| = 2^6$, 5^2 or 7^2, then G has a strongly regular orbit on $\mathfrak{P}(\Omega)$.

Proof. (1) We first consider $|\Omega| = 2^6$, hence $S \leq GL(6, 2)$. Let $g \in G^{\#}$. If $s(g) = 0$ or $s(g) = |\mathbf{C}_V(g)| \mid 2^4$, then Lemma 5.1 yields $n(g) \leq (|\Omega| + s(g))/2 \leq 40$. If $s(g) = 2^5$, then g centralizes a hyperplane of V. Observe that the centralizer in $GL(6, 2)$ of a hyperplane is elementary abelian of order 32. Thus $n(g) = 32 + 32/2 = 48$. We set $G_0 = \{g \in G \mid n(g) = 48\}$ and $G_1 = G^{\#} \backslash G_0$. As V has exactly 63 hyperplanes, it follows that $|G_0| \leq 63 \cdot 31$. By Corollary 3.6, $|G| \leq (2^6)^{13/4}/2$. Recalling the definition of the sets $\mathfrak{T}(X)$, we obtain $|\mathfrak{T}(G^{\#})| \leq |\mathfrak{T}(G_0)| + |\mathfrak{T}(G_1)| \leq 2^{48} \cdot 63 \cdot 31 + 2^{40} \cdot (2^6)^{13/4}/2 < 2^{59} + 2^{59} = 2^{60}$. Now $|\{\Delta \subseteq \Omega \mid |\Delta| \neq 32\}| \geq \binom{64}{33} \geq 2^{60} > |\mathfrak{T}(G^{\#})|$, and G has a strongly regular orbit on $\mathfrak{P}(\Omega)$.

(2) Let now $|\Omega| = 7^2$. Since $|\Omega|$ is odd, we only have to establish the existence of a regular orbit. To do so note that Lemma 5.1 yields $n(g) \leq 28$ for all $g \in G^{\#}$. Since S acts irreducibly on V, Theorem 2.11 implies $|S| \leq 144$. Therefore

$$|\mathfrak{T}(G^{\#})| \leq 2^{28} \cdot |G| \leq 2^{28} \cdot (144 \cdot 49) < 2^{42} < 2^{49} = |\mathfrak{P}(\Omega)|$$

and we have settled case (2).

(3) Suppose finally $|\Omega| = 5^2$. We first note that S is isomorphic to a subgroup of $GL(2, 5)$ of order dividing 32, 48 or 96 (cf. Theorem 2.11).

We define $G_r = \{g \in G \mid o(g) = r\}$ and $G_2^i = \{g \in G_2 \mid s(g) = i\}$ for $r = 2, 3, 5$ and $i = 1, 5$. Then each $\Delta \subseteq \Omega$ is stabilized by an element in $X := G_2^1 \cup G_2^5 \cup G_3 \cup G_5$.

Observe the following values of $n(g)$:

$g \in$	$n(g)$
G_2^1	13
G_2^5	15
G_3	9
G_5	5 (because $5 \nmid \lvert S \rvert$).

If $g \in G_2^1$, then g is a central involution even in $GL(2,5)$ and hence each conjugate of S contains at most one such g. Thus $\lvert G_2^1 \rvert \le 25$. Furthermore, $g \in G_2^5$ is contained in five conjugates of S, and $G_5 = V^{\#}$. Altogether we obtain

$$\lvert \{\Delta \subseteq \Omega \mid \operatorname{stab}_G(\Delta) \ne 1\} \rvert \le \lvert \mathfrak{T}(X) \rvert$$
$$\le 2^{13} \lvert G_2^1 \rvert + 2^{15} \lvert G_2^5 \rvert + 2^9 \lvert G_3 \rvert + 2^5 \lvert G_5 \rvert$$
$$\le 2^{13} \cdot 25 + 2^{15} \cdot 5 \cdot 96 + 2^9 \cdot 25 \cdot 96 + 2^5 \cdot 24 < 2^{25} = \lvert \mathfrak{P}(\Omega) \rvert,$$

which completes the proof. $\qquad\qquad\qquad\qquad\qquad\qquad\qquad\qquad\qquad$ $\square$

Whereas in the previous lemma we only needed bounds for the order of S, the proof of the next lemma relies on the actual structure of S.

5.5 Lemma. *If $\lvert \Omega \rvert = 2^4$, 3^3 or 2^5, then G has a strongly regular orbit on $\mathfrak{P}(\Omega)$.*

Proof. (1) Let $\lvert \Omega \rvert = 2^5$. Then Corollary 2.13(a) implies that S is a subgroup of $\Gamma(2^5)$. In order to guarantee a regular orbit, we may assume that $S = \Gamma(2^5)$ and $G = A\Gamma(2^5)$. Recall the permutation actions of S on Ω and V are isomorphic. If $\Delta \subseteq \Omega$ with $\lvert \Delta \rvert = 3$, then Δ is stabilized by some element $g \in G$ of prime order. Since $\lvert G \rvert = 2^5 \cdot 31 \cdot 5$, indeed g must centralize some $\delta \in \Delta$ and consequently g is conjugate to an element of S. If $x \in S^{\#}$, then $o(x)$ is 31 or 5 and $\lvert \mathbf{C}_V(x) \rvert$ is 1 or 2 (respectively). Since the actions of S on V and Ω are permutation isomorphic, the orbit pattern for $\langle x \rangle$ on Ω is $(1, 31)$ or $(1, 1, 5, 5, 5, 5, 5, 5)$, respectively. Thus Δ is stabilized by no element of S or G.

(2) Assume now that $|\Omega| = 2^4$. By Corollary 2.15 it is now sufficient to consider the cases $S = \Gamma(2^4)$ and $S = \Gamma(2^2) \mathrm{wr}\, Z_2$.

We first assume that $S = \Gamma(2^4)$. Let $\Omega_7 = \{\Delta \subseteq \Omega \mid |\Delta| = 7\}$. We may assume that each $\Delta \in \Omega_7$ is fixed by some element $g \in G$ of prime order. Since $o(g) \nmid |\Delta|$, g centralizes some $\delta \in \Omega_7$ and so g is conjugate to some element of S. If $x \in S$ has order 3 or 5, then $\mathbf{C}_V(x) = \{0\}$ and hence $\langle x \rangle$ has an orbit pattern $(1, 3, 3, 3, 3, 3)$ or $(1, 5, 5, 5)$ on Ω. Thus x stabilizes 10 elements of Δ_7 if $o(x) = 3$ and none if $o(x) = 5$. An element $y \in S$ of order 2 satisfies $|\mathbf{C}_V(y)| = 4$ and y stabilizes $\binom{6}{3} 4 + \binom{6}{2} 4 = 140$ elements of Ω_7. Furthermore such a y belongs to 4 conjugates of S as $|\mathbf{C}_V(g)| = 4$. Since S has exactly one subgroup of order 3 and five of order 2 (all conjugate in S) and since S has 16 conjugates in G, $\binom{16}{7} = |\Omega_7| \leq 16 \cdot 10 + (16 \cdot 5 \cdot 140 / 4) = 2960$, a contradiction.

We now consider the case where $S = \Gamma(2^2) \mathrm{wr}\, Z_2$ as a linear group on V. Then $V = V_1 \oplus V_2$, where $V_i \cong A(2^2)$ and Z_2 permutes the V_i. Since $A\Gamma(2^2) \cong S_4$, we conclude that $G = A\Gamma(2^2) \mathrm{wr}\, Z_2 \cong S_4 \mathrm{wr}\, Z_2$. Again Ω and V are permutation isomorphic and we may thus identify Ω with $\{(i, j) \mid i, j = 1, \ldots, 4\}$. Let $\Delta = \{(1, 1), (1, 2), (1, 3), (2, 2), (2, 4), (3, 1)\}$ and choose $g \in \mathrm{stab}_G(\Delta)$. Then $g = (g_1, g_2) \in S_4 \times S_4$, since three different entries appear as first coordinates in Δ, but four different ones as second coordinates. Now the entry i ($i = 1, \ldots, 4$) occurs exactly $4 - i$ times as first coordinate in Δ, which clearly implies that $g_1 = 1$. Similarly, g_2 has to fix the sets $\{1, 2\}$ and $\{3, 4\}$. But as $g_1 = 1$, we also see that $g_2 = 1$, and Δ generates a strongly regular orbit on $\mathfrak{P}(\Omega)$, because $|\Delta| \neq |\Omega|/2$.

(3) In the case $|\Omega| = 3^3$, we may proceed similarly to case (2). By Corollary 2.13(b) namely we have to investigate the possibilities $S = \Gamma(3^3)$ and $S = \Gamma(3)$ wr S_3.

Suppose that $S = \Gamma(3^3)$. We let $\Omega_{11} = \{\Delta \subseteq \Omega \mid |\Delta| = 11\}$. Each element $\Delta \in \Omega$ is stabilized by some element $g \in G$ of prime order. Since $o(g) \neq 11$, g centralizes some $\delta \in \Delta$ and so g is conjugate to some element of S. If $x \in S$ has order 2 or 13, then $\mathbf{C}_V(x) = \{0\}$ and x has exactly one trivial orbit on Ω. Hence x stabilizes $\binom{13}{5}$ elements of Ω_{11} if $o(x) = 2$ and none if $o(x) = 13$. If $y \in S$ with $o(g) = 3$, then $|\mathbf{C}_V(y)| = 3$. So y has 3 fixed points in Ω, y stabilizes $\binom{8}{3} \cdot 3$ elements of Ω_{11}, and y belongs to 3 conjugates of S. Since S has one subgroup of order 2 and 13 subgroups of order 3 and since S has 27 conjugates in G, $\binom{27}{11} = |\Omega_{11}| \leq 27 \cdot \binom{13}{5} + 27 \cdot 13 \cdot \binom{8}{3} = 27 \cdot 2015$, a contradiction. Hence S has a strongly regular orbit on $\mathfrak{P}(\Omega)$.

Suppose finally that $S = \Gamma(3)$ wr $S_3 \cong Z_2$ wr S_3. Then we have $G = S_3$ wr S_3 and Ω can be identified with $\{(i, j, k) \mid i, j, k = 1, 2, 3\}$, where the base group $S_3 \times S_3 \times S_3$ acts componentwise and the S_3 outside permutes the coordinates. Let $\Delta = \{(1,1,3), (1,2,1), (1,2,3), (2,2,2), (2,2,3)\} \subseteq \Omega$ and let $g \in \text{stab}_G(\Delta)$. Comparing the occurrence pattern of the entries in the distinct components, it easily follows that $g = (g_1, g_2, g_3) \in S_3 \times S_3 \times S_3$, and then that $g_i = 1$ ($i = 1, 2, 3$). This completes the proof of the lemma. $\qquad\square$

5.6 Theorem (Gluck). *Let G be a primitive solvable permutation group on a finite set Ω with point stabilizer S. Then G has a regular orbit on $\mathfrak{P}(\Omega)$, unless one of the following cases occurs:*

(1) $|\Omega| = 3$ and $G \cong S_3$;

(2) $|\Omega| = 4$ and $G \cong A_4$ or S_4;

(3) $|\Omega| = 5$ and $G \cong F_{10}$ or F_{20};

(4) $|\Omega| = 7$ and $G \cong F_{42}$;

(5) $|\Omega| = 8$ and $G \cong A\Gamma(2^3)$;

(6) $|\Omega| = 9$ and G is the semi-direct product of $Z_3 \times Z_3$ with D_8, SD_{16}, $SL(2,3)$ or $GL(2,3)$.

In the exceptional cases, no regular orbit exists. Also if (Ω, G) is non-exceptional and if $|\Omega| \neq 2$, then there even exists a strongly regular orbit on $\mathfrak{P}(\Omega)$.

Proof. By Proposition 5.2 and Lemmas 5.4 and 5.5, G has a strongly regular orbit on $\mathfrak{P}(\Omega)$ provided that $|\Omega| \geq 81$ or $|\Omega| = 2^4$, 2^5, 2^6, 3^3, 5^2 or 7^2. If $|\Omega| = 2$, then $G = Z_2$ and G has a regular, but no strongly regular orbit. If $|\Omega|$ is an odd prime, then a regular orbit automatically is strongly regular, and Lemma 5.3 tells us that the exceptions are precisely given by (1), (3) and (4) above. Let next $|\Omega| = 2^2$. Then $G \cong A_4$ or S_4 and in both cases no regular orbit on $\mathfrak{P}(\Omega)$ exists. Thus we still have to discuss the cases $|\Omega| = 2^3$ and 3^2.

If $|\Omega| = 2^3$, Corollary 2.13(a) yields $S \leq \Gamma(2^3)$. If equality holds, then every subset $\Delta \subseteq \Omega$ has a non-trivial stabilizer, because $\binom{8}{i} < 168 = |A\Gamma(2^3)| = |G|$ for $i = 0, 1, \ldots, 8$. Otherwise we have $|S| = 7$, since S acts irreducibly on the minimal normal subgroup V of G. As V is an elementary abelian 2-group, every subset $X \subseteq V$ of cardinality three has a trivial stabilizer in G and thus generates a strongly regular orbit.

Let finally $|\Omega| = 3^2$. Then S is an irreducible subgroup of $GL(2,3)$. By Theorem 2.11, S is a subgroup of D_8, or a subgroup of SD_{16}, or is isomorphic to Q_8, $SL(2,3)$ or $GL(2,3)$. If $S = GL(2,3)$, $SL(2,3)$ or SD_{16}, then $\binom{9}{i} < |G|$ for all i and no regular orbit exists. Since S acts irreducibly, the remaining possibilities for S are D_8, Q_8, Z_8 and Z_4. Let first $S = D_8$, and consider $SD_{16} \cong T \in \mathrm{Syl}_2(GL(2,3))$ with $S \leq T$. Note that both S and T have 5 involutions. By the Sylow Theorems, it follows that every element of prime order in VT is in VS. As each $X \subseteq V$ has a non-trivial stabilizer in VT, it follows that $G = VS$ as well has no regular orbit on $\mathfrak{P}(\Omega)$. In each of the cases $S = Q_8$, Z_8 and Z_4, the group S acts fixed-point-freely on V and contains exactly one involution. Consequently $G = VS$ contains exactly 9 involutions and their orbit pattern on V is $(1, 2, 2, 2, 2)$. The orbit pattern of the 8 elements of G of order 3 is $(3, 3, 3)$. Thus $\mathrm{stab}_G(\Delta)$ is a 2-group whenever $|\Delta| = 4$. Each involution in G stabilizes $\binom{4}{2}$ subsets of order 4 of Ω. Since $\binom{9}{4} > 9 \cdot \binom{4}{2}$, some $X \subseteq \Omega$ with $|X| = 4$ is in a regular orbit of G on $\mathfrak{P}(\Omega)$. $\square$

We draw two Corollaries from Theorem 5.6. The first will be needed in §17, and the second is an essential ingredient of §9.

5.7 Corollary. *Let G be a solvable permutation group on a finite set Ω.*

(a) *There exists a subset $\Delta \subseteq \Omega$ such that* $\mathrm{stab}\,_G(\Delta)$ *is a $\{2, 3\}$-group. Here, Δ can be chosen to have non-empty intersection with every orbit of G on Ω.*

(b) *If $|G|$ is odd, then there exists a regular orbit of G on $\mathfrak{P}(\Omega)$.*

Proof. (a) Let $\Omega_1, \ldots, \Omega_n$ be the orbits of the action of G on $\mathfrak{P}(\Omega)$. If we can find subsets $\Delta_i \subseteq \Omega_i$ $(i = 1, \ldots, n)$ such that $\mathrm{stab}\,_G(\Delta_i)/\mathbf{C}_G(\Omega_i)$ is a $\{2, 3\}$-group, then obviously $\Delta = \Delta_1 \cup \cdots \cup \Delta_n$ satisfies the desired condition. Note that we may assume that $\Delta_i \neq \varnothing$, since otherwise $\Delta_i = \Omega_i$ can be taken. Hence we may assume G to be transitive.

We first suppose that G acts primitively on Ω. We in fact prove the existence of $\Delta \subseteq \Omega$, $|\Delta| \neq |\Omega|/2$, such that $\mathrm{stab}\,_G(\Delta)$ is a $\{2, 3\}$-group. This is certainly clear if G has a strongly regular orbit on $\mathfrak{P}(\Omega)$. Therefore we have to consider the exceptional cases of Theorem 5.6. If $|\Omega| = 2$, 3, 4 or 9, we can take $\Delta = \Omega$. If $|\Omega| = 5$, 7 or 8, any subset Δ with $|\Delta| = 1$, 1 or 2 (respectively) works.

If G is imprimitive, let H denote the point stabilizer of $\alpha \in \Omega$. We fix a subgroup J such that $H < J < G$ and H is maximal in J. If we choose $\{g_1 = 1, g_2, \ldots, g_t\}$ as right coset representatives of J in G, we set $J_i = J^{g_i}$ and let Δ_i be the J_i-orbit of αg_i $(i = 1, \ldots, t)$. It then follows that $\Omega = \Delta_1 \dot{\cup} \ldots \dot{\cup} \Delta_t$, stab $_G(\Delta_i) = J_i$ and G transitively permutes the Δ_i. Let $K = \bigcap_i J_i$. Then G/K faithfully permutes the Δ_i and hence $\{1,\ldots,t\}$. Since $t < |\Omega|$, we may choose by induction $s \geq 1$ such that stab $_{G/K}(\{1,\ldots,s\})$ is a $\{2,3\}$-group. Since J_i primitively permutes Δ_i, the previous paragraph yields the existence of $\Xi_i \subseteq \Delta_i$, $|\Xi_i| \neq |\Delta_i|/2$, such that stab $_{J_i}(\Xi_i)/\mathbf{C}_{J_i}(\Delta_i)$ is a $\{2,3\}$-group. In particular, stab $_K(\Xi_i)/\mathbf{C}_K(\Delta_i)$ is a $\{2,3\}$-group, because $K \leq J_i$. We may clearly assume that $|\Xi_i| > |\Delta_i|/2$ if and only if $i \leq s$. Thus $\Xi := \bigcup_{i=1}^t \Xi_i \subseteq \Omega$ is non-empty. Since $\bigcap_i \mathbf{C}_K(\Delta_i) = 1$,

$$\text{stab }_K(\Xi) \leq \prod_i \text{stab }_K(\Xi_i)/\mathbf{C}_K(\Delta_i)$$

is a $\{2,3\}$-group. By the choice of s, also

$$\text{stab }_G(\Xi)/\text{stab }_K(\Xi) \cong K \cdot \text{stab }_G(\Xi)/K \leq \text{stab }_{G/K}(\{1,\ldots,s\})$$

is a $\{2,3\}$-group, and thus the same holds for stab $_G(\Xi)$.

(b) By Theorem 5.6, every primitive permutation group of odd order has a strongly regular orbit on $\mathfrak{P}(\Omega)$. But then a similar induction argument as in (a) yields assertion (b). □

5.8 Corollary. *Let G be a primitive solvable permutation group on a finite set Ω. Let q be a prime divisor of $|G|$, and assume that for all $\Delta \subseteq \Omega$, stab $_G(\Delta)$ contains a Sylow q-subgroup of G. Then one of the following cases occurs.*

(i) *$|\Omega| = 3$, $q = 2$ and $G \cong D_6$;*

(ii) *$|\Omega| = 5$, $q = 2$ and $G \cong D_{10}$;*

(iii) *$|\Omega| = 8$, $q = 3$ and $G \cong A\Gamma(2^3)$.*

Proof. We first eliminate case (6) of Theorem 5.6. Here we must have $q = 2$ and we fix $Q \in \text{Syl}_q(G)$. Since $\mathbf{C}_V(Q) = 1$, Q has exactly one fixed point on Ω. Since Q must stabilize a set of size j for all $j \leq 9$, the possible orbit sizes of Q are $(1,2,2,2,2)$ and $(1,2,2,4)$. Let us denote by z the number of subsets of Ω of size 4 that are fixed by Q. Observe that

$t \leq 6$. Since $|\mathrm{Syl}_2(G)| \cdot t \geq \binom{9}{4}$, and since $|\mathrm{Syl}_2(G)| \mid 3^3$, it follows that $|\mathrm{Syl}_2(G)| = 27$ and $t \geq 5$. Thus the orbit sizes of Q are $(1, 2, 2, 2, 2)$ and Q is elementary abelian. This is a contradiction, because $GL(2,3)$ does not contain an elementary abelian subgroup of order 8. We next eliminate case (4) of Theorem 5.6. If $q = 3$, the orbit sizes of $Q \in \mathrm{Syl}_3(G)$ are $(1, 3, 3)$, and thus no subset $\Delta \subseteq \Omega$ of cardinality two can be stabilized. If on the other hand $q = 2$, $Q \in \mathrm{Syl}_2(G)$ has orbit sizes $(1, 2, 2, 2)$. Now the number of subsets $\Delta \subseteq \Omega$ such that $|\Delta| = 3$ and such that $Q \subseteq \mathrm{stab}_G(\Delta)$ equals 3. Since $|\mathrm{Syl}_2(G)| = 7$, we obtain $7 \cdot 3 < 35 = \binom{7}{3}$, a contradiction.

The remaining cases of Theorem 5.6 which do not appear among (i)–(iii) can be easily ruled out by considering orbit sizes of Sylow q-subgroups.

□

§6   Solvable Doubly Transitive Permutation Groups

In the 1950s, Huppert classified the solvable doubly transitive permutation groups on a set Ω. Such a group G is certainly primitive ([Hu, II, 1.9]) and hence contains a unique minimal normal subgroup V that acts regularly on Ω. In particular, $|V| = |\Omega| = q^n$ for a prime q. Furthermore, $VG_\alpha = G$, $V \cap G_\alpha = 1$ and G_α acts faithfully on V. Since G is doubly transitive, G_α acts transitively on $V^\#$. Now Huppert's result, which has many uses as we shall see later, states that G may be identified as a subgroup of $A\Gamma(q^n)$ or $q^n = 3^2$, 5^2, 7^2, 11^2, 23^2 or 3^4. Huppert's original proof did not use Zsigmondy's prime theorem, but that was later modified in [HB, chap. XII]. Another approach is given in Passman's book [Pa 2]. We present a different proof which exploits the Zsigmondy prime theorem fully.

6.1 Definition. Let $a > 1$ and n be positive integers. A prime p is called a *Zsigmondy prime divisor* for $a^n - 1$ if $p \mid a^n - 1$ but $p \nmid a^j - 1$ for $1 \leq j < n$. (Note that this is dependent upon a and n and not just on $a^n - 1$.) If p is a Zsigmondy prime divisor for $a^n - 1$, then n is the order of a module p.

Hence $n \mid p - 1$.

6.2 Theorem. *Let $a > 1$ and n be positive integers. Then there exists a Zsigmondy prime divisor for $a^n - 1$ unless*

(i) $n = 2$ and $a = 2^k - 1$ for some $k \in \mathbb{N}$, or

(ii) $n = 6$ and $a = 2$.

Proof. See e.g. [HB, IX, 8.3]. A short, elementary proof is also given by Lüneburg in [Lü 1]. □

6.3 Proposition. *Assume that G is a solvable subgroup of $GL(n, q)$, q a prime power. Suppose that $p \mid |G|$ where p is a Zsigmondy prime divisor of $q^n - 1$. Let $P \in \mathrm{Syl}_p(G)$ and V be the corresponding G-module. Then*

(i) *G acts irreducibly and quasi-primitively on V, and*

(ii) *P is cyclic.*

Proof. We may assume that $n \geq 2$ and $p > 2$. Let $x \in P$ denote an element of order p. Since $p \nmid q^j - 1$ for all $j < n$, $\langle x \rangle$ and hence G act irreducibly on V. If V is not quasi-primitive, we may choose $C \trianglelefteq G$ such that $V_C = V_1 \oplus \cdots \oplus V_m$ for non-zero C-modules V_i permuted faithfully by G/C. Since $m \leq n < p$, P fixes each V_i, whence $P \leq C$ and $x \in C$. Now $\langle x \rangle$ and hence C act irreducibly on V. Thus $m = 1$ and $C = G$. This proves (i).

Applying (i) with $P = G$, we have that P acts irreducibly and quasi-primitively on V. Thus every normal abelian subgroup of P is cyclic. Since $p > 2$, P is cyclic, by Corollary 1.3. □

If G is a transitive subgroup of $GL(n, q)$, then $|G|$ is divisible by all Zsigmondy prime divisors of $q^n - 1$. In all but a few cases, $|\mathbf{F}(G)|$ will be divisible by a Zsigmondy prime divisor of $q^n - 1$, and as Lemma 6.4 shows, this forces solvable G to be a subgroup of $\Gamma(q^n)$. Lemma 6.7 will handle the exceptional cases where $|G|$ is divisible by a Zsigmondy prime divisor of $q^n - 1$, but $|\mathbf{F}(G)|$ is not.

6.4 Lemma. *Assume that $G \leq GL(n, q)$ is solvable (q a prime power). Suppose that $p \mid |\mathbf{F}(G)|$ for a Zsigmondy prime divisor p of $q^n - 1$. Let $P \in \mathrm{Syl}_p(G)$. Then*

(i) $G \leq \Gamma(q^n)$.

(ii) *When $1 \neq P_0 \leq P$, then $\mathbf{F}(G) = \mathbf{C}_G(P_0) \geq P$.*

(iii) *$\mathbf{F}(G)$ and $G/\mathbf{F}(G)$ are cyclic, $|\mathbf{F}(G)| \mid q^n - 1$ and $|G/\mathbf{F}(G)| \mid n$.*

Proof. We may assume that $n > 1$ and $p > 2$, and we choose $P_1 \leq P$ with $|P_1| = p$. The hypotheses of the lemma imply that $P \cap \mathbf{F}(G) \neq 1$, and thus $P_1 \leq \mathbf{F}(G)$, because P is cyclic (Proposition 6.3). Also $P_1 \trianglelefteq G$ and $P_1 \leq \mathbf{Z}(\mathbf{F}(G))$. By Proposition 6.3 and Lemma 2.9, the natural module V of $GL(n, q)$ is an irreducible P_1-module and $\mathbf{C}_G(P_1)$ is cyclic. Thus $\mathbf{F}(G) = \mathbf{C}_G(P_1) \geq P$ and $\mathbf{F}(G)$ is cyclic. Part (ii) now follows.

Since P_1 acts irreducibly on V, Theorem 2.1 implies that $G \leq \Gamma(q^n) =: \Gamma$. Now $p \mid |\Gamma_0|$, where $\Gamma_0 \trianglelefteq \Gamma$ is the cyclic subgroup of multiplications of order $q^n - 1$. Thus $p \mid |\mathbf{F}(\Gamma)|$ and the arguments of the last paragraph apply to Γ as well. It follows that $P_1 \leq P \leq \mathbf{F}(\Gamma) = \mathbf{C}_\Gamma(P_1)$, $\mathbf{F}(\Gamma)$ is cyclic and $\mathbf{F}(\Gamma) = \Gamma_0$. Also $\Gamma/\mathbf{F}(\Gamma)$ is cyclic of order n. Because

$$\mathbf{F}(\Gamma) \cap G \subseteq \mathbf{F}(G) = \mathbf{C}_\Gamma(P_1) \cap G \leq \mathbf{F}(\Gamma) \cap G,$$

we have that $\mathbf{F}(G) = G \cap \mathbf{F}(\Gamma)$ and the result follows. $\square$

The key to finding the exceptional cases of Huppert's Theorem is describing those solvable groups $G \leq GL(n, q)$ whose Fitting factor group $G/\mathbf{F}(G)$ is divisible by a Zsigmondy prime divisor of $q^n - 1$. Before we do so, we stop and study the structure of semi-linear groups in more detail.

6.5 Lemma. *Let $\Gamma = \Gamma(q^n)$ and $\Gamma_0 = \Gamma_0(q^n)$ for a prime power q.*

(a) *If $n = 2$ and $q \in \mathfrak{M}$, then $\Gamma = R \times S$ where R is a non-abelian 2-group and S is cyclic of odd order.*

(b) *If $q^n = 2^6$, then $\Gamma_0 = \mathbf{F}(\Gamma)$. Let S and T be the unique subgroups of Γ_0 of order 3 and 7, respectively. Then $|\mathbf{C}_\Gamma(S)/\Gamma_0| = 3$ and $|\mathbf{C}_\Gamma(T)/\Gamma_0| = 2$.*

(c) *In all other cases, $\Gamma_0 = \mathbf{F}(\Gamma)$. For each Zsigmondy prime divisor p of $q^n - 1$ and each non-trivial p-subgroup P_0 of Γ, $\Gamma_0 = \mathbf{C}_\Gamma(P_0) \geq P_0$.*

Proof. Recall that Γ has a cyclic subgroup $B = \langle \beta \rangle$ of order n consisting of field automorphisms, $\Gamma = \Gamma_0 B$ and $\Gamma_0 \cap B = 1$. If $U \leq \Gamma_0$, then $\mathbf{C}_\Gamma(U) = \Gamma_0 \mathbf{C}_B(U)$. For $\alpha \in B$ of order t, it holds that $|\mathbf{C}_{\Gamma_0}(\alpha)| = q^{n/t} - 1$.

If $n = 2$ and $q \in \mathfrak{M}$, $\mathbf{C}_{\Gamma_0}(\beta)$ has order $q - 1$ and thus contains the Hall $2'$-subgroup of Γ. Part (a) now follows.

Assume next $q^n = 2^6$. Since $|\mathbf{C}_{\Gamma_0}(\beta^2)| = 3$ and $|\mathbf{C}_{\Gamma_0}(\beta^3)| = 7$, we have that $|\mathbf{C}_\Gamma(S)/\Gamma_0| = 3$ and $|\mathbf{C}_\Gamma(T)/\Gamma_0| = 2$. Clearly $\Gamma_0 \leq \mathbf{F}(\Gamma)$, and equality follows, since β^2 does not act trivially on $T \in \mathrm{Syl}_7(\Gamma_0)$ and β^3 does not act trivially on the Sylow 3-subgroup of Γ_0.

If (a) or (b) does not apply, then $q^n - 1$ has a Zsigmondy prime divisor p, by Theorem 6.2. Therefore $p \mid |\Gamma_0|$, whence $p \mid |\mathbf{F}(\Gamma)|$. By Lemma 6.4, $\mathbf{F}(\Gamma) = \mathbf{C}_\Gamma(P_0)$ is cyclic and contains P_0. Since each subgroup of Γ which properly contains Γ_0 is non-abelian, $\Gamma_0 = \mathbf{F}(\Gamma)$ follows. $\qquad\square$

6.6 Corollary. *Suppose that $G \leq \Gamma(q^n)$ and $p \mid |G|$ for a Zsigmondy prime divisor p of $q^n - 1$. Then $p \mid |\mathbf{F}(G)|$ and Lemma 6.4 applies.*

Proof. Let $P \in \mathrm{Syl}_p(G)$ and write Γ for $\Gamma(q^n)$. By Lemma 6.5 (c), Γ has a unique Sylow p-subgroup, and therefore

$$P \leq \mathbf{O}_p(\Gamma) \cap G \leq \mathbf{F}(\Gamma) \cap G \leq \mathbf{F}(G). \qquad\square$$

We let $F_1(G) \leq F_2(G) \leq \ldots$ denote the ascending Fitting series, i.e. $F_1(G) = \mathbf{F}(G)$ and $F_{i+1}(G)/F_i(G) = \mathbf{F}(G/F_i(G))$.

6.7 Lemma. *Suppose that G is a solvable subgroup of $GL(n,q)$ (q a prime power) and p is a Zsigmondy prime divisor of $q^n - 1$. Assume also that $p \mid |G/\mathbf{F}(G)|$. Then $p \nmid |\mathbf{F}(G)|$ and*

(i) *$n = p - 1 = 2^m$ for an integer $m = 2^k \geq 1$;*

(ii) *$\mathbf{F}(G) = ET$ where $E \trianglelefteq G$ is an extra-special 2-group of order 2^{2m+1}, T is cyclic, $T = \mathbf{Z}(G)$ and $T \cap E = \mathbf{Z}(E)$;*

(iii) *$T \leq \mathbf{Z}(GL(n,q))$ and $|T| \mid q - 1$;*

(iv) *$F_2(G)/\mathbf{F}(G)$ is cyclic of order p, and $G/F_2(G)$ is a cyclic 2-group with $|G/F_2(G)| \mid 2m$.*

Proof. By Lemma 6.4 and Corollary 6.6, $p \nmid |\mathbf{F}(G)|$ and $G \not\leq \Gamma(q^n)$. In particular, $n > 1$ and $p > 2$. Let V be the corresponding G-module of order q^n and let $P \in \mathrm{Syl}_p(G)$. By Proposition 6.3, P is cyclic and V is an irreducible quasi-primitive G-module. Since $G \not\leq \Gamma(q^n)$, Corollary 2.3 implies that $\mathbf{F}(G)$ is non-abelian. Set $F = \mathbf{F}(G)$, $F_2 = F_2(G)$ and $P_0 \leq P$ with $|P_0| = p$.

Let A be a normal abelian subgroup of G. Then A is cyclic, because V is quasi-primitive. If $P \not\leq \mathbf{C}_G(A)$, then every faithful $\varphi \in \mathrm{Irr}\,(AP)$ has degree divisible by p. Note that $(q, |AP|) = 1$ by the choice of p and since $A \leq F$. Then Lemma 2.4 implies that $p \leq \dim_{GF(q)}(V) = n$, a contradiction to $n \mid p - 1$. Hence $P \leq \mathbf{C}_G(A)$ for all abelian $A \trianglelefteq G$.

Since $\mathbf{C}_G(F) \leq F$, we may choose a prime r and a Sylow r-subgroup R of F with $P_0 \not\leq \mathbf{C}_G(R)$. By the last paragraph, R is non-abelian. Every normal abelian subgroup of G is cyclic and we apply Theorem 1.9 to conclude that $R = ES$ with E, $S \trianglelefteq G$ such that $E \cap S$ is the unique subgroup Z of $\mathbf{Z}(R)$ of order r, E is extra-special or $E = Z$, and S has a cyclic subgroup $U \trianglelefteq G$ with $|S \colon U| \leq 2$. By the last paragraph, $P \leq \mathbf{C}_G(U)$. Clearly, P centralizes S/U and $p \neq 2$. Thus P centralizes S. Since V is an irreducible P-module (see Proposition 6.3), $\mathbf{C}_G(P)$ is cyclic by Lemma 2.9 and hence S is cyclic. As R is non-abelian, $E > Z$.

Let $H = EP$. Then V_H is irreducible and quasi-primitive, by Proposition 6.3. Now P_0 acts faithfully on E/Z and we may choose a chief factor E_1/Z of H such that P_0 and P act faithfully on E_1/Z. Furthermore, applying Theorem 1.9 to H, we may assume that $Z = \mathbf{Z}(E_1)$ and E_1 is extra-special, say $|E_1/Z| = r^{2l}$. Since P centralizes Z, P is a cyclic irreducible subgroup of $Sp(2l, r)$. Consequently, $|P| \mid r^l + 1$ (see [Hu, II, 9.23]). Now $|E/Z| = r^{2m}$ for an integer $m \geq l$ and Corollary 2.6 implies that $\dim(V) = tr^m \dim(W)$ for an irreducible S-submodule W of V and an integer t. Then

$$n \leq p - 1 \leq |P| - 1 \leq r^l \leq r^m \leq tr^m \dim(W) = \dim(V) = n.$$

Equality must hold throughout. Thus P has order p, $p = r^m + 1$ is a Fermat prime, $r = 2$ and m is a power of 2. Also $\dim(V) = n = 2^m$ and $l = m$, i.e. E/Z is a chief factor even in G.

We chose R to be a Sylow r-subgroup of F not centralized by $P_0 = P$ and proved that $r = 2$. Thus P centralizes $S_1 \in \mathrm{Hall}_{2'}(F)$. Let $T = SS_1 \trianglelefteq G$ and note that $T \leq \mathbf{C}_G(P)$. Thus T is cyclic by Lemma 2.9, $F = ET$ and $E \cap T = Z = \mathbf{Z}(E)$. Also $T \leq \mathbf{Z}(GL(V))$ by Lemma 2.10 (iii), as $n = \dim(V) = |E/Z|^{1/2}$. Parts (i), (ii) and (iii) now follow.

Since $p = 2^m + 1$, p is a Zsigmondy prime divisor of $2^{2m} - 1$. If also $p \nmid |F_2/F|$, then (ii) applied to the faithful action of G/F on E/Z implies that $\mathbf{O}_2(G/F) \neq 1$, a contradiction because E/Z has characteristic 2. Hence $p \mid |F_2/F|$, and Lemma 6.4 implies that F_2/F and G/F_2 are cyclic with $|G/F_2| \mid 2m$. By Proposition 6.3, F_2/F acts irreducibly on E/Z. Again employing [Hu, II, 9.23], we have that $|F_2/F| \mid 2^m + 1 = p$. Hence $G/F_2 \leq \mathrm{Aut}(Z_p)$ which is a cyclic 2-group of order 2^m. But we also know that $|G/F_2| \mid 2m$. This proves (iv). $\qquad\square$

6.8 Theorem. *Let V be a vector space of dimension n over $GF(q)$, q a prime power. Suppose that G is a solvable subgroup of $GL(V)$ that transi-*

tively permutes the elements of $V^{\#}$. Then $G \leq \Gamma(q^n)$, or one of the following occurs:

(a) $q^n = 3^4$, $\mathbf{F}(G)$ is extra-special of order 2^5, $|F_2(G)/\mathbf{F}(G)| = 5$ and $G/F_2(G) \leq Z_4$.

(b) $q^n = 3^2$, 5^2, 7^2, 11^2 or 23^2. Here $\mathbf{F}(G) = QT$, where $T = \mathbf{Z}(G) \leq \mathbf{Z}(GL(V))$ is cyclic, $Q_8 \cong Q \trianglelefteq G$, $T \cap Q = \mathbf{Z}(Q)$ and $Q/\mathbf{Z}(Q) \cong \mathbf{F}(G)/T$ is a faithful irreducible $G/\mathbf{F}(G)$-module. We also have one of the following entries:

q^n	$|T|$	$G/\mathbf{F}(G)$
3^2	2	Z_3 or S_3
5^2	2 or 4	Z_3
5^2	4	S_3
7^2	2 or 6	S_3
11^2	10	Z_3 or S_3
23^2	22	S_3

Proof. We may assume that $n > 1$. Since G acts transitively on $V^{\#}$, V is an irreducible quasi-primitive G-module and $q^n - 1 \mid |G|$.

Suppose first that $n = 2$. We may assume that $G \not\leq \Gamma(q^2)$. Then Theorem 2.11 implies that $\mathbf{F}(G) = QT$ where $Q_8 \cong Q \trianglelefteq G$, $T = \mathbf{Z}(G) \leq \mathbf{Z}(GL(V))$ is cyclic, $|T| \mid q - 1$, $T \cap Q = \mathbf{Z}(Q)$, $G/\mathbf{F}(G)$ is isomorphic to Z_3 or S_3 and $G/\mathbf{F}(G)$ acts faithfully on $Q/\mathbf{Z}(Q)$. Now $|G| = |G/\mathbf{F}(G)| \, |\mathbf{F}(G)/T| \, |T|$ divides $24(q - 1)$. But $q^2 - 1 \mid |G|$ and so $q + 1 \mid 24$. Since $\mathbf{O}_2(G) \neq 1$, q is odd, and $q = 3$, 5, 7, 11 or 23. Counting yields conclusion (b) or that $q^n = 5^2$, $|T| = 2$ and $G/\mathbf{F}(G) \cong S_3$. In this case, $|G| = 24 \cdot 2$. Fix $v \in V^{\#}$ so that $|\mathbf{C}_G(v)| = 2$. Also $\mathbf{F}(G) \cap \mathbf{C}_G(v) = 1$. Now $\mathbf{F}(G) \trianglelefteq S \trianglelefteq G$ where $|G/S| = 2$ and S also acts fixed-point-freely on $V^{\#}$. Now let $Z = \mathbf{Z}(GL(V))$ so that $|Z| = 4$ and $Z \cap G = T$ has order 2. Set $G_1 = ZG$, then $|G_1| = 2|G|$ and $G_1/S \cong Z_2 \times Z_2$. Also set $H = \mathbf{C}_{G_1}(v)$. Since $|G_1| = 4 \cdot 24$, we have that $|H| = 4$, $H \cap S = 1$ and $HS = G_1$. Thus $H \cong Z_2 \times Z_2$. Since $q \nmid |H|$, H acts faithfully on the one-dimensional space $V/\langle v \rangle$. Thus H is cyclic, a contradiction. We may now assume that $n > 2$.

If secondly $q = 2$ and $n = 6$, then $63 \mid |G|$. Since V is a quasi-primitive G-module, Corollary 2.15 implies that $G \leq \Gamma(2^6)$. In the remaining cases, Theorem 6.2 allows us to choose a Zsigmondy prime divisor p of $q^n - 1$. By Lemma 6.4, we may also assume that $p \nmid |\mathbf{F}(G)|$. Applying Lemma 6.7, we have that

   (i) $n = p - 1 = 2^m$ for an integer m;

   (ii) $\mathbf{F}(G) = ET$ where $E \trianglelefteq G$ is an extra-special group of order 2^{2m+1}, $T = \mathbf{Z}(G) \leq \mathbf{Z}(GL(V))$ and $E \cap T = \mathbf{Z}(E)$;

   (iii) $F_2(G)/\mathbf{F}(G)$ is cyclic of order p;

   (iv) $G/F_2(G)$ is a cyclic 2-group with order dividing $2m$.

In particular, $|G| = |G/F_2(G)| \, |F_2(G)/\mathbf{F}(G)| \, |\mathbf{F}(G)/T| \, |T|$ indeed divides $2 \cdot m \cdot p \cdot 2^{2m} \cdot (q-1)$ and p is the only odd prime dividing $|G|/(q-1)$. By the transitive action, $(q^n - 1)/(q-1) \mid |G|/(q-1)$. If $n = 2^m \geq 8$, then $q^{n/2} - 1$ has an odd Zsigmondy prime divisor $p_0 \neq p$ which must divide $|G|/(q-1)$, a contradiction. By (i) just above, we are left with $n = 4$. Then $p = 5$ and $m = 2$. Now $(q^4 - 1)/(q-1)$ divides $2^6 \cdot 5$. Thus $q^3 \leq 2^6 \cdot 5$ and $q = 3$. Because $T \leq \mathbf{Z}(GL(V))$, $|T| \mid q - 1 = 2$ and $\mathbf{F}(G)$ is extra-special of order 2^5. Observe that conclusion (a) of the theorem is satisfied. $\qquad\square$

An example due to Bucht is presented in [HB, XII, 7.4] of a solvable group $G \leq GL(4, 3)$ such that $|G| = 2^7 \cdot 5$, G has subgroups $G_1 \geq G_2$ of index 2 and 4 with G_2 (and hence G_1 and G) transitive on $V^{\#}$. Furthermore, G_2 (and hence G_1 and G) is not a subgroup of $\Gamma(3^4)$. For each of the other exceptional values of q^n and corresponding values of $|G|$ given in Theorem 6.8, there is a solvable subgroup $G \leq GL(n, q)$ with $G \nleq \Gamma(q^n)$, but G transitive on $V^{\#}$. We refer the reader to Huppert's original paper [Hu 2].

Theorem 6.8 can now be stated in terms of doubly transitive permutation groups. To avoid redundancy, we do not list the structure of the exceptional cases, which again do exist. The proofs of this section were derived by Wolf with encouragement from P. Sin.

6.9 Theorem (Huppert). *If G is a solvable doubly transitive permutation*

group on Ω, then $|\Omega| = q^n$ for a prime q and $G \leq A\Gamma(q^n)$ unless $q^n = 3^2, 5^2,$
$7^2, 11^2, 23^2$ or 3^4. In the non-exceptional cases, the unique minimal normal
subgroup of G is $A(q^n)$.

Proof. Since G is 2-fold transitive, G is in fact primitive (see [Hu, II, 1.9]).
Then, by solvability, G has a unique minimal normal subgroup V that acts
regularly on Ω, $VG_\alpha = G$ ($\alpha \in \Omega$), $V \cap G_\alpha = 1$ and the actions of G_α on
Ω and V are permutation isomorphic. Thus G_α transitively permutes the
elements of $V^\#$. If $q^n \neq 3^2$, 5^2, 7^2, 11^2, 23^2 or 3^4, then $G_\alpha \leq \Gamma(q^n)$, by
Theorem 6.8, and $G \leq A\Gamma(q^n)$, by [Hu, II, 3.5]. $\qquad\qquad\qquad\square$

Suppose that $G \leq GL(n,p)$ is solvable and irreducible (p a prime). Let
V be the corresponding natural module, and r be the number of orbits of G
on $V^\#$. Theorem 6.8 states that when $r = 1$, then $G \leq \Gamma(p^n)$ or p^n is one
of six values. In [Sa 1], S. Saeger shows that if G is primitive (as a linear
group) and if $r \leq p^{n/2}/(12n + 1)$, then $G \leq \Gamma(p^n)$ or p^n is one of 17^4, 19^4,
7^6, 5^8, 7^8, 13^8, 7^9, 3^{16} or 5^{16}. Of course, when $r = 1$, it is easy to see that
G is a primitive linear group. However, the inequality cannot be met for
small values of p^n, including the exceptional values in Theorem 6.8.

§7   Regular Orbits of Sylow Subgroups of Solvable
Linear Groups

In this section, we return to the study of regular orbits of a p-group P.
But this time we consider P as a Sylow p-subgroup of some solvable linear
group in characteristic p, and present a remarkable result due to A. Espuelas
[Es 1]. Our proof however is different from Espuelas' one; whereas he uses
tensor induction, we rely on the methods developed so far. We shall also
use a result which admittedly does not lie at the surface, but which is often
helpful when studying indecomposable modules for p-nilpotent groups over
arbitrary fields of characteristic p, namely:

7.1 Proposition. *Let G be p-nilpotent and V a finite-dimensional inde-composable $\mathcal{K}[G]$-module, char $(\mathcal{K}) = p > 0$. If*

$$0 = V_0 < V_1 < \cdots < V_n = V$$

is a composition series of V, then all composition factors V_i/V_{i-1} ($i = 1, \ldots, n$) are mutually isomorphic.

Proof. Since V is indecomposable, V belongs to a block of $\mathcal{K}[G]$ and all its composition factors belong to the same block (see [HB, VII, 12.1]). Since G is p-nilpotent, each block of $\mathcal{K}[G]$ only contains one irreducible $\mathcal{K}[G]$-module (see [HB, VII, 14.9]) and the assertion follows. $\qquad \square$

The next lemma, which will as well become important in Chapter III, has some connections to the Hall–Higman results (see [HB, chap IX] and the remarks following the lemma). Our techniques are elementary and work for arbitrary fields.

7.2 Lemma. *Suppose that $Z := \mathbf{Z}(E) \leq E \trianglelefteq H$ with $|H : E| = p$, $p \nmid |E|$ and E/Z is an abelian q-group for primes p and q. For $P \in \mathrm{Syl}_p(H)$, assume that $P \not\leq \mathbf{C}_H(E)$. Let V be a finite-dimensional H-module with char $(V) \nmid |E|$ such that V_E is faithful and homogeneous. Then*

(i) $\dim(\mathbf{C}_V(P)) \leq \dim(V)/2$; *or*

(ii) $p = 2$, $P \leq \mathbf{C}_H(Z)$ *and* $\dim(\mathbf{C}_V(P)) \leq ((q+1)/(2q)) \cdot \dim(V)$.

Proof. We argue by induction on $|H|$. Note that $P \not\leq \mathbf{C}_H(E)$ implies $E \neq 1$. Since V_E is homogeneous and $Z = \mathbf{Z}(E)$, V_Z is homogeneous. Thus Z acts fixed-point-freely on V and is cyclic.

If $Z \not\leq \mathbf{Z}(H)$, we may choose $1 \neq Y \leq Z$ with $Y \trianglelefteq H$ and YP a Frobenius group. Since $\mathbf{C}_V(Y) = 0$, Lemma 0.34 implies that $\dim(\mathbf{C}_V(P)) \leq (\dim(V))/p$. We may thus assume that $Z \leq \mathbf{Z}(H)$ and $E > Z$. Let $L/Z = [E/Z, P]$, recall that V_E is homogeneous and write $V_L = V_1 \oplus \cdots \oplus V_n$

for homogeneous components V_i that are transitively permuted by E/L. Now V_i^g again is a homogeneous L-module ($g \in P$, $i = 1, \ldots, n$), and thus P permutes the V_i. Since P centralizes E/L, Glauberman's Lemma 0.14(a,b) implies that P fixes each V_i. Set $C_i = \mathbf{C}_L(V_i)$, $i = 1, \ldots, n$, and suppose that $[P, L] \leq C_j$ for some j. Since $E/Z = L/Z \times \mathbf{C}_{E/Z}(P)$ and $E/L \cong \mathbf{C}_{E/Z}(P)$ transitively permutes the V_i, we then have $[P, L] \leq \bigcap_{i=1}^{n} C_i = 1$. This implies that P centralizes L, E/L and hence E, a contradiction. Thus $[P, L/C_i] \neq 1$ for all i. If $L < E$, we apply the inductive hypothesis to the action of PL/C_i on V_i ($i = 1, \ldots, n$) and the conclusion of the lemma holds. We may therefore assume that $L = E$.

Let M/Z be a minimal normal subgroup of H/Z. First assume that V_M is not homogeneous and write $V_M = W_1 \oplus \cdots \oplus W_l$ ($l > 1$) for homogeneous components W_i that are transitively permuted by E/M. Again P permutes the W_i. Since $\mathbf{C}_{E/M}(P) = 1$, Glauberman's Lemma 0.14(a,b) implies that P fixes exactly one W_j and permutes the others. Thus $p \mid l-1$ and $\dim(\mathbf{C}_V(P)) \leq (1 + (l-1)/p) \cdot \dim(W_1)$. For p odd, it suffices to show that $1 + (l-1)/p \leq l/2$ or equivalently that $(1/(l-1)) + (2/p) \leq 1$. This holds as $l - 1 \geq p \geq 3$. For $p = 2$, it suffices to show that $1 + (l-1)/2 \leq l \cdot (q+1)/(2q)$ or equivalently that $q \leq l$. But this follows, since the q-group E/M transitively permutes the l homogeneous components W_i. Thus the result holds when V_M is not homogeneous and we may assume that V_M is homogeneous.

Since $[M/Z, P] \neq 1$, P acts non-trivially on M. Thus if $M < E$, the result follows from the inductive hypothesis applied to MP. Hence E/Z is a chief factor of H. Since P centralizes $Z = \mathbf{Z}(E)$, but does not centralize E, we also have $E' \neq 1$.

Let $0 \neq x \in \mathbf{C}_V(P)$. As $\mathbf{C}_E(x) \cong \mathbf{C}_E(x)Z/Z$ is abelian, $\mathbf{C}_E(x)Z$ is an abelian normal subgroup of H. Thus the last paragraph implies $\mathbf{C}_E(x) \leq Z$ and $\mathbf{C}_E(x) = 1$. Let $P_0 \in \mathrm{Syl}_p(H)$ with $P_0 \neq P$. Then $1 \neq [P, P_0] \leq E$ and $[P, P_0]$ centralizes $\mathbf{C}_V(P) \cap \mathbf{C}_V(P_0)$. Consequently $\mathbf{C}_V(P) \cap \mathbf{C}_V(P_0) = 0$ and $\dim(\mathbf{C}_V(P)) \leq (\dim(V))/2$ follows. $\square$

As indicated before, Lemma 7.2 can also be proved more heavy-handedly by Hall–Higman techniques. The advantage of this approach however is that it clarifies the module structure of V_P in the case where $P \leq \mathbf{C}_H(Z)$, E is extra-special and $[E/Z, P] = E/\mathbf{Z}$. We may also assume that V is an indecomposable H-module and that the underlying field $\mathcal{F}$ is a splitting-field. In this situation the proofs of [Hu, V, 17.13] and [HB, IX, 2.6] yield:

(1) char$(\mathcal{F}) \neq p$, $\chi_P = m\rho + \delta\mu$, where χ is the character of V, ρ is the regular character of $\mathcal{F}[P]$, μ is a linear character, $m \in \mathbb{N}$ and $\delta = \pm 1$.

(2) char$(\mathcal{F}) = p$, $V_P = m \cdot \mathcal{F}[P] \oplus W$, where W is an indecomposable $\mathcal{F}[P]$-module (possibly $W = 0$) and $m \in \mathbb{N}$.

If $p > 2$, the estimate of Lemma 7.2 immediately follows. If $p = 2$, the only critical case is where $\mu = 1_P$ or $W \cong \mathcal{F}$. Then $q \leq \dim(V) = 2m + 1$, $\dim(\mathbf{C}_V(P)) = m + 1$ and we obtain

$$\dim(\mathbf{C}_V(P)) = m + 1 \leq ((q+1)/(2q)) \cdot (2m+1) = ((q+1)/(2q)) \cdot \dim(V).$$

7.3 Theorem (Espuelas). *Let G be a solvable group, p an odd prime, $P \in \mathrm{Syl}_p(G)$ and $\mathbf{O}_p(G) = 1$. Let V be a finite and faithful G-module with* char$(V) = p$. *Then P has a regular orbit on V.*

Proof. We proceed by induction on $|G|\,|V|$.

Step 1. $G = \mathbf{O}^{p'}(G)$ is p-nilpotent with nilpotent p-complement $F = \mathbf{F}(G)$.

Proof. Since $P \leq \mathbf{O}^{p'}(G)$ and $\mathbf{O}_p(\mathbf{O}^{p'}(G)) = 1$, induction yields $G = \mathbf{O}^{p'}(G)$. We also have $\mathbf{O}_p(FP) = 1$ and thus, again by induction, $G = FP$ is p-nilpotent with p-complement F.

Step 2. V is an irreducible G-module.

Proof. We first decompose $V = I_1 \oplus \cdots \oplus I_n$ into indecomposable G-modules I_j and pick an irreducible G-submodule $M_j \leq I_j$ $(j = 1, \ldots, n)$. Since G

is p-nilpotent, Proposition 7.1 implies that each G-composition factor of I_j is isomorphic to M_j. Consequently, we obtain for the p'-group F that $\mathbf{C}_F(M_j) = \mathbf{C}_F(I_j)$ and thus $\bigcap_{j=1}^n \mathbf{C}_F(M_j) = \bigcap_{j=1}^n \mathbf{C}_F(I_j) = \mathbf{C}_F(V) = 1$. Since $\mathbf{O}_p(G) = 1$, G acts faithfully on the completely reducible module $M_1 \oplus \cdots \oplus M_n$ and the inductive hypothesis implies $V = M_1 \oplus \cdots \oplus M_n$. If $n > 1$, then $P/\mathbf{C}_P(M_j) \le G/\mathbf{C}_G(M_j)$ has a regular orbit on M_j, generated by $v_j \in M_j$ $(j = 1, \ldots, n)$. Since then $v = v_1 + \cdots + v_n$ generates a regular orbit for P on V, Step 2 holds.

Step 3. V is quasi-primitive.

Proof. Suppose there is $M \trianglelefteq G$ such that $V_M = V_1 \oplus \cdots \oplus V_m$ with $m > 1$ homogeneous components V_i that are transitively permuted by G. It follows from Clifford's Theorem that V_i is an irreducible $\mathbf{N}_G(V_i)$-module and hence $\mathbf{O}_p(\mathbf{N}_G(V_i)/\mathbf{C}_G(V_i)) = 1$. Note further that

$$\mathbf{N}_P(V_i)/\mathbf{C}_P(V_i) = \mathbf{N}_P(V_i)/(\mathbf{C}_G(V_i) \cap \mathbf{N}_P(V_i))$$

$$\cong \mathbf{N}_P(V_i)\mathbf{C}_G(V_i)/\mathbf{C}_G(V_i) \le \mathbf{N}_G(V_i)/\mathbf{C}_G(V_i).$$

Since $m > 1$, the inductive hypothesis shows that $\mathbf{N}_P(V_i)/\mathbf{C}_P(V_i)$ has a regular orbit on V_i $(i = 1, \ldots, m)$.

Also P permutes the subspaces $V_1, \ldots, V_m$ (possibly intransitively). Since $p > 2$, the exceptional cases of Lemma 4.3 cannot occur, and P has a regular orbit on V. This completes Step 3.

As an immediate consequence, we obtain

Step 4. Let $Z = \mathbf{Z}(F)$. Then $V_Z \cong W \oplus \cdots \oplus W$ for an irreducible Z-module W, Z acts fixed-point-freely on W and $|Z| \mid |W| - 1$.

Step 5. $P \le \mathbf{C}_G(Z)$, i.e. $Z = \mathbf{Z}(G)$.

Proof. Set $C = \mathbf{C}_P(Z)$ and assume $C < P$. Note that $\mathbf{O}_p(ZP) = C$ and $\mathbf{F}(ZP/C) = ZC/C$. Let $0 = V_0 < V_1 < \cdots < V_t = V$ be a composition series of V considered as a ZP-module. Then each V_j/V_{j-1} is an irreducible and faithful ZP/C-module. By Step 4,

$$(V_j/V_{j-1})_{ZC/C} \cong (V_j/V_{j-1})_Z \cong W \oplus \cdots \oplus W,$$

and it follows from Lemma 2.2 and Corollary 2.3 that $(V_j/V_{j-1})_Z \cong W$ $(j = 1, \ldots, t)$ and that ZP/C is a subgroup of a semi-linear group. In particular $P/C \cong (ZP/C)/(ZC/C)$ is cyclic.

Since $C < P$ and $\mathbf{O}_p(FC) = 1$, we may apply the inductive hypothesis to the action of FC on V, and we thus find $v \in V$ generating a regular C-orbit on V. Choose $j \in \{1, \ldots, t\}$ such that $v \in V_j \setminus V_{j-1}$. Since V_j/V_{j-1} is irreducible as a Z-module, V_j/V_{j-1} is spanned by $\{v^z + V_{j-1} \mid z \in Z\}$ as a vector space. By the previous paragraph, P/C is cyclic and we can thus find $y \in Z$ such that $\mathbf{C}_{P/C}(v^y + V_{j-1}) = 1$. Since v generates a regular C-orbit and since $[Z, C] = 1$, v^y as well generates a regular C-orbit. Thus $\mathbf{C}_P(v^y) \le \mathbf{C}_C(v^y) = 1$, and Step 5 holds.

Step 6. Each Sylow-subgroup of F is extra-special and non-abelian and $|V| = |W|^e$, where $|F/Z| = e^2 > 1$.

Proof. By Step 3, every normal abelian subgroup A of G is cyclic. Since $\mathbf{O}^{p'}(G) = G$, we conclude $F = \mathbf{O}^p(G) \le G' < \mathbf{C}_G(A)$ and A is central in F. By Step 5, $P \le \mathbf{C}_G(Z) \le \mathbf{C}_G(A)$. We now apply Corollary 1.10 with $P \le \mathbf{C}_G(Z)$. Because $\mathbf{O}^{p'}(G) = G$, each Sylow-subgroup of F is extra-special and non-abelian. Since V_F is homogeneous and $G/F \cong P$ is a p-group, [HB, VII, 9.19] implies that V_F is irreducible. Therefore, $|V| = |W|^e$ follows from Corollary 2.6.

Step 7. $|P| \le (e^{16/5})/2$.

Proof. Since $\mathbf{O}_p(G) = 1$ and $P \le \mathbf{C}_G(Z)$, P acts faithfully and completely reducibly on F/Z (possibly over different finite fields). Therefore Theorem 3.3 applies and $|P| \le (|F/Z|^{8/5})/2 = (e^{16/5})/2$.

Step 8. $|\mathbf{C}_V(g)| \le |V|^{1/2}$ for all $g \in P^{\#}$.

Proof. We pick $1 \neq h \in \langle g \rangle$ with $h^p = 1$. Then there exists $Q \in \mathrm{Syl}_q(F)$ such that $h \notin \mathbf{C}_G(Q)$, and the hypotheses of Lemma 7.2 are satisfied for $H = Q\langle h \rangle$. Since $p > 2$, we obtain $|\mathbf{C}_V(G)| \leq |\mathbf{C}_V(h)| \leq |V|^{1/2}$.

Step 9. We may assume that

(i) $|P| > |V|^{1/2}$, and

(ii) $e^{32} \geq 2^{10}|W|^{5e}$.

Proof. Since we may assume that P has no regular orbit on V, we have that $V = \bigcup_{g \in P\#} \mathbf{C}_V(g)$ and

$$|V| \leq \sum_{g \in P\#} |\mathbf{C}_V(g)| \leq (|P| - 1)|V|^{1/2},$$

using Step 8. Part (i) follows. By Steps 6 and 7, $|V| = |W|^e$ and $|P| \leq e^{16/5}/2$. Thus $e^{32} \geq 2^{10}|W|^{5e}$, proving (ii).

Step 10. Conclusion.

Proof. Since $e^{32} \leq 2^{10} \cdot 7^{5e}$ for all integers $e \geq 2$, it follows from Step 9 (ii) that $|W| < 7$. Since $\mathrm{char}(W) = p$ is odd, we have that $|W| = p$ is 3 or 5. Since $|Z| \mid |W| - 1$, Z has order 2 or 4. Thus F is a 2-group and $e > 1$ must be a power of two. It easily follows now from Step 9 (ii) that e and p are as in the following table:

e	$p = |W|$	$|V|$	$n := \log_2(e^2)$	$|GL(n,2)|_p$
2	3	3^2	2	3^1
4	3	3^4	4	3^2
4	5	5^4	4	5^1
8	3	3^8	6	3^4

Note that $|V| = |W|^e$ by Step 6. Also $|F/Z|$ has order $e^2 = 2^n$. Since P acts faithfully on $|F/Z|$, we have $|P| \leq |GL(n,2)|_p$. Thus, in all cases, $|P| \leq |V|^{1/2}$. This contradicts Step 9 (i), completing the proof. $\square$

7.4 Remark. As follows from Example 7.5 (a) below, Theorem 7.3 does not hold for $p = 2$. If one however requires in addition that P does not involve a copy of D_8, then P has a regular orbit on V. Whereas the proof in the quasi-primitive case runs similarly to the above (and does not rely on the assumption about D_8), in the primitive case Lemma 4.3 cannot be applied any longer. It has to be replaced by a result similar to Theorem 4.8. For details, we refer to Espuelas' paper [Es 1].

7.5 Examples. (a) Let $W = Z_2 \times Z_2$ be the faithful irreducible S_3-module over $GF(2)$ and $G = S_3$ wr Z_2. Then $V := W^G$ is a faithful irreducible $GF(2)[G]$-module, $\mathbf{O}_2(G) = 1$ and $D_8 \cong Z_2$ wr Z_2 is a Sylow-2-subgroup of G. Furthermore, the orbit sizes of G on V are $(1, 6, 9)$ and D_8 has no regular orbit on V.

(b) Let V be a 3-dimensional vector space over $GF(p)$. We consider

$$g = \begin{bmatrix} 1 & 0 & 0 \\ 0 & 1 & 1 \\ 0 & 0 & 1 \end{bmatrix}, \quad h = \begin{bmatrix} 1 & 0 & 1 \\ 0 & 1 & 0 \\ 0 & 0 & 1 \end{bmatrix} \quad \in GL(3, p).$$

Then $P = \langle g \rangle \times \langle h \rangle \cong Z_p \times Z_p$. Let $v = (x, y, z) \in V$. If $x = 0$, then v is fixed by $h \in P^{\#}$; if $x \neq 0$, then $gh^{-y/x} \in P^{\#}$ fixes v. Therefore P does not have a regular orbit on V. In particular, the hypothesis that $\mathbf{O}_p(G) = 1$ is necessary in Theorem 7.3.

In his thesis, W. Carlip [Ca 1, 2] replaced the Sylow subgroup in Theorem 7.3 by a nilpotent Hall subgroup H. Under the assumption that both $|G|$ and p are odd, he proves the existence of a regular H-orbit.

We remark that a faithful module action of G on a finite-dimensional $\mathcal{F}$-vector space V always has a regular orbit, provided that $|\mathcal{F}| = \infty$. Namely V then cannot be written as the union of a finite number of proper subspaces, and therefore $\bigcup_{g \in G^{\#}} \mathbf{C}_V(g) < V$.

§8   Short Orbits of Linear Groups of Odd Order

In this section, we are looking in quite the other direction, namely we try to find short orbits $\neq \{0\}$ for a solvable group G which acts faithfully on a finite vector space V. If $|G||V|$ is odd and V carries a symplectic G-invariant bilinear form, T. Berger [Be 1] gave an upper estimate for such an orbit and we present his result as Theorem 8.4. In §16, we shall use this theorem to bound the derived length of a (solvable) group of odd order in terms of its different character degrees.

We start with the following number theoretical lemma.

8.1 Lemma. Let p and r be distinct odd primes and let $n \in \mathbb{N}$ such that $r \mid n$. Then $(r - 1)r(n + 1) \leq 2(p^n + 1)/(p^{n/r} + 1)$.

Proof. We set $a = n/r$ and have to show that $(r - 1)r(ar + 1) \leq 2(p^{ar} + 1)/(p^a + 1)$. Observe first that $(r - 1)r(ar + 1) = ar^3 + r^2(1 - a) - r \leq ar^3$. Note further

$$2(p^{ar} + 1)/(p^a + 1) = 2(1 - (-p^a)^r)/(1 - (-p^a)) = 2\sum_{j=0}^{r-1}(-p^a)^j$$

$$\geq 2(p^{a(r-1)} - p^{a(r-2)}) = 2p^{a(r-2)}(p^a - 1) \geq 2p^{a(r-2)}a.$$

It thus suffices to show that

$$r^3 \leq 2p^{a(r-2)}. \tag{8.1}$$

If $r \geq 7$, then $r^3 \leq 2 \cdot 3^{r-2} \leq 2p^{a(r-2)}$ and (8.1) holds. If $r = 5$, then

$$r^3 = 5^3 \leq 2 \cdot 7^{5-2} \leq 2p^{a(r-2)} \qquad \text{for } p \geq 7, \text{ and}$$

$$r^3 = 5^3 \leq 2 \cdot 3^{2(5-2)} \leq 2p^{a(r-2)} \quad \text{for } a \geq 2.$$

Thus (8.1) holds when $r = 5$, unless $p = 3$ and $a = 1$. Check that the lemma is also valid then. Let finally $r = 3$. By (8.1), we may assume that $3^3 > 2p^a$. Since $p \geq 5$, this implies $a = 1$ and therefore

$$(r - 1)r(ar + 1) = 2 \cdot 3 \cdot 4 \leq 2(5^3 + 1)/(5 + 1) \leq 2(p^{ar} + 1)/(p^a + 1). \qquad \square$$

In the proof of Theorem 8.4 we shall need that a certain group extension splits. A criterion is provided by the following result of Gaschütz.

8.2 Lemma. *Let A be an abelian normal subgroup of G. Suppose that for each prime p and $P \in \mathrm{Syl}_p(G)$, P splits over $P \cap A$. Then G splits over A, i.e. there exists $H \leq G$ such that $G = AH$ and $A \cap H = 1$.*

Proof. We proceed by induction on $|A|$, and choose a prime divisor p of $|A|$ and $P \in \mathrm{Syl}_p(G)$. Since $P \cap A$ is abelian and $(|P|, |G : P|) = 1$, Gaschütz's Theorem ([Hu, I, 17.4]) asserts that G splits over $P \cap A$, i.e. there exists $K \leq G$ such that $G = (P \cap A)K$ and $(P \cap A) \cap K = 1$.

Let $B \in \mathrm{Hall}_{p'}(A)$. Then $B \leq K$ and induction yields $H \leq K$ such that $K = BH$ and $B \cap H = 1$. Obviously H then is the required complement for A in G. $\qquad\qquad\square$

8.3 Lemma. *Let E be an extra-special group of order p^{2n+1} ($n \in \mathbb{N}$), and V a faithful irreducible $\mathcal{F}[E]$-module for a field $\mathcal{F}$. If $g \in E \setminus \mathbf{Z}(E)$ is an element of order p, then $\dim_{\mathcal{F}}(\mathbf{C}_V(g)) = (1/p) \cdot \dim_{\mathcal{F}} V$.*

Proof. Let $\mathcal{K}$ be an algebraically closed field extension of $\mathcal{F}$. Then $V \otimes_{\mathcal{F}} \mathcal{K} = W_1 \oplus \cdots \oplus W_m$ for irreducible Galois-conjugate $\mathcal{K}[E]$-modules W_i and we may thus assume that $\mathcal{F}$ is algebraically closed. Since $\mathrm{char}(\mathcal{F}) \neq p$ must hold, it is no loss to assume that $\mathcal{F} = \mathbb{C}$. We denote by χ the character afforded by V. Then $\chi(1) = p^n$ and χ vanishes off $\mathbf{Z}(E)$ ([Hu, V, 16.14]). Therefore $\chi_{\langle g \rangle} = p^{n-1} \cdot \rho$, where ρ is the regular character of $\langle g \rangle$, and $\dim_{\mathcal{F}}(\mathbf{C}_V(g)) = [\chi_{\langle g \rangle}, 1_{\langle g \rangle}] = p^{n-1}$. $\qquad\qquad\square$

8.4 Theorem (Berger). *Let G be a (solvable) group of odd order, p an odd prime and V a symplectic $GF(p)[G]$-module with respect to the non-singular symplectic G-invariant form $(\ ,\ )$. We set $\dim(V) = 2n$. Then there exists an element $v \in V^{\#}$ such that $|G : \mathbf{C}_G(v)| \leq (p^n + 1)/2$.*

Proof. We proceed by induction on $|G| + \dim(V)$, and may clearly assume that V is faithful.

Step 1. V is irreducible.

Proof. If not, we choose an irreducible submodule W of V of smallest possible dimension. Set $\dim(W) = m$. Since $(\ ,\ )$ is G-invariant, the subspace $\{w \in W \mid (w, w') = 0$ for all $w' \in W\}$ is a submodule of W. Thus the form $(\ ,\ )$ is either trivial or non-singular on W. In the first case, W is totally isotropic, and [Hu, II, 9.11] implies that $m \leq \dim(V)/2 = n$. Since $|W^{\#}| = p^m - 1$ is even but $|G|$ is odd, there are at least two different G-orbits on $W^{\#}$, and we find a vector $w \in W^{\#}$ such that

$$|G : \mathbf{C}_G(w)| \leq (p^m - 1)/2 \leq (p^n + 1)/2.$$

The assertion holds in this case. We may thus assume that $(\ ,\ )$ is non-singular on W. But then $m = 2l$ (for some $l \in \mathbb{N}$), and since $l < n$, the inductive hypothesis implies the existence of some $w \in W^{\#}$ such that

$$|G : \mathbf{C}_G(w)| \leq (p^l + 1)/2 \leq (p^n + 1)/2.$$

This completes Step 1.

Step 2. V is quasi-primitive.

Proof. Suppose not, and choose $N \trianglelefteq G$ such that $V_N = V_1 \oplus \cdots \oplus V_t$ with homogeneous components V_i and $t > 1$. Set $H = \mathbf{N}_G(V_1)$. By Clifford's Theorem, V_1 is an irreducible H-module, and we argue as in the last step that the form $(\ ,\ )$ is either trivial or non-singular on V_1. Since G transitively permutes the V_i, the G-invariant form $(\ ,\ )$ simultaneously is either trivial or non-singular on each V_i.

Set $V_j^{\perp} = \{v \in V \mid (v, v_j) = 0$ for all $v_j \in V_j\}$. For $v \in V$, we consider the map $f_v \in V_j^* := \mathrm{Hom}\,_{GF(p)}(V_j, GF(p))$, defined by

$$f_v(v_j) = (v, v_j),\ v_j \in V_j.$$

Then

$$v \mapsto f_v, \; v \in V,$$

induces a G-isomorphism between $V/V_j^{\perp}$ and the dual space V_j^*. Since V_N is completely reducible, there exists an N-module U_j such that $V_N = V_j^{\perp} \oplus U_j$. Thus $U_j \cong V_j^*$ is homogeneous and consequently $U_j = V_{\pi(j)}$ for a permutation $\pi \in S_t$. If (,) is trivial on each V_j, then $V_j \subseteq V_j^{\perp}$ and $\pi(j) \neq j$ for all $j = 1, \ldots, t$. Therefore all V_j occur in dual pairs $(V_j, V_{\pi(j)})$. On the other hand, $t = |G : H|$ is odd, a contradiction.

We may thus assume that (,) is non-singular on each V_j. We set $\dim(V_1) = 2m$. By induction, there is $v \in V_1^{\#}$ such that $|H : \mathbf{C}_H(v)| \leq (p^m + 1)/2$. Since $\mathbf{C}_G(v) = \mathbf{C}_H(v)$, we obtain

$$|G : \mathbf{C}_G(v)| = |G : H| \, |H : \mathbf{C}_H(v)| \leq t(p^m + 1)/2,$$

and we are done provided that $t(p^m + 1) \leq p^n + 1$. Since $t = n/m$ and $p \geq 3$, it otherwise would follow that

$$t > (p^n + 1)/(p^m + 1) \geq p^n/(2p^m) = p^{tm-m}/2 \geq p^{t-1}/2 \geq 3^{t-1}/2,$$

which contradicts $t \geq 3$. This proves Step 3.

Step 3. $F := \mathbf{F}(G)$ is non-abelian.

Proof. Suppose that F is abelian. By Step 2, V_F is homogeneous and Corollary 2.3 yields a labelling of the points of V such that $G \leq \Gamma(p^{2n})$ and $F \leq \Gamma_0(p^{2n})$. By Lemma 2.2, V_F is irreducible. Since F acts symplectically on V, [Hu, II 9.23] implies that $|F| \mid p^n + 1$.

Let F_1 be the Hall subgroup of $\Gamma_0 := \Gamma_0(p^{2n})$ corresponding to the odd prime divisors of $p^n + 1$. Then $G \cap F_1 = F$. We set $G_1 = GF_1$ and claim that G_1 splits over F_1. To establish the claim, we let $S \in \mathrm{Syl}_s(G_1)$ for a prime number s. By Lemma 8.2, it suffices to show that S splits over $S \cap F_1$. We may therefore assume that $s \mid |F_1|$, and so $s \nmid |\Gamma_0/F_1|$, by the definition of F_1. Certainly, $S\Gamma_0$ splits over Γ_0, i.e. there exists $U \leq S\Gamma_0$

such that $S\Gamma_0 = \Gamma_0 U$ and $\Gamma_0 \cap U = 1$. Now $S \in \mathrm{Syl}_s(S\Gamma_0)$ and so there exists $g \in \Gamma_0$ such that $V := U^g \le S$. In particular, $S\Gamma_0 = \Gamma_0 V$ and $\Gamma_0 \cap V = 1$. This implies $S = \Gamma_0 V \cap S = (\Gamma_0 \cap S)V = (F_1 \cap S)V$ and $V \cap (S \cap F_1) = V \cap F_1 \le V \cap \Gamma_0 = 1$. Thus G_1 splits over F_1, and there exists $H_1 \le G_1$ such that $G_1 = F_1 H_1$ and $F_1 \cap H_1 = 1$.

As $H_1 \cong G_1/F_1 \cong G/(G \cap F_1) = G/F$ is cyclic, we may write $H_1 = \langle T \rangle$ for a semi-linear transformation

$$T = T_{a,\sigma} : x \mapsto a x^\sigma, \quad (a \in GF(p^{2n})^\#, \ \sigma \in \mathrm{Gal}\,(GF(p^{2n}/GF(p)))).$$

Let $t = o(\sigma)$. Then $T^t(x) = a \cdot a^\sigma \ldots a^{\sigma^{t-1}} \cdot x^{\sigma^t} = (a \cdot a^\sigma \ldots a^{\sigma^{t-1}})x$ for all $x \in GF(p^{2n})$. Consequently, since $F_1 \le \Gamma_0$ and $F_1 = \mathbf{F}(G_1)$, we obtain $T^t \in \Gamma_0 \cap H_1 \le \mathbf{C}_{G_1}(F_1) \cap H_1 = F_1 \cap H_1 = 1$. Thus $a \cdot a^\sigma \ldots a^{\sigma^{t-1}} = 1$ and a is contained in the kernel of the norm map from $GF(p^{2n})^\#$ to $GF(p^{2n/t})^\#$. By Hilbert's Theorem 90 (see [Ja 1, Theorem 4.28]), there is $v \in GF(p^{2n})^\# = V^\#$ with $a = v \cdot (v^{-1})^\sigma$. But then $T(v) = av^\sigma = v$ and $H_1 \le \mathbf{C}_{G_1}(v)$. Thus

$$|G : \mathbf{C}_G(v)| \le |G_1 : \mathbf{C}_{G_1}(v)| \le |G_1 : H_1| = |F_1| \le (p^n + 1)/2,$$

and Step 3 is complete.

Step 4. Conclusion.

Proof. Since F is non-abelian of odd order, Corollary 1.10 yields an extra-special normal subgroup E of G with exponent r and order $|E| = r^{2t+1}$ (r an odd prime, $t \in \mathbb{N}$). Set $C = \mathbf{C}_G(\mathbf{Z}(E))$. Then $|G : C| \mid r - 1$, and C fixes the non-singular symplectic form on $\bar{E} := E/\mathbf{Z}(E)$. By Corollary 2.6, $r^t \mid \dim_{GF(p)} V = 2n$. Now $|C| + \dim(\bar{E}) = |C| + 2t < |G| + \dim(V)$ and the inductive hypothesis yields an element $1 \ne \bar{x} \in \bar{E}$ ($x \in E$) such that $|C : \mathbf{C}_C(\bar{x})| \le (r^t + 1)/2$. Set $H = \mathbf{C}_C(\bar{x})$ and $M = \langle x, \mathbf{Z}(E) \rangle$. We apply Lemma 1.5 to M and $H/\mathbf{C}_H(M)$ (in the role of E and A there). Since $\mathbf{C}_H(M) = \mathbf{C}_H(x)$, it follows that $|H : \mathbf{C}_H(x)| \mid r$.

Now $p \ne r$, and we have $V = \mathbf{C}_V(x) \oplus [V, x]$. Let $v \in \mathbf{C}_V(x)$ and

$wx - w \in [V, x]$. Then

$$(v, wx - w) = (vx^{-1}, w) - (v, w) = (v, w) - (v, w) = 0.$$

This shows that the form $(,)$ is still non-singular when restricted to $\mathbf{C}_V(x)$. As V_E is homogeneous, Lemma 8.3 implies that $\dim_{GF(p)} \mathbf{C}_V(x) = (1/r) \cdot \dim_{GF(p)} V = 2n/r$. We now apply induction to the action of $L := \mathbf{C}_G(x)$ on $\mathbf{C}_V(x)$, and obtain $|L : \mathbf{C}_L(v)| \leq (p^{n/r} + 1)/2$ for some $v \in \mathbf{C}_V(x)^\#$. To finish the proof, we gather what we have so far, namely

$$|G : \mathbf{C}_G(v)| \leq |G : L| \cdot |L : \mathbf{C}_L(v)|$$

$$\leq |G : C| \cdot |C : H| \cdot |H : \mathbf{C}_H(x)| \cdot |L : \mathbf{C}_L(v)|$$

$$\leq (r - 1) \cdot (r^t + 1)/2 \cdot r \cdot (p^{n/r} + 1)/2$$

$$\leq (r - 1) \cdot (n + 1) \cdot r \cdot (p^{n/r} + 1)/4,$$

because $r^t \mid n$. Now Lemma 8.1 applies, and $|G : \mathbf{C}_G(v)| \leq (p^n + 1)/2$, which was to be shown. $\qquad\qquad\qquad\qquad\qquad\qquad\qquad\qquad\qquad\qquad\quad \Box$

8.5 Example. If $|G|$ is even or $p = 2$, then the assertion of Theorem 8.4 definitely does not hold. The symplectic group $Sp(2n, p)$ namely contains a cyclic irreducible subgroup S of order $p^n + 1$, the so-called *Singer cycle*. Since S acts fixed-point-freely on the natural $GF(p)$-module V of dimension $2n$, all S-orbits $\neq \{0\}$ have length $p^n + 1$.

We show how to construct the Singer cycle inside $Sp(2n, p)$.
(1) Let $q = p^n$, $V_0 = GF(q^2)$, and fix some $a \in GF(q^2) \setminus GF(q)$. Set

$$(v, w) = a \cdot (vw^q - v^q w) + (a \cdot (vw^q - v^q w))^q, \quad v, w \in V_0.$$

Let $s \in \Gamma_0(q^2)$ denote an element of order $q + 1$. Then $(,)$ is a symplectic s-invariant $GF(q)$-bilinear form on V_0. We claim that $(,)$ is non-singular. Since $\langle s \rangle$ acts irreducibly on V_0, it suffices to show that $(,)$ does not vanish on V_0. Choose $x \in GF(q^2) \setminus GF(q)$. Then

$$(x, 1) = a \cdot (x - x^q) + a^q \cdot (x^q - x) = (a - a^q) \cdot (x - x^q) \neq 0.$$

(2) Consider V_0 as a $GF(p)$-vector space V of dimension $2n$, and set

$$[v, w] = \text{tr}_{GF(q)/GF(p)}(v, w), \quad v, w \in V.$$

It is then easy to check that $[\,,\,]$ is a symplectic s-invariant $GF(p)$-bilinear form on V, which is non-degenerate. Clearly, $\langle s \rangle$ of order $p^n + 1$ acts irreducibly on V.

In Huppert's paper [Hu 3], Singer cycles are also constructed in the orthogonal and unitary groups.

8.6 Remark. Odd order symplectic groups in odd characteristic also behave well with respect to long orbits, as A. Espuelas [Es 2] showed. His results runs as follows. Assume that the hypotheses of Theorem 8.4 hold. Then G has at least two regular orbits on V.

Chapter III

MODULE ACTIONS WITH LARGE CENTRALIZERS

§9 Sylow Centralizers — the Imprimitive Case

In the next two sections, we study a situation where a solvable group G acts faithfully and irreducibly on a finite vector space V and each $v \in V$ is centralized by a Sylow p-subgroup (for a fixed prime divisor p of $|G|$). The basic thrust is to show that the examples given in 9.1 and 9.4 are essentially the only ones. If V is a quasi-primitive G-module, we show in Section 10 that $G \leq \Gamma(V)$ (compare with Example 9.1) or $G \leq GL(2,3)$ and $|V| = 3^2$. In this section we assume that G is imprimitive and $\mathbf{O}^{p'}(G) = G$. Choose $C \trianglelefteq G$ maximal such that V_C is not homogeneous and write $V_C = V_1 \oplus \cdots \oplus V_n$ for homogeneous components V_i of V_C. The main result (Theorem 9.3) employs Gluck's result in Section 5 to show that $n = 3, 5$ or 8, $G/C \cong D_6$, D_{10} or $A\Gamma(2^3)$ and p is 2, 2 or 3 (respectively). Furthermore, C transitively permutes the non-zero vectors of V_i for each i. Then Huppert's results of Section 6 apply and $C/\mathbf{C}_C(V_i) \leq \Gamma(V_i)$ unless $|V_i|$ is one of six values. The remainder of this section, somewhat technical, exploits these facts to give detailed information about the normal structure of G.

9.1 Example. Let q, p be primes and n an integer such that $p \nmid q^n - 1$. Let V be an n-dimensional vector space over $GF(q)$. Suppose that $H \trianglelefteq \Gamma(V) = \Gamma(q^n)$ and $p \mid |H|$. If $v \in V$, then $\mathbf{C}_H(v)$ contains a Sylow p-subgroup of H (and of course of $\mathbf{O}^{p'}(H)$). Also, $\mathbf{O}^{p'}(H)$ acts irreducibly on V.

Proof. Since $\Gamma := \Gamma(V)$ acts transitively on $V^{\#}$, we have for $v \in V^{\#}$ that $|\Gamma : \mathbf{C}_\Gamma(v)| = q^n - 1$. Thus $\mathbf{C}_\Gamma(v)$ contains a Sylow p-subgroup of Γ. Since $H \trianglelefteq \Gamma$, we have $P \cap H \in \mathrm{Syl}_p(H)$ for all $P \in \mathrm{Syl}_p(\Gamma)$. Consequently, $\mathbf{C}_H(v)$ contains a Sylow p-subgroup of H for all $v \in V$. It remains to show that $L := \mathbf{O}^{p'}(H)$ acts irreducibly on V.

The last paragraph implies that for all $v \in V$, $\mathbf{C}_L(v)$ contains a Sylow p-subgroup of L, and therefore $\mathbf{O}_p(L)$ acts trivially on V. Thus $\mathbf{O}_p(L) = 1$ and L has at least two Sylow p-subgroups. For $v \in V^\#$, $\mathbf{C}_\Gamma(v) \lesssim \Gamma/\Gamma_0(V)$ and is cyclic. Thus $\mathbf{C}_L(v)$ contains a unique Sylow p-subgroup. If $V = W_1 \oplus W_2$ for L-submodules $W_i \neq 0$, we may choose $w_i \in W_i^\#$, and $P_i \in \mathrm{Syl}_p(L)$ with $P_1 \neq P_2$, and P_i the Sylow p-subgroup of $\mathbf{C}_L(w_i)$. Then $w_1 + w_2$ is not centralized by a Sylow p-subgroup of L. This contradiction implies that $L = \mathbf{O}^{p'}(H)$ acts irreducibly on V. $\square$

If π is a non-empty set of prime divisors of n, each of which is coprime to $q^n - 1$, then $\mathbf{O}^{\pi'}(\Gamma(V))$ acts irreducibly on V, and each $v \in V$ is centralized by a Hall π-subgroup of $\mathbf{O}^{\pi'}(\Gamma(V))$. This follows immediately from the above example.

9.2 Lemma. *Assume that G acts faithfully on a finite vector space V and $2 \nmid |G : \mathbf{C}_G(v)|$ for all $v \in V$. If $|G|$ is even, then $\mathrm{char}(V) = 2$.*

Proof. View V as a multiplicative group and form the semi-direct product $H = VG$. Since G acts faithfully on V, we may choose $v \in V$ and an involution $t \in G$ such that $v^t \neq v$. Let $y = v^{-1}v^t$, so that $y \in V^\#$. Now $y^t = (v^t)^{-1}v = (v^{-1}v^t)^{-1} = y^{-1}$ and $t \in \mathbf{N}_G(\langle y \rangle)$. The hypotheses imply that $\mathbf{N}_G(\langle y \rangle)/\mathbf{C}_G(y)$ has odd order and $t \in \mathbf{C}_G(y)$. Thus $y^{-1} = y^t = y$. Since $y \neq 1$, $\mathrm{char}(V) = 2$ follows. $\square$

9.3 Theorem. *Suppose that a solvable group G acts faithfully and irreducibly on a finite vector space V and each $v \in V$ is centralized by a Sylow p-subgroup of G (p a fixed prime). Furthermore, assume that $C \trianglelefteq G$, $p \mid |G/C|$, that $V_C = V_1 \oplus \cdots \oplus V_n$ for C-invariant subspaces V_i, and that G/C primitively and faithfully permutes $\{V_1, \ldots, V_n\}$. Then*

(a) *$n = 3, 5$ or 8, and $p = 2, 2$ or 3 (respectively);*

(b) *G/C is isomorphic to D_6, D_{10} or $A\Gamma(2^3)$ (respectively); and*

(c) *$C/\mathbf{C}_C(V_i)$ acts transitively on $V_i^\#$ for each i.*

Proof. Let $1 \leq t \leq n$ and let $u_i \in V_i^{\#}$. Set $u = (u_1, \ldots, u_t, 0, \ldots, 0) \in V$. Then a Sylow p-subgroup P of G centralizes u, and so P and PC/C stabilizes $\{V_1, \ldots, V_t\}$. Likewise, every $\Delta \subseteq \{V_1, \ldots, V_n\}$ is stabilized by a Sylow p-subgroup of G/C. Since $p \mid |G/C|$, parts (a) and (b) follow from Corollary 5.8.

For part (c), first assume that $n = 8$, $G/C \cong A\Gamma(2^3)$ and $p = 3$. Observe that a Sylow 3-subgroup of G/C has orbits of size 1, 1, 3 and 3 on $\{V_1, \ldots, V_8\}$ and that $\text{stab}_{G/C}\{V_1, V_2, V_3\} \cong Z_3$ is a Sylow 3-subgroup of G/C. Let $x, w \in V_1^{\#}$, $y \in V_2^{\#}$ and $z \in V_3^{\#}$. Now $(x, y, z, 0, \ldots, 0)$ and $(w, y, z, 0, \ldots, 0)$ are centralized by Sylow 3-subgroups Q_1 and Q_2 of G (respectively). Then $Q_1 C/C = \text{stab}_{G/C}\{V_1, V_2, V_3\} = Q_2 C/C$. In particular, there exist $a, b \in Q_1 C$ such that $x^a = y$, $y^a = z$, $z^a = x$, $w^b = y$, $y^b = z$ and $z^b = w$. Now ab^{-1} stabilizes each of V_1, V_2 and V_3. Since only the trivial element of $\text{stab}_{G/C}\{V_1, V_2, V_3\} \cong Z_3$ stabilizes each of V_1, V_2 and V_3, we have that $ab^{-1} \in C$. Now $x^{ab^{-1}} = w$. Hence C transitively permutes the non-identity elements of V_1. Part (c) now follows in the case when $n = 8$.

To prove (c) when $n = 3$ or 5, $G/C \cong D_{2n}$ and $p = 2$, observe that $\text{stab}_{G/C}\{V_1, V_2\} \cong Z_2$. If $x, w \in V_1^{\#}$, choose $y \in V_2^{\#}$ and consider the vectors $(x, y, 0, \ldots, 0)$ and $(w, y, 0, \ldots, 0)$ in V. Arguments like those above show that C is transitive on $V_1^{\#}$. $\qquad\square$

Note that if C is chosen maximal with respect to $C \trianglelefteq G$ and V_C non-homogeneous, then G/C faithfully and primitively permutes the homogeneous components V_i of V_C (see Lemma 0.2).

9.4 Example. Let p be 2 or 3 and let q^m be a prime power such that $p \nmid q^m - 1$. If $p = 2$, let n be 3 or 5 and $H = D_{2n}$. If $p = 3$, let $n = 8$ and $H = A\Gamma(2^3)$. Observe that H is a primitive permutation group on n letters, and let $G = \Gamma(q^m) \text{ wr } H$ (w.r.t. this permutation action). Then G acts faithfully and irreducibly on a vector space V of dimension mn over $GF(q)$. We let π be the set of prime divisors t of $|G|$ which do not divide $q^m - 1$ or $|H|/p$, (in particular, each t in π must divide m or equal p).

We claim that each $v \in V$ is centralized by a Hall π-subgroup of G. Now G has a normal subgroup $G_0 \cong \Gamma(q^m) \times \cdots \times \Gamma(q^m)$, $G_0 H = G$ and $G_0 \cap H = 1$. Also $V = V_1 \oplus \cdots \oplus V_n$ for irreducible G_0-modules V_i that are permuted by H. Fix $v \in V^{\#}$. Without loss of generality, $v = (v_1, \ldots, v_l, 0, \ldots, 0)$ for non-zero $v_i \in V_i$ $(i = 1, \ldots, l)$. Let $X = \{(w_1, \ldots, w_l, 0, \ldots, 0) \mid w_i \neq 0\}$. Observe that G_0 transitively permutes the elements of X, and X is the G_0-orbit of v. In particular, $|G_0 : \mathbf{C}_{G_0}(v)| = (q^m - 1)^l$.

Suppose $z = (z_1, \ldots, z_n) \in V$ and $\Delta = \{j \mid z_j \neq 0\}$. Then v and z are G-conjugate if and only if $\{1, \ldots, l\}$ and Δ are H-conjugate. Hence $|G : \mathbf{C}_G(v)| = (q^m - 1)^l \cdot |H : \text{stab}_H(\{1, \ldots, l\})|$. Since $n = 3, 5$ or 8, and H is D_6, D_{10} or $A\Gamma(2^3)$, p does not divide the index of any set-stabilizer (i.e. the converse of Corollary 5.8 holds). Thus $|G : \mathbf{C}_G(v)|$ is a π'-number.

Theorem 9.3 gives us important information when V is an imprimitive G-module, $\mathbf{O}^{p'}(G) = G$, and each $v \in V$ is centralized by a Sylow p-subgroup of G. We will apply this in Sections 10 and 12, and hence we will need more specific information. The remainder of this section will study this situation in more detail, although we first give a general proposition.

9.5 Proposition. *Assume that* $C_i \trianglelefteq C$ *and* $\bigcap_i C_i = 1$. *Let* $F_i/C_i = \mathbf{F}(C/C_i)$, *let* r *be a prime and* $R_i/C_i \in \text{Syl}_r(F_i/C_i)$. *Set* $F = \mathbf{F}(C)$ *and let* R *be the Sylow* r-*subgroup of* F. *Then*

(a) $F = \bigcap_i F_i$;

(b) $R = \bigcap_i R_i$; *and*

(c) *If* $r \nmid |C/F_i|$ *for all* i *and* $D_i = \mathbf{C}_C(R_i/C_i)$, *then* $R \in \text{Syl}_r(C)$ *and* $\bigcap_i D_i = \mathbf{C}_C(R)$.

Proof. (a) Let $H = \bigcap_i F_i$. Then $HC_i \leq F_i$ for all i and the final term H^∞ of the descending central series of H lies in C_i. Thus $H^\infty \leq \bigcap_i C_i = 1$ and H is a normal nilpotent subgroup of C, whence $H \leq F$. But FC_i/C_i is a normal nilpotent subgroup of C/C_i and thus $F \leq F_i$ for all i. Therefore $F = H$, proving (a).

(b) Now $RC_i/C_i \leq \mathbf{O}_r(C/C_i) = R_i/C_i$ for all i and so $R \leq \bigcap_i R_i$. If $S = \bigcap_i R_i$, then $S \trianglelefteq C$ and $S/S \cap C_i \cong SC_i/C_i \leq R_i/C_i$. So $S/S \cap C_i$ is an r-group. Since $\bigcap C_i = 1$, $S \leq \mathbf{O}_r(C) = R$, proving (b).

(c) Assume $r \nmid |C/F_i|$ for all i. Then $r \nmid |C/R_i|$ for all i, whence by (b), $C/R = C/(\bigcap_i R_i) \leq \prod_i C/R_i$ is an r'-group. Hence $R \in \mathrm{Syl}_r(C)$, and $RC_i \leq R_i$ implies that $RC_i = R_i$. Therefore $\mathbf{C}_C(R)$ centralizes R_i/C_i and so $\mathbf{C}_C(R) \leq \bigcap_i D_i$. On the other hand, $[\bigcap_i D_i, R] \leq [D_j, R] \leq C_j$ for all j. Hence $\bigcap_i D_i \leq \mathbf{C}_C(R)$, and (c) follows. $\square$

9.6 Notation. Throughout the remainder of Section 9, we will be assuming that G satisfies the hypotheses of Theorem 9.3. We will let C_i denote $\mathbf{C}_C(V_i)$ and let $F_i/C_i = \mathbf{F}(C/C_i)$. Also, write $|V_i| = q^m$ for a prime q and an integer m. Also F will denote $\mathbf{F}(C)$. Recall that by Theorem 9.3, $p \in \{2,3\}$ and $\bigcap_i C_i = 1$.

Next is a corollary to Theorem 9.3.

9.7 Corollary.

(a) If $p = 2$, then $q = 2$.

(b) If $p = 3$, then $q = 3$ or m is odd.

(c) If $q^m \neq 3^2$ or 3^4, then F_i/C_i and C/F_i are cyclic groups whose orders divide $(q^m - 1)$ and m (respectively).

(d) If $q^m \neq 3^2$, 3^4 or 2^6, then there is a Zsigmondy prime divisor r of $q^m - 1$, and if $R_i/C_i \in \mathrm{Syl}_r(C/C_i)$, then $F_i = \mathbf{C}_C(R_i/C_i) \geq R_i$.

(e) If $q^m = 2^6$, and $R_i/C_i \in \mathrm{Syl}_7(C/C_i)$, then $\mathbf{C}_C(R_i/C_i)/F_i$ has order at most 2 and $F_i \geq R_i$.

Proof. Since C acts transitively on $V_i^{\#}$, we have that $|C : \mathbf{C}_C(x)| = q^m - 1$ for each $x \in V_i^{\#}$. The hypothesis on centralizers implies that $p \nmid q^m - 1$. Parts (a) and (b) follow immediately. (Of course, (a) is also a consequence of Lemma 9.2.)

We may now assume that $q^m \neq 3^2$ or 3^4. Since C/C_i acts transitively on $V_i^{\#}$, we may apply Theorem 6.8. By parts (a) and (b) above, none of the exceptional six values of q^m can occur here. We thus conclude that $C/C_i \leq \Gamma(q^m)$. If, in addition, $q^m \neq 2^6$, the Zsigmondy Prime Theorem 6.2 together with (a) and (b) imply that $q^m - 1$ has a Zsigmondy prime divisor r. By transitivity, $r \mid |C/C_i|$. Parts (c) (except for the case $q^m = 2^6$) and (d) now follow from Corollary 6.6 and Lemma 6.4.

To complete the proof, we can assume that $q^m = 2^6$ and must prove (c) and (e). By transitivity, $3^2 \cdot 7 \mid |C/C_i|$, and by the last paragraph, $C/C_i \leq \Gamma(2^6) =: \Gamma$. Let S and T be the unique subgroups of $\Gamma_0 := \Gamma_0(2^6)$ of order 3 and 7 (respectively). Then $ST \leq F_i/C_i$, $T = R_i/C_i$ and Lemma 6.5 (b) yields $F_i/C_i \leq \mathbf{C}_{C/C_i}(ST) \leq \mathbf{C}_\Gamma(ST) = \Gamma_0$. Therefore $F_i/C_i = C/C_i \cap \Gamma_0$ and parts (c) and (e) follow, since $|\mathbf{C}_\Gamma(T)/\Gamma_0| = 2$. $\qquad\square$

For an abelian p-group P, the *rank* of P is m when $p^m = |\{x \in P \mid x^p = 1\}|$. For an abelian group A, $\mathrm{rank}\,(A) = \max\{\mathrm{rank}\,(P) \mid P \in \mathrm{Syl}_p(A)\}$.

9.8 Lemma. *Assume that $q^m \neq 3^2$ or 3^4. Then*

(a) F *and* C/F *are abelian of rank at most* n.

(b) *The exponent of* C/F *divides* m.

(c) *If* $q^m \neq 2^6$, *there exists a Zsigmondy prime divisor* r *of* $q^m - 1$; *and for every such* r *and* $R \in \mathrm{Syl}_r(C)$, *we have* $1 \neq R \leq F = \mathbf{C}_C(R)$.

(d) *If* $q^m = 2^6$ *and* $R \in \mathrm{Syl}_7(C)$, *then* $1 \neq R \leq F \leq \mathbf{C}_C(R)$ *and* $\mathbf{C}_C(R)/F$ *is a 2-group.*

Proof. By Proposition 9.5, $F = \bigcap_i F_i$ and thus $C/F \leq \prod_{i=1}^n C/F_i$. But each C/F_i is cyclic and $|C/F_i| \mid m$, by Corollary 9.7 (c). Hence C/F is abelian, $\mathrm{rank}\,(C/F) \leq n$ and $\exp(C/F) \mid m$. Now $F/(F \cap C_i) \cong FC_i/C_i \leq F_i/C_i$. Since $\bigcap_i (F \cap C_i) = 1$, we have $F \leq \prod_{i=1}^n F_i/C_i$ and F is abelian of rank at most n, by Corollary 9.7 (c). This proves (a) and (b).

If $q^m \neq 2^6$, Corollary 9.7 (d) yields the existence of a Zsigmondy prime divisor r of $q^m - 1$. If $q^m = 2^6$, we let $r = 7$. In all cases, $r \mid |C/C_i|$ by Corollary 9.7, but $r \nmid |C/F_i|$. Parts (c) and (d) now follow by Corollary 9.7 (d), (e) and Proposition 9.5. $\square$

9.9 Corollary. *Assume that $q^m \neq 3^2$ or 3^4. Suppose that $H \leq F$, $H \trianglelefteq G$ and $\mathbf{C}_G(H) \not\leq C$. Then H is cyclic.*

Proof. Let L/C be the minimal normal subgroup of G/C and note that L/C transitively permutes the V_i and hence also the $H \cap C_i$. Since we have $\mathbf{C}_G(H) \not\leq C$, $L \leq C \cdot \mathbf{C}_G(H)$. Thus $\mathbf{C}_G(H)$ permutes the $H \cap C_i$ transitively. In particular, $H \cap C_1 = H \cap C_2 = \cdots = H \cap C_n = 1$, with the last equality holding because $\bigcap_i C_i = 1$. Then $H \cong HC_1/C_1 \leq FC_1/C_1 \leq F_1/C_1$. Now F_1/C_1 is cyclic by Corollary 9.7 (c). Thus H is cyclic. $\square$

9.10 Lemma. *Assume that $\mathbf{O}^{p'}(G) = G$, $p \nmid |C|$ and $q^m \neq 3^2$ or 3^4. Then*

(a) $[G, C] = C$.

(b) *If $1 \neq S/F \in \mathrm{Syl}_s(C/F)$ for some prime s, then $C/F = \mathbf{C}_{G/F}(S/F)$.*

(c) *If $1 \neq S/F \in \mathrm{Syl}_s(C/F)$ and $p = 3$, then $s = 2$ or $\mathrm{rank}\,(S/F) \geq 7$.*

Proof. Recall that we have $C < L \leq K \leq G$ with $|L/C| = n$, L/C a chief factor of G, and $|G/K| = p$. If $p = 2$, then $L = K$. Also L/C transitively permutes the V_i, and hence the C_i and F_i.

(a) We may assume that there exists $A \trianglelefteq G$ with $A \leq C$, $C/A \leq \mathbf{Z}(G/A)$ and $|C/A|$ prime. If the nilpotent group L/A is non-abelian, then $|L/A|$ is a prime power and $C/A = \mathbf{Z}(L/A) = \Phi(L/A) = (L/A)'$. Hence L/A is extra-special and $|L/C|$ is a square, a contradiction as $|L/C| = n \in \{3, 5, 8\}$. Hence L/A is abelian. Since L/C is an irreducible faithful G/L-module, we may write $L/A = U/A \times C/A$ with $U \trianglelefteq G$ and U/A G-isomorphic to L/C (note $(|G/L|, |L/C|) = 1$). If $(|G/L|, |C/A|) = 1$, then G has a factor group isomorphic to $L/U \cong C/A$, contradicting the hypothesis that $\mathbf{O}^{p'}(G) = G$.

Thus $|C/A| \mid |G/L|$. Since $p \nmid |C|$, in fact $|C/A| \mid |K/L|$, whence $p = 3$ and $|K/L| = 7 = |C/A| = |L/U|$. Then $|G/U| = 3 \cdot 7^2$, $|\mathbf{Z}(G/U)| = 7$ and $\mathbf{O}^7(G/U) < G/U$, a contradiction to $\mathbf{O}^{3'}(G) = G$. This proves (a).

(b) We now assume that $1 \neq S/F \in \mathrm{Syl}_s(C/F)$. Since C/F is abelian (by Lemma 9.8), $C/F \leq \mathbf{C}_{G/F}(S/F)$. Since we wish to prove that $C/F = \mathbf{C}_{G/F}(S/F)$ and since L/C is the unique minimal normal subgroup of G/C, we assume that $L/F \leq \mathbf{C}_{G/F}(S/F)$. Now $F \leq F_i \cap S \leq S$ for each i. Since L transitively permutes the F_i and likewise the $F_i \cap S$, and since L/F centralizes S/F, we have that $F_1 \cap S = F_2 \cap S = \cdots = F_n \cap S$. But $\bigcap_i (F_i \cap S) = (\bigcap_i F_i) \cap S = F$. Thus $F_1 \cap S = F$ and consequently $S/F = S/(F_1 \cap S) \cong SF_1/F_1 \leq C/F_1$. By Corollary 9.7 (c), S/F must be cyclic and so $G/\mathbf{C}_G(S/F)$ is abelian. Since $(G/L)' = K/L$ and $L \leq \mathbf{C}_G(S/F)$, we now have that $S/F \leq \mathbf{Z}(K/F)$.

Let $P \in \mathrm{Syl}_p(G)$ and note that $|P| = p$ because $p \nmid |C|$. We claim that P centralizes S/F. First observe that P stabilizes some V_i, without loss of generality $P \leq \mathbf{N}_G(V_1)$. Set $H = PC$, hence $H/C \cong P$ has order p. By the last paragraph, S/F is H-isomorphic to a subgroup of C/F_1. For the claim, it therefore suffices to show that P centralizes C/F_1. We may thus assume that $\mathbf{O}_p(H/C_1) = 1$ and that $F_1/C_1 = \mathbf{F}(H/C_1)$. In particular, H/C_1 acts faithfully on V_1 because C/C_1 does. Also, H/C_1 transitively permutes $V_1^{\#}$. Since $|V_1| \neq 3^2$ nor 3^4, it follows from Corollary 9.7 (a), (b) and Theorem 6.8 that $H/C_1 \leq \Gamma(q^m)$. In particular, $(H/C_1)/\mathbf{F}(H/C_1) \cong H/F_1$ is cyclic, and P centralizes C/F_1, as desired. We have established our claim that $P \leq \mathbf{C}_G(S/F)$.

By the last two paragraphs, S/F is centralized by $KP = G$. Let U/F be the Hall s'-subgroup of C/F. Since C/F is abelian and $S/F \neq 1$, we have that $[G, C] \leq U < C$, contradicting part (a). This contradiction arises because we assumed that $C/F > \mathbf{C}_{G/F}(S/F)$.

(c) We now assume that $p = 3$ and that $1 \neq S/F \in \mathrm{Syl}_s(C/F)$ for a prime $s \geq 3$. By (b), we have $C/F = \mathbf{C}_{G/F}(S/F)$, i.e. G/C acts faithfully on the abelian group S/F. Since L/C is a 2-group and $s \neq 2$, L/C acts faithfully on $\Omega_1(S/F) = \{g \in S/F \mid g^s = 1\}$. In fact we may choose $Y \leq X \leq \Omega_1(S/F)$ such that X/Y is a chief factor of G and L/C acts faithfully on X/Y. Since L/C is the only minimal subgroup of G/C, X/Y is a faithful G/C-module and $\mathbf{C}_{X/Y}(L/C) = 0$. But K/C is a Frobenius group of order $2^3 \cdot 7$ and so $\dim(X/Y) \geq 7$ by Lemma 0.34. Consequently, $|\Omega_1(S/F)| \geq s^7$ and $\mathrm{rank}\,(S/F) \geq 7$.   □

9.11 Corollary. *Assume that* $p = 2$, $\mathbf{O}^{2'}(G) = G$ *and* $2 \nmid |C|$. *If* $q^m \neq 2^6$, *let* r *be a Zsigmondy prime divisor of* $q^m - 1$. *If* $q^m = 2^6$, *then let* $r = 7$. *If* R *is the Sylow* r-*subgroup of* F *and* $C > F$, *then* $\mathrm{rank}\,(R) = n$.

Proof. Now $1 \neq C/F$ has odd order and $\exp(C/F) \mid m$, by Lemma 9.8 (b). So m is divisible by an odd prime. As $q^m = 2^m$ (see Lemma 9.2) and $q^m \neq 2^2$ nor 2^4, r is not 3 or 5. Consequently $r \nmid |G/C|$, because $G/C \cong D_6$ or D_{10}. Since $|C/F|$ is odd, Lemma 9.8 (c), (d) implies that $R \in \mathrm{Syl}_r(G)$ and $\mathbf{C}_C(R) = F$.

Now $C < L < G$ with $|G/L| = 2$ and $|L/C| = 3$ or 5. Since $C > F$, it follows from Lemma 9.10 (b) that L/F is non-abelian. Since C/F acts faithfully on R, so must L/F. Now $r \nmid |L/F|$ and thus L/F acts faithfully and completely reducibly on $\Omega_1(R) = \{x \in R \mid x^r = 1\}$. Thus there exists an irreducible L/F-module $X \leq \Omega_1(R)$ such that $L/\mathbf{C}_L(X)$ is non-abelian. In particular, $F \leq \mathbf{C}_L(X) \leq C$, and $L/\mathbf{C}_L(X)$ has a normal abelian subgroup $C/\mathbf{C}_L(X)$ of prime index $n \in \{3,5\}$. Consequently, $\dim(X) \geq n$ because $r \nmid |L/\mathbf{C}_L(X)|$ and every absolutely irreducible faithful $L/\mathbf{C}_L(X)$-module in characteristic r has degree n. Thus $|\Omega_1(R)| \geq r^n$ and $n \geq \mathrm{rank}\,(R)$. By Lemma 9.8 (a), $\mathrm{rank}\,(R) = n$.   □

9.12 Lemma. *Suppose that* $p = 3$, *that* $E \trianglelefteq G$ *and* E *is an abelian* s-*group for a prime* s, $s \nmid |L/F|$. *Assume that* $\mathbf{C}_G(E) \leq C$. *Also assume that* $q^m \neq 3^2$ *or* 3^4. *Then* $\mathrm{rank}\,(E) \geq 7$.

Proof. Set $B = \mathbf{C}_G(E)$. The hypotheses and Lemma 9.8 imply that $E \leq F \leq B \leq C$. Note that K/C is a Frobenius group of order $2^3 \cdot 7$ with Frobenius kernel L/C of order 2^3.

We claim there exist subgroups $B < J \lhd H \leq G$ such that J/B is abelian, H/B is non-abelian, $|H/J| = 7$ and $s \nmid |J/B|$. If L/B is abelian, we let $J = L$ and $H = K$. Since $s \nmid |L/F|$ and K/C is non-abelian, the claim holds in this case. We thus assume that L/B is non-abelian. Set $J = C$ and $H/C \in \mathrm{Syl}_7(G/C)$. Then $s \nmid |J/B|$. We need just show that H/B is non-abelian. If not, then $H \leq \mathbf{C}_G(C/B)$. Since L/C is the unique minimal normal subgroup of G/C, we have that $\mathbf{C}_G(C/B)$ contains L and $LH = K$. Because $B < L'B \leq C$ and $C/B \leq \mathbf{Z}(K/B)$, we may choose $B \leq D < C$ such that L/D is non-abelian and C/D is cyclic. Since L/C is a chief factor of K, it follows that $C/D = \mathbf{Z}(L/D) = \mathbf{Z}(K/D)$. Furthermore, every normal abelian subgroup of K/D must be contained in $C/D = \mathbf{Z}(K/D)$. By Corollary 1.4, $|L : C| = |\mathbf{F}(K/D) : \mathbf{Z}(K/D)|$ is a square. Since $|L : C| = 2^3$, this contradiction establishes the claim.

Now let $1 = E_0 < E_1 < \cdots < E_t = E$ with each E_{i+1}/E_i an irreducible H/B-module. Since $1 \neq (H/B)' \leq C/B$ and $s \nmid |(H/B)'|$, it follows that for some j, $(H/B)' \neq \mathbf{C}_{H/B}(E_{j+1}/E_j)$. Thus $T := H/\mathbf{C}_H(E_{j+1}/E_j)$ is non-abelian. Now T has an abelian normal subgroup of index 7. Thus $7 \mid \dim(E_{j+1}/E_j)$ and $\mathrm{rank}(E) \geq 7$. $\qquad\square$

§10   Sylow Centralizers — the Primitive Case

We continue to study the situation where V is an irreducible G-module and every element is centralized by a Sylow p-subgroup of G, but our emphasis now will be on when V is quasi-primitive. Actually, we just assume that V is pseudo-primitive. Recall that V is called pseudo-primitive if V_N is homogeneous for all characteristic subgroups N of G. So the results of the last section still come into play, specifically in Lemma 10.1. If V is pseudo-primitive and $\mathbf{C}_G(v)$ contains a Sylow p-subgroup of G (for all $v \in V$ and a fixed prime divisor p of G), then $G \leq \Gamma(V)$ or $|V| \in \{3^2, 2^6\}$ (see Theorem 10.5).

10.1 Lemma. *Suppose that a solvable group $G \neq 1$ acts irreducibly and faithfully on a finite vector space V such that each $v \in V$ is centralized by a Sylow p-subgroup of G. Assume that $\mathbf{O}^{p'}(G) = G$ and V is pseudo-primitive. Then either*

   (i) *Every normal abelian subgroup of G is cyclic; or*

   (ii) *$|V| = 2^6$, $p = 2 = |G : \mathbf{F}(G)|$, $\mathbf{F}(G)$ is an extra-special 3-group of exponent 3 and order 3^3, and $\mathbf{Z}(\mathbf{F}(G)) = \mathbf{Z}(G)$.*

Proof. The hypotheses imply that $\mathbf{O}_p(G) \leq \mathbf{C}_G(V) = 1$, $p \nmid |\mathbf{F}(G)|$, G/G' is a p-group and thus $\mathbf{F}(G) \leq G'$. Every characteristic abelian subgroup of G is cyclic, central in G', and thus contained in the center Z of $\mathbf{F}(G)$. We set $F = \mathbf{F}(G)$. If $Z = F$, conclusion (i) holds. We may assume via Corollary 1.4 that $F = E \cdot Z$ where E is a direct product of extra-special groups, $Z \cap E = \mathbf{Z}(E)$, and $|F/Z| = e^2$ for an integer $e > 1$. (Note that Z has a different meaning than in Cor. 1.4 and that we do not assume that E is normal in G.)

We can assume that there exists a non-cyclic abelian normal subgroup $A \trianglelefteq G$. Of course, $A \leq F$, and V_A is not homogeneous. By Proposition 0.2, there exists a normal subgroup $A \leq C \trianglelefteq G$ such that V_N is not homogeneous for all normal subgroups N of G with $A \leq N \leq C$. Moreover, $V_C = U_1 \oplus \cdots \oplus U_n$ for C-invariant U_i that G/C faithfully and primitively permutes. In particular, $F \not\leq C$ by the hypotheses of the theorem. By Theorem 9.3, $G/C \cong D_6$, D_{10} or $A\Gamma(2^3) =: J$, $n = 3$, 5, 8 and $p = 2$, 2 or 3, respectively. Thus G/C has a unique minimal normal subgroup L/C and a unique maximal normal subgroup K/C; of course $K = L$ when $p = 2$. Since $F \not\leq C$, we have that $FC = L$ and $F \cap C = \mathbf{F}(C)$. Consequently, $F/\mathbf{F}(C) \cong L/C$ is a chief factor of G with order $n = 3$, 5 or 2^3 (respectively).

If $C_i = \mathbf{C}_C(U_i)$ and $F_i/C_i = \mathbf{F}(C/C_i)$, then $\bigcap_i C_i = 1$ and $\bigcap_i F_i = \mathbf{F}(C)$, by Proposition 9.5. Since $F/\mathbf{F}(C) \cong L/C$ as G-modules, $F/\mathbf{F}(C)$ transitively permutes the C_i and likewise the F_i. But $\mathbf{F}(C) \le F_i \le C$ and $F/\mathbf{F}(C)$ centralizes $C/\mathbf{F}(C)$. Thus $F_1 = \cdots = F_n = \mathbf{F}(C)$ and $C/\mathbf{F}(C) = C/F_1$.

We claim that $\mathbf{F}(C)$ is abelian, or $p = 3$ and $|U_i| = 3^2$. We may assume by Lemma 9.8 (a) and Corollary 9.7 (a) that $|U_i| = 3^4$ and $p = 3$. It follows from Theorem 9.3 (c) that C/C_1 transitively permutes the elements of $U_1^{\#}$. Consequently, Theorem 6.8 implies that either $C/C_1 \le \Gamma(3^4)$, or $C/\mathbf{F}(C) = C/F_1$ has order 5, 10 or 20. In the first case, $|C/C_1|$ is divisible by the Zsigmondy prime divisor 5 of $3^4 - 1$, and by Corollary 6.6, F_1/C_1 and hence $\mathbf{F}(C)$ are abelian. In the second case, $L/F \cong C/\mathbf{F}(C)$ has a normal Sylow 5-subgroup T/F of order 5. Now G/L is non-abelian of order $3 \cdot 7$, and $\mathrm{Aut}\,(T/F) \le Z_4$, contradicting $\mathbf{O}^{3'}(G) = G$. This establishes the claim.

Recall that $F/\mathbf{F}(C) \cong L/C$ is a chief factor of order $n = 3$, 5 or 2^3. Also F is non-abelian and Z is cyclic. By the last paragraph, $\mathbf{F}(C)$ is abelian or $p = 3$ and $n = 2^3$. Thus $Z \le \mathbf{F}(C)$. Since $A \le \mathbf{F}(C) \le C$, V does not restrict homogeneously to $\mathbf{F}(C)$ (see second paragraph) and the hypotheses imply that $\mathbf{F}(C)$ is not characteristic in G. Since $F/\mathbf{F}(C)$ is irreducible, it follows that there exists $M \trianglelefteq G$ with $Z \le M \le \mathbf{F}(C)$ such that $F/M = X_1 \oplus X_2$ for irreducible G-modules X_i of order n. Also $G/\mathbf{C}_G(X_2) \cong G/\mathbf{C}_G(X_1) = G/\mathbf{C}_G(L/C) = G/L$. Since $K/L = (G/L)'$, it follows that $G/\mathbf{C}_G(F/M) \lesssim G/L \times G/L$ and $|G'/\mathbf{C}_{G'}(F/M)| \mid |K/L|^2$.

Since $F = EZ$ for a direct product of extra-special groups E and $Z = \mathbf{Z}(F)$, an abelian subgroup B of F with $Z \le B$ must satisfy $|B/Z| \le |F/B|$ (see [Hu, III, 13.7]). Since $Z \le M \le \mathbf{F}(C) \le F$ and $|\mathbf{F}(C)/M| = |F/\mathbf{F}(C)|$, it follows that $M = Z$ whenever $\mathbf{F}(C)$ is abelian. By the next to the last paragraph, $M = Z$ unless possibly $p = 3$ and $|U_i| = 3^2$. For the moment, assume that $M = Z$. By the last paragraph, $|G'/\mathbf{C}_{G'}(F/Z)| \mid |K/L|^2$. By Theorem 1.12, G/F acts faithfully and completely reducibly on $F/\Phi(G)$.

Then G'/F acts faithfully on $F/(Z\Phi(G))$ and on F/Z. Thus $|G'/F| \mid |K/L|^2$ except possibly when $M > Z$, $p = 3$ and $|U_i| = 3^2$. In this exceptional case, $n = 8$ and $|V| = |U_i|^8 = 3^{16}$. Now $e = |F : Z|^{1/2} > |F : M|^{1/2} = 8$ and $e \mid \dim(V)/\dim(V_0)$ for an irreducible Z-submodule V_0 of V. Since $|V| = 3^{16}$, we have that $e = 16$, $|V_0| = 3$, $|M/Z| = 4$ and $|Z| = 2$. Now $Z = \mathbf{Z}(G) \leq \Phi(G)$ and $F/\Phi(G)$ is a completely reducible faithful G/F-module whose irreducible constituents are G-isomorphic to X_1, X_2 or a submodule Y of M/Z. Note that $G/\mathbf{C}_G(Y)$ has order 1 or 3 because $\mathbf{O}^{p'}(G) = G$. Recalling that $|K/L|$ is 1 or 7 for $p = 2$ or 3 (respectively), we summarize:

$$|G' : F| = 1 \quad \text{when } p = 2;$$

$$|G' : F| \mid 7^2 \quad \text{when } p = 3.$$

Also $|F/Z| = n^2$, except possibly when $|F/Z| = 4 \cdot n^2$, $p = 3$, $|U_i| = 3^2$ and $Z = \mathbf{Z}(G)$ has order 2.

First suppose that $p = 3$, so that $n = 8$. We set $|U_i| = q^m$ for a prime q and an integer m. Since V_Z is homogeneous and $\mathbf{Z} \leq C$, we have that $|Z| \mid q^m - 1$. Now $|\mathrm{Syl}_3(G)| = |G' : \mathbf{C}_{G'}(P)|$ for $P \in \mathrm{Syl}_3(G)$, because G/G' is a 3-group. Thus

$$|\mathrm{Syl}_3(G)| \leq 7^2 \cdot 8^2 \cdot |Z| \leq 7^2 \cdot 2^6 \cdot (q^m - 1), \text{ or}$$
$$|\mathrm{Syl}_3(G)| \leq 7^2 \cdot 8^2 \cdot 4 = 7^2 \cdot 2^8 \text{ and } q^m = 3^2. \tag{10.1}$$

Now P permutes the U_i in orbits of size 1, 1, 3 and 3. Next we let $X_0 = \{(u_1, \ldots, u_8) \mid u_i \in U_i$, exactly six u_i are non-zero$\}$. Then P centralizes at most $(q^m - 1)^2$ elements of X_0. The hypotheses imply that each $x \in X_0$ is centralized by some Sylow 3-subgroup. Hence

$$|\mathrm{Syl}_3(G)| \cdot (q^m - 1)^2 \geq |X_0| = \binom{8}{2} \cdot (q^m - 1)^6. \tag{10.2}$$

Combining (10.1) and (10.2), we get

$$7^2 \cdot 2^6 \geq 7 \cdot 2^2 \cdot (q^m - 1)^3, \text{ or}$$
$$7^2 \cdot 2^8 \geq 7 \cdot 2^2 \cdot (q^m - 1)^4 = 7 \cdot 2^{14} \text{ when } q^m = 3^2.$$

The second equation is nonsense and the first only holds when $q^m \leq 5$. Note that 2 divides $|F/Z|$ and hence $|Z|$. Thus q^m is 3 or 5, and $Z \leq \mathbf{Z}(G)$ because $\mathbf{O}^{3'}(G) = G$. Since $m = 1$, we have $C = F_1$ and $L/F \cong C/\mathbf{F}(C) = C/F_1 = 1$ (see the third paragraph of the proof). Thus $G/L = G/F$ is non-abelian of order $3 \cdot 7$ and $F/Z = X_1 \oplus X_2$ with each X_i a faithful G/F-module of order 2^3. Then $|\mathbf{C}_{F/Z}(P)| = 2^2$. Since $Z \leq \mathbf{Z}(G)$, $|\mathrm{Syl}_3(G)| \leq |K/F| \cdot |(F/Z) : \mathbf{C}_{F/Z}(P)| = 7 \cdot 2^4$. This contradicts (10.2). Hence $p \neq 3$.

Now $p = 2$ and $G/G' = G/F$ is a 2-group. Also $G/C \cong D_{2n}$ with $n = 3$ or 5 and $P \in \mathrm{Syl}_2(G)$ fixes exactly one U_i and permutes the others in pairs. Now $\mathbf{F}(C)$ is abelian with index n in F, $|F/Z| = n^2$ and $n \mid |Z|$. By Lemma 9.2, $|U_i| = 2^m$ for an integer m.

We claim that $L = F$. Assume not, so that $L/F \cong C/\mathbf{F}(C) = C/F_1$ is a non-trivial 2-group. Since C/C_i transitively permutes the elements of $U_i^{\#}$, Theorem 6.8 implies that $C/C_i \leq \Gamma(2^m)$. Since $|\Gamma(2^m)| = m(2^m - 1)$ and $2 \mid |C/F_1|$, m is even. Now a Sylow 2-subgroup Q of C centralizes at most $2^{m/2}$ elements of U_i for each i. Thus Q centralizes at most $(2^{m/2} - 1)^n$ elements of $X := \{(u_1, \ldots, u_n) \mid u_i \in U_i, \text{ all } u_i \text{ non-zero}\}$. Since each element of X is centralized by a Sylow 2-subgroup of C, we have that $|\mathrm{Syl}_2(C)|(2^{m/2} - 1)^n \geq |X| = (2^m - 1)^n$. On the other hand, $|\mathrm{Syl}_2(C)| \leq |\mathbf{F}(C)| = n \cdot |Z| \leq n \cdot (2^m - 1)$, with the last inequality because $|Z| \mid |U_i| - 1$. Thus $n \geq (2^{m/2} + 1)^n / (2^m - 1)$. Since $n = 3$ or 5, this is a contradiction. Therefore, $L = F$ has index 2 in G and $C = \mathbf{F}(C)$. As $|F|$ is odd, we also have $|P| = 2$.

Now P inverts the module F/Z of order n^2. Since the Sylow n-subgroup N of F is non-abelian, P does not induce a fixed-point-free automorphism of N. Hence P centralizes the cyclic group $\mathbf{Z}(N)$. Write $Z = \mathbf{Z}(N) \times S$ for a cyclic $\{2, n\}'$-group S. Now $F = N \times S$ and $\mathbf{C}_S(P) = 1$, as $\mathbf{O}^{2'}(G) = G$. Without loss of generality, P stabilizes U_1. Applying Lemma 0.34 to the action of SP on U_1, we conclude that $\mathbf{C}_{U_1}(P) = 2^{m/2}$ or $S = 1$. For the set $X := \{(u_1, \ldots, u_n) \mid u_i \in U_i, \text{ all } u_i \text{ non-zero}\}$, we have that $|\mathbf{C}_X(P)| \leq (2^m - 1)^{(n+1)/2}$, and even $|\mathbf{C}_X(P)| \leq (2^{m/2} - 1)(2^m - 1)^{(n-1)/2}$ provided that $S \neq 1$. Now $|\mathrm{Syl}_2(G)| \leq |F : \mathbf{C}_F(P)| = n^2|S| \leq n^2|Z|/n \leq n(2^m - 1)$. Since each element of X is centralized by a Sylow 2-subgroup of G, it holds that

$$n(2^m - 1)(2^m - 1)^{(n+1)/2} \geq |X| = (2^m - 1)^n, \quad \text{and}$$

$$n(2^m - 1)(2^m - 1)^{(n-1)/2}(2^{m/2} - 1) \geq |X| = (2^m - 1)^n \quad \text{if } S \neq 1.$$

If $n = 5$, the first inequality implies that $2^m \leq 6$, a contradiction because $n \mid |Z| \mid 2^m - 1$. Thus $n = 3$. Should $S \neq 1$, then the second inequality implies that $2^m = 4$, whence $|Z| = 3$ and $S = 1$, a contradiction. Thus $S = 1$, i.e. $F = N$. Now P centralizes at most $2^m - 1$ elements of $Y :=$ $\{(u_1, u_2, u_3) \mid u_i \in U_i,$ exactly one u_i is zero$\}$. Since $|\mathrm{Syl}_2(G)| = 3^2$, we have that $3^2 \cdot (2^m - 1) \geq |Y| = 3 \cdot (2^m - 1)^2$, and therefore $2^m = 4$. This means that $|V| = 2^6$, $|Z| = 3$ and F is extra-special of order 3^3. That F has exponent 3 follows from Theorem 1.2.   □

10.2 Lemma. *Suppose that V is a finite faithful irreducible G-module, that $\mathbf{O}^{p'}(G) = G \neq 1$ and $\mathbf{O}^p(G)$ is nilpotent (p a prime). Furthermore assume that $p \nmid |G : \mathbf{C}_G(v)|$ for all $v \in V$ and that V is pseudo-primitive. Then one of the following assertions occurs:*

(i) $\mathbf{O}^p(G)$ *is a cyclic p'-group;*

(ii) $|V| = 3^2$, $p = 3 = |G : \mathbf{O}^p(G)|$ *and* $G \cong SL(2,3)$;

(iii) $|V| = 2^6$, $\mathbf{O}^p(G)$ *is extra-special of order 3^3 and exponent 3, $p = 2 = |G : \mathbf{O}^p(G)|$ and $\mathbf{Z}(\mathbf{O}^p(G)) = \mathbf{Z}(G)$.*

Proof. By Lemma 10.1, we may assume that every normal abelian subgroup of G is cyclic. Let $F = \mathbf{F}(G)$. Since $\mathbf{O}_p(G) \leq P$ for all $P \in \mathrm{Syl}_p(G)$, the hypotheses imply that $\mathbf{O}_p(G) \leq \mathbf{C}_G(V) = 1$. Thus $p \nmid |F|$, $F = \mathbf{O}^p(G)$ and $F \leq G'$. If A is a normal abelian subgroup of G, then A is cyclic and thus central in $G' \geq F$. Let $Z = \mathbf{Z}(F)$. By Corollary 1.10, F/Z is a completely reducible G-module and $|F/Z| = e^2$ for an integer e. (Note that Z has a different meaning than in Cor. 1.10.) By Corollary 2.6, $\dim(V) = t \cdot e \cdot \dim(W)$

for an irreducible Z-submodule W of V and an integer t. Since V_Z is homogeneous, $|Z| \mid |W| - 1$. If $e = 1$, then $F = \mathbf{O}^p(G)$ is cyclic and conclusion (i) holds. We thus assume that $e \geq 2$, i.e. $F > Z$.

Let $v \in V^{\#}$ and observe that $\mathbf{C}_Z(v) = 1$ and $Z\mathbf{C}_F(v) \trianglelefteq F$. Since v is centralized by a Sylow p-subgroup P_0 of G, even $Z\mathbf{C}_F(v) \trianglelefteq FP_0 = G$. Now $Z\mathbf{C}_F(v)$ is an abelian normal subgroup of G. Hence $Z\mathbf{C}_F(v) \leq Z$ and $\mathbf{C}_F(v) = 1$. Consequently, F acts fixed-point-freely on V. Since F is nilpotent and non-abelian, it follows that $F = Q \times S$ with a cyclic group S of odd order and Q a quaternion group (see [Hu, V, 8.7]). By the first paragraph, $Q/\mathbf{Z}(Q)$ is elementary abelian and thus $Q \cong Q_8$ (cf. Proposition 1.1). Since $\mathbf{O}^{p'}(G) = G$, it follows that for $P \in \mathrm{Syl}_p(G)$, $\mathbf{C}_S(P) = 1$, $\mathbf{C}_Q(P) = \mathbf{Z}(Q)$ and $p = 3$. In particular, then $|\mathrm{Syl}_3(G)| = |F : \mathbf{C}_F(P)| = |F : \mathbf{Z}(Q)| = 4 \cdot |S|$. For $v \in V^{\#}$ we have seen that $\mathbf{C}_F(v) = 1$ and thus $\mathbf{C}_G(v) \in \mathrm{Syl}_3(G)$. Hence $|V^{\#}| = |\mathrm{Syl}_3(G)| \cdot |(\mathbf{C}_V(P))^{\#}|$ with $P \in \mathrm{Syl}_3(G)$. Letting $|V| = q^n$ and $|\mathbf{C}_V(P)| = q^m$ for a prime q, we have $(q^n - 1)/(q^m - 1) = 4 \cdot |S|$; in particular $m \mid n$.

We claim that $|G/F| = 3$. Assume not. Because $G/\mathbf{C}_G(Q/\mathbf{Z}(Q)) = 3$ and G/F is a 3-group, we may choose $J \trianglelefteq G$ such that J has a cyclic normal subgroup J_0 of index 3, $J_0 \leq S$, and J is a Frobenius group. Every $v \in V$ is centralized by a Sylow 3-subgroup of J because $J \trianglelefteq G$. Let $P_1 \in \mathrm{Syl}_3(J)$. Then $|\mathbf{C}_V(P_1)| = |V|^{1/3}$ (see Lemma 0.34). Hence $|J_0| = |\mathrm{Syl}_3(J)| \geq |V|^{2/3}$ and $|Z| \geq |V|^{2/3}$. This contradicts Corollary 2.6 which implies that $|Z| \leq |V|^{1/2}$. Hence $|G/F| = 3$.

If $S \neq 1$, then SP is a Frobenius group and $q^m = |\mathbf{C}_V(P)| = |V|^{1/3}$, by Lemma 0.34. Thus $|\mathrm{Syl}_3(G)| = 4|S| = (q^{3m} - 1)/(q^m - 1) = 1 + q^m + q^{2m}$, a contradiction because the right hand side is odd. Thus $S = 1$. Now $4 = 1 + q^m + \cdots + q^{n-m}$. Hence $|V| = q^n = 3^2$ and $Q_8 \leq G \leq GL(2,3)$. Since $\mathbf{O}^{3'}(G) = G$, it follows that $G \cong SL(2,3)$. $\qquad\square$

10.3 Example. Suppose that $F = \mathbf{F}(G)$ is extra-special of order 3^3 and exponent 3, that $|G : F| = 2$, $\mathbf{Z}(G) = \mathbf{Z}(F)$ and $\mathbf{O}^{2'}(G) = G$. Then G has a unique faithful irreducible module V over $GF(2)$. Furthermore

  (a) $|V| = 2^6$ and $2 \mid |\mathbf{C}_G(v)|$ for all $v \in V$.

  (b) There exists $y \in V$ with $|\mathbf{C}_G(y)| = 2$.

  (c) Suppose that G char $G_0 \leq GL(V)$ with G_0 solvable and V a pseudo-primitive G_0-module. Assume that each $v \in V$ is centralized by a Sylow 2-subgroup of G_0. Then $G_0 = G$.

Proof. Observe that in characteristic 2, G has two absolutely irreducible and faithful representations, both of degree 3 (see [Hu, V, 17.13]). Hence over $GF(2)$, G has either exactly one faithful irreducible representation necessarily of degree 6, or exactly two faithful irreducible representations both of degree 3. If V is a faithful irreducible $GF(2)[G]$-module, then $|\mathbf{Z}(G)| \mid |V| - 1$. Thus $|V| = 2^6$ and V is unique. The same argument shows that V_F is irreducible.

Since G/F inverts $F/\mathbf{Z}(G)$, we may choose $C \trianglelefteq G$ with C elementary abelian of order 3^2. Then V_C is not homogeneous. But V_F is irreducible and so $V_C = V_1 \oplus V_2 \oplus V_3$ for homogeneous components V_i of V_C that are transitively and faithfully permuted by $G/C \cong S_3$. For $i \neq j$, $\mathbf{C}_C(V_i) \cap \mathbf{C}_C(V_j) = 1$. Let $Y = \{(v_1, v_2, v_3) \mid v_i \in V_i, \text{ exactly two } v_i \text{ non-zero}\}$. Then $\mathbf{C}_C(y) - 1$ and consequently $\mathbf{C}_F(y) - 1$ for all $y \in Y$. In particular, $\mathbf{C}_G(y)$ has order 1 or 2, and $y \in Y$ is centralized by at most one Sylow 2-subgroup of G. If $P \in \mathrm{Syl}_2(G)$, then P fixes one V_i and interchanges the other two. Thus P centralizes 3 elements of Y. Since $|Y| = 3^3$ and $|\mathrm{Syl}_2(G)| = |F : \mathbf{Z}(G)| = 3^2$, we have $|\mathbf{C}_G(y)| = 2$ for all $y \in Y$. This establishes assertion (b).

Say P fixes V_1, so that $\mathrm{stab}_G(V_1) = CP$. Now $C/\mathbf{C}_C(V_1) \cong \mathbf{Z}(G)$, and since $[\mathbf{Z}(G), P] = 1$, it follows that $CP/\mathbf{C}_{CP}(V_1)$ has a normal Sylow 2-subgroup. As V_1 is an irreducible $GF(2)[CP]$-module, $P \leq \mathbf{C}_G(V_1)$.

We next establish assertion (a), and let $v = (v_1, v_2, v_3) \in V$. If two v_i are zero, say $v_1 \neq 0$, we then choose $P_1 \in \mathrm{Syl}_2(G)$ that fixes V_1. By the last paragraph, $P_1 \leq \mathbf{C}_G(V_1)$, and so $P_1 \leq \mathbf{C}_G(v)$. If however all v_i are non-zero, then $y = (v_1, v_2, 0) \in Y$. We choose $P_2 \in \text{Syl}(G)$ with $P_2 = \mathbf{C}_G(y)$. Since P_2 fixes V_3, the last paragraph implies that $P_2 \leq \mathbf{C}_G(V_3)$ and consequently $P_2 \leq \mathbf{C}_G(v)$. Part (a) follows.

We finally prove (c) by induction on $|G_0/G|$. For $G \leq H$ char G_0, we have that V_H is pseudo-primitive and irreducible, and that each $v \in V$ is centralized by a Sylow 2-subgroup of H. We may assume that G_0/G is a q-group for a prime q. By Lemma 10.2, we may also assume that $q \neq 2$. Now let $M/F \in \text{Syl}_q(G_0/F)$, so that $|G_0/M| = 2$ and M char G_0. Either $q = 3$ and M is a 3-group, or $q > 3$ and M/F centralizes both F/Z and Z. In all cases, M is nilpotent. Apply Lemma 10.2 for a contradiction. $\square$

The hypothesis that $\mathbf{O}^{p'}(G) = G$ in the next theorem is more for convenience. We remove it in Corollary 10.5 (but we must also allow the conclusion $G = GL(2,3)$ when $|V| = 3^2$).

10.4 Theorem. *Let G be a solvable group acting completely reducibly and faithfully on a finite vector space V such that $p \nmid |G : \mathbf{C}_G(v)|$ for all $v \in V$ (p a fixed prime). Assume that $\mathbf{O}^{p'}(G) = G \neq 1$ and that V is pseudo-primitive. Then V is an irreducible G-module and one of the following occurs:*

(i) $\mathbf{O}^p(G)$ *is a cyclic p'-group and $G \leq \Gamma(V)$;*

(ii) $G \cong SL(2,3)$, $p = 3$ *and* $|V| = 3^2$; *or*

(iii) $\mathbf{O}^p(G)$ *is extra-special of order 3^3 and exponent 3, $p = 2 = |G : \mathbf{O}^p(G)|$, $\mathbf{Z}(G) = \mathbf{Z}(\mathbf{O}^p(G))$ and $|V| = 2^6$.*

Proof. We argue by induction on $|G||V|$. If V is not irreducible, we may write $V = X \oplus Y$ for faithful G-modules X and Y, because V is homogeneous. Applying the inductive hypothesis, we may assume that G and X satisfy the conclusion of the theorem. If $\mathbf{O}^p(G)$ is cyclic or $G \cong SL(2,3)$, then $\mathbf{C}_G(x) \in \text{Syl}_p(G)$ for all $x \in X^{\#}$. If $\mathbf{O}^p(G)$ is extra-special of order 3^3, we may choose $x \in X$ such that $\mathbf{C}_G(x) \in \text{Syl}_p(G)$ (see Example 10.3). Choose $y \in Y$ such that $\mathbf{C}_G(x)$ does not centralize y. Then the vector

$(x, y) \in V$ is not centralized by a Sylow p-subgroup of G, a contradiction. Hence V is an irreducible G-module.

We let $K = \mathbf{O}^p(G)$, $F = \mathbf{F}(G)$ and $Z = \mathbf{Z}(F)$. Since $\mathbf{O}^{p'}(G) = G$, $K \leq G'$. If $S \trianglelefteq G$ and $P \in \mathrm{Syl}_p(G)$, then $P \cap S \in \mathrm{Syl}_p(S)$. Thus $p \nmid |S : \mathbf{C}_S(v)|$ for all $v \in V$. If S is also characteristic in G, then $\mathbf{O}^{p'}(S)$ and V satisfy the hypotheses of the theorem or $p \nmid |S|$. Since G is faithful on V, $\mathbf{O}_p(G) = 1$ and $p \nmid |F|$. Thus $F \leq K \leq G'$.

Step 1. V is an irreducible K-module.

Proof. Let V_0 be an irreducible K-submodule of V, and let $v \in V_0^{\#}$. Now v is centralized by a Sylow p-subgroup P_0 of G. Then V_0 is invariant under $KP_0 = G$, and $V_0 = V$ follows.

Step 2. We may assume that

(a) $F < K$;

(b) $Z < F$; and

(c) Every normal abelian subgroup of G is cyclic and contained in Z and $Z = \mathbf{Z}(K)$.

Proof. (a) if $F = K$, it follows from Lemma 10.2 that either K is a cyclic p'-group or conclusion (ii) or (iii) of the theorem hold. As V_K is irreducible, $G \leq \Gamma(V)$ in the first case by Theorem 2.1.

(b) If $F = Z$, then F is cyclic and therefore $K \leq G' \leq \mathbf{C}_G(F) = F$. Thus $F = K$, contradicting (a).

(c) By Lemma 10.1, we may assume that each normal abelian subgroup B of G is cyclic, and thus central in $G' \geq F$. Hence $B \leq Z \leq \mathbf{Z}(G')$. Since $F \leq K \leq G'$ and $Z = \mathbf{Z}(F)$, part (c) follows.

Step 3.

    (a) $F/Z = H_1/Z \times \cdots \times H_m/Z$ for chief factors H_i/Z of G/Z with $m \geq 1$.

    (b) $Z = \mathbf{Z}(H_i)$ and $H_i = ZF_i$ for an extra-special group $F_i \trianglelefteq G$.

    (c) $|H_i : Z| = f_i^2$ for a prime power $f_i = q_i^{n_i} > 1$.

    (d) If W is an irreducible Z-submodule of V, then $\dim(V) = te \cdot \dim(W)$ for an integer t and $e = f_1 \cdots f_m > 1$.

    (e) Let $C_i = \mathbf{C}_G(H_i/Z)$. If $C_i \leq H \leq \mathbf{C}_G(Z)$, then H/C_i is isomorphic to a subgroup of $Sp(2n_i, q_i)$.

Proof. By Step 2, every normal abelian subgroup of G is cyclic and central in F. This step thus is a consequence of Corollary 1.10 and Corollary 2.6. (Note again that Z has a different meaning than in Cor. 1.10).

Step 4. If N is characteristic in G and $p \mid |N|$, then $N = G$. In particular, $p \nmid |K|$, $K = G'$ and G/K is elementary abelian. Also by Step 2 (a), $\mathbf{O}_p(G/F) = 1$.

Proof. Assume that N is a proper characteristic subgroup of G and $p \mid |N|$. Without loss of generality, $N = \mathbf{O}^{p'}(N)$. Let $L = \mathbf{O}^p(N)$. By the inductive hypothesis, V is an irreducible N-module. By the same argument as in Step 1, V is an irreducible L-module. Suppose that L is cyclic. Then $L \leq \mathbf{Z}(G')$ and $L \leq Z$. But by Step 3 (d), V_Z is not irreducible, and so L is not cyclic. Applying the inductive hypothesis to N, we may conclude either

$$|V| = 3^2, \ p = 3 \quad \text{and} \quad N \cong SL(2,3),$$

or

$$|V| = 2^6, \ p = 2 \quad \text{and} \quad N \text{ has the structure described in}$$

$$\text{conclusion (iii).}$$

In the first case, $N \trianglelefteq G \leq GL(2,3)$. Since $N < G$, we have $|G/N| = 2$, contradicting $\mathbf{O}^{p'}(G) = G$. In the second case, $N = G$ by Example 10.3.

Step 5. F/Z is a faithful G/F-module, i.e. $\bigcap_i C_i = F$.

Proof. If not, then $D := \bigcap_i C_i > F$. By Corollary 1.10, F/Z is a faithful G'/F-module and hence $G'/F \cap D/F = 1$. Since G/G' is a p-group, D/F is a non-trivial p-group. Then $\mathbf{O}_p(G/F) \neq 1$, contradicting Step 4.

Step 6. Let H_i/Z be a chief factor of G as in Step 3 with $|H_i/Z| = f_i^2$ and $C_i = \mathbf{C}_G(H_i/Z)$. Then

(a) $C_i < G$.

(b) $K \leq C_i$ if and only if $f_i = 2$. Also $p = 3$ in this case.

(c) At most one f_i equals 2.

Proof. If $C_i = G$, then H_i/Z is a central chief factor of G/Z. In this case, $|H_i/Z|$ is prime, a contradiction. Thus $C_i < G$, proving (a).

If $f_i = 2$, then $G/C_i \leq S_3$. As $p \nmid |F|$ and $\mathbf{O}^{p'}(G) = G$, we have $|G/C_i| = 3 = p$ and $K \leq C_i$. To prove (b) and (c), we may assume for some $j \in \{1, \ldots, m\}$ that $K \leq C_i$ if and only if $i \leq j$. Let $H = H_1 \cdots H_j$ and $B = \mathbf{C}_K(H) \trianglelefteq G$. Now H/Z is central in K and thus K/B is isomorphic to a subgroup of $\mathrm{Aut}(H)$ that acts trivially on both H/Z and Z. By Lemma 1.5, $|K/B| \leq |H/Z|$. But $B \cap H = \mathbf{Z}(H) = Z$ and hence $K = BH$. For $v \in V^{\#}$, v is centralized by a Sylow p-subgroup P_1 of G and thus $Z \cdot \mathbf{C}_H(v) \trianglelefteq HBP_1 = KP_1 = G$. Since $\mathbf{C}_H(v) \cap Z = \mathbf{C}_Z(v) = 1$ and H/Z is abelian, $Z \cdot \mathbf{C}_H(v)$ is an abelian normal subgroup of G and by Step 2 (c), contained in Z. Hence H acts fixed-point-freely on V. By [Hu, V, 8.7], a Sylow-subgroup of H is cyclic or isomorphic to Q_8. Thus $|H/Z| = 2^2$, $j = 1$ and $f_1 = 2$. This step is complete.

Step 7.

(a) If $f_i = 2$, then $p = 3$.

(b) f_i is not 2^2, 2^4 or 3.

(c) If $f_i = 2^3$, then $p = 3$ and $P \not\leq \mathbf{C}_G(Z)$ ($P \in \mathrm{Syl}_3(G)$).

(d) If $f_i = 2^5$, then $p = 5$ and $|K/K \cap C_i| \leq 2^5 \cdot 3^5$.

(e) If $f_i = 3^2$, then $p = 2$ and $|K/K \cap C_i| \leq 5$, or $p = 5$ and $K/K \cap C_i$ is extra-special of order 2^5.

(f) If $f_i = 3^3$, then $p = 2$ and $|K/K \cap C_i| \leq 3^2 \cdot 13$.

(g) If $f_i = q_i$, then $p \leq 3$.

Proof. Part (a) is immediate from Step 6 and is restated here for convenience. Assume that $f_i \neq 2$. By Step 6, $C_i = \mathbf{C}_G(H_i/Z)$ does not contain K. It thus follows from Step 4 that $KC_i/C_i = (G/C_i)'$ is a non-trivial p'-group and G/KC_i is a p-group. Since G/C_i acts irreducibly and faithfully on H_i/Z, parts (b), (d), (e) and (g) are now immediate consequences of Lemma 2.16.

(c) By Lemma 2.16, we may assume that $|G/C_i| = 3^2 \cdot 7$ and $p = 3$. Thus G/C_i has a cyclic normal subgroup I/C_i of order 21 that must act irreducibly on H_i/Z. If $P \leq \mathbf{C}_G(Z)$, then I/C_i acts symplectically and irreducibly on H_i/Z. Since I/C_i is cyclic, $|I/C_i| \mid 2^3 + 1$ (by [Hu, II, 9.23]), a contradiction.

(f) In this case, Lemma 2.16 yields $p = 2$ and $|KC_i/C_i|$ divides $3^2 \cdot 13^2$ or $7 \cdot 13$. Since $K \leq \mathbf{C}_G(Z)$, in fact $K/K \cap C_i \cong KC_i/C_i \leq Sp(6,3)$ and so $13^2 \nmid |K/K \cap C_i|$. Therefore $|K/K \cap C_i| \leq 3^2 \cdot 13$.

Step 8.

(a) $e \geq 5$.

(b) If $p > 3$ and $e < 48$, then one of the following occurs:

(1) $e = f_1 = 5^2$;

(2) $e = f_1 = 3^2$, $p = 5$ and K/F is extra-special of order 2^5; or

(3) $e = f_1 = 2^5$, $p = 5$ and $|K/F| \leq 2^5 \cdot 3^5$.

Proof. Now $e = f_1 \cdots f_m$ with each $f_i > 1$ and $e > 1$ (see Step 3). By Steps 7 (b) and 6 (c), $e \neq 3$ or 4. If $e = 2$, Steps 6 (b) and 5 yield $F = C_1 = K$, contradicting Step 2 (a). This establishes (a). Assume that $p \geq 5$, so by Step 7 (g), no f_i is prime. Since $e = \prod_i f_i < 48$ and $F = \bigcap_i C_i$, part (b) follows from Step 7.

Step 9. Let $P \in \mathrm{Syl}_p(G)$ and W be an irreducible Z-submodule of V. Then

(a) $|\mathrm{Syl}_p(G)| \cdot |\mathbf{C}_V(P)| \geq |V|$;

(b) $|V| = |W|^{et}$; and

(c) $|Z| \mid |W| - 1$.

Proof. Part (a) follows from the hypothesis that every vector is centralized by a Sylow p-subgroup. Part (b) is just a restatement of Step 3 (d). Since V_Z is homogeneous and faithful and since Z is cyclic, part (c) follows.

Step 10.

(a) $|K/F| \leq e^{9/2}/2$.

(b) If $p = 2$, then $|K/F| \leq e^3/2$.

(c) If $p = 3$, then $|K/F| \leq e^4/2$.

(d) If $p = 3$ and $f_1 = 2$, then $|K/F| \leq e^4/2^5$.

Proof. Since $K \trianglelefteq G$, we have H_i/Z is a completely reducible and faithful $K/K \cap C_i$ module. By Theorem 3.5, $|K/K \cap C_i| \leq (f_i^2)^{9/4}/2 = f_i^{9/2}/2$. Since $\bigcap_i (K \cap C_i) = F$ (by Step 5), $|K/F| \leq \prod_i (f_i^{9/2}/2) \leq e^{9/2}/2$. Noting that $p \nmid |K/F| \cdot |F/Z|$, parts (b) and (c) similarly follow from Theorem 3.5.

If $p = 3$ and $f_1 = 2$, then $K \leq C_1$ by Step 6 (b). For $i > 1$, $|K/K \cap C_i| \leq f_i^4/2$ by Theorem 3.5. Since $e > 2$ (by Step 8 (a)), it follows that $m \geq 2$ and $\bigcap_{i=2}^m (K \cap C_i) = K \cap C_2 \cap \cdots \cap C_m = F$. Hence

$$|K/F| \leq \prod_{i=2}^m (f_i^4/2) \leq (e/2)^4/2 = e^4/2^5.$$

Step 11. $P \leq \mathbf{C}_G(Z)$.

Proof. Assume not. Since $K \leq \mathbf{C}_G(Z)$ and $\mathbf{C}_G(Z)$ is a proper characteristic subgroup of G, Step 4 implies that $\mathbf{C}_G(Z) = K$ is a p'-group. We may thus choose $P_0 \leq P$ with $|P_0| = p$ and $P_0 \not\leq \mathbf{C}_G(Z)$. There exists a Sylow subgroup Z_0 of Z with $Z_0 P_0$ a Frobenius group. Applying Lemma 0.34, we obtain $|\mathbf{C}_V(P_0)| = |V|^{1/p}$. Thus $p \mid \dim(V)$ and $|\mathbf{C}_V(P)| \leq |V|^{1/p}$. Since $p \nmid e$, Step 9 yields that

$$|\mathrm{Syl}_p(G)| \geq |W|^{te(p-1)/p} \quad \text{and} \quad p \mid t \cdot \dim(W), \tag{10.3}$$

where W is an irreducible Z-submodule of V. Recall that $|Z| \mid |W| - 1$. Also $|\mathrm{Syl}_p(G)| \leq |K| = |K/F| \cdot |F/Z| \cdot |Z|$.

First assume that $p \geq 5$. By the first paragraph and Step 10,

$$e^{13/2}|W|/2 \geq |\mathrm{Syl}_p(G)| \geq |W|^{4et/5},$$

and so $e^{65} \geq |W|^{8te-10} \cdot 2^{10}$. Note that $|W| \geq 3$ because $|Z| \mid |W| - 1$. If $t \geq 5$, then $e^{65} \geq 3^{40e-10} \cdot 2^{10}$ and $e < 5$, contradicting Step 8 (a). So $t < 5 \leq p$, and (10.3) implies that $|W| \geq 32$. Then $e^{65} \geq 2^{40e-40}$ and $e < 5$, again a contradiction. We thus assume that $p \leq 3$.

We next consider the case $p = 3$. By the first paragraph and Step 10 (c),

$$e^6|W|/2 \geq |\mathrm{Syl}_3(G)| \geq |W|^{2te/3}, \quad \text{and so} \tag{10.4 a}$$

$$e^{18} \geq |W|^{2te-3} \cdot 2^3. \tag{10.4 b}$$

If $t \geq 3$, then $e^{18} \geq 3^{6e-3} \cdot 2^3$ and $e < 5$, contradicting Step 8 (a). By (10.3), $3 \mid \dim(W)$. If $|W| \geq 64$, then (10.4 b) implies $e < 5$, a contradiction. If $|W| = 27$, then (10.4 b) implies that $e < 8$ and so $5 \leq e \leq 7$. This is a contradiction since each prime divisor of e divides $|W| - 1$. Hence $|W| = 8$ and (10.4 b) implies that $e < 14$. Since $|Z| \mid |W| - 1$, $|Z| = 7 = e$. Now F/Z is a faithful irreducible G/F-module of order 7^2. Since $\mathbf{O}^{3'}(G) = G$, it follows from Theorem 2.11 that $|G/F| \leq 96$ and $|K/F| \leq 32$. Then $|\mathrm{Syl}_3(G)| \leq |K/F| \cdot |F/Z| \cdot |Z| \leq 32 \cdot 7^3$, and (10.4 a) yields that $2^5 \cdot 7^3 \geq 8^{14/3} = 2^{14}$, a contradiction.

To conclude this step, we consider the case $p = 2$. By the first paragraph and Step 10 (b),

$$e^5|W|/2 \geq |\mathrm{Syl}_2(G)| \geq |W|^{te/2}, \quad \text{and so} \tag{10.5 a}$$

$$e^{10} \geq 4|W|^{te-2}. \tag{10.5 b}$$

Since $p = 2$, also $\mathrm{char}\,(V) = 2$ by Lemma 9.2, and thus $|W| \geq 4$. If $t \geq 2$, $e^{10} \geq 4^{2e-1}$ and $e < 9$. But $2 \nmid e$ and Step 8 (a) implies that $e = 5$ or 7. Each prime divisor of e divides $|W| - 1$. Thus $e = 5$ and $|W| \geq 16$, or $e = 7$ and $|W| \geq 8$. In both cases, (10.5 b) gives a contradiction because $t \geq 2$. Hence $t = 1$ and $\dim(W)$ is even by (10.3).

If $|W| \geq 2^6$, then (10.5 b) implies that $e < 6$ and so $e = 5$ by Step 8 (a). In this case, $5 \mid |W| - 1$ and so $|W| \geq 2^8$, contradicting (10.5 b). Hence $|W| = 2^2$ or 2^4. Now every prime divisor of e divides $|Z|$ and therefore $|W| - 1$. Thus e is a $\{3,5\}$-number and when $|W| = 4$, e is a 3-power. Recall that no f_i equals 3 (by Step 7). Using (10.5 b), we have only the following cases and $e = f_1$ in all:

$|W|$	$e = f_1$	$|Z|$	$|K/F|$
4	27	3	at most $3^2 \cdot 13$
4	9	3	5
16	9	3 or 15	5
16	5	5 or 15	3.

Note that the order of K/F is determined by Step 7 in the first 3 cases and by Theorem 2.11 in the last case, as $2 \nmid |K|$. By (10.5 a),

$$|W|^{e/2} \leq |\mathrm{Syl}_2(G)| \leq |K| = |K/F| \cdot e^2 \cdot |Z|.$$

This rules out the first case. In the remaining cases, $|K/F|$ has prime order. Since $\mathbf{O}_2(G/F) = 1$ and G/K is elementary abelian (see Step 4), we have that G/F is a Frobenius group of order $2|K/F|$ that acts faithfully and irreducibly on F/Z. By Lemma 0.34, $|\mathbf{C}_{F/Z}(P)| = |F/Z|^{1/2} = e$. Thus $|W|^{e/2} \leq |\mathrm{Syl}_2(G)| \leq |K/F| \cdot e \cdot |Z|$, which gives a contradiction in each remaining case.

Step 12. Recall that $P \in \mathrm{Syl}_p(G)$. We have

(a) $|\mathbf{C}_V(P)| \leq |V|^{2/3}$ and $|\mathrm{Syl}_p(G)| \geq |W|^{te/3}$; and

(b) If $p \neq 2$, then $|\mathbf{C}_V(P)| \leq |V|^{1/2}$ and $|\mathrm{Syl}_p(G)| \geq |W|^{te/2}$.

Proof. Let $P_0 \leq P$ with $|P_0| = p$, and choose a Sylow subgroup Q of F such that $P_0 \not\leq \mathbf{C}_G(Q)$. Since V_Q is homogeneous, Lemma 7.2 applied to QP_0

implies that

$$|\mathbf{C}_V(P)| \leq |\mathbf{C}_V(P_0)| \leq \begin{cases} |V|^{1/2} & \text{if } p \neq 2, \\ |V|^{2/3} & \text{if } p = 2. \end{cases}$$

Since $|\mathrm{Syl}_p(G)| \cdot |\mathbf{C}_V(P)| \geq |V|$ and $|V| = |W|^{te}$ by Step 9, this Step follows.

Step 13. $p < 5$.

Proof. Assume that $p \geq 5$. By Steps 12 (b), 11 and 10 (a), we have

$$|\mathrm{Syl}_p(G)| \geq |W|^{te/2} \quad \text{and} \tag{10.6}$$

$|\mathrm{Syl}_p(G)| \leq |K/F| \cdot |F/Z| \leq e^{13/2}/2$. Thus

$$e^{13} \geq 4|W|^e. \tag{10.7}$$

But $|W| \geq 3$ and hence $e < 48$. Every prime divisor of e divides $|Z|$, which in turn divides $|W| - 1$. If $e = 5^2$, then $|W| \geq 11$ contradicting (10.7). Thus Step 8 implies that

$$e = 2^5, \ p = 5 \ \text{and} \ |K/F| \leq 2^5 \cdot 3^5, \ \text{or}$$

$$e = 3^2, \ p = 5 \ \text{and} \ K/F \ \text{is extra-special of order } 2^5.$$

In the first case, (10.6) implies that $|W|^{16} \leq |\mathrm{Syl}_5(G)| \leq 2^{15} \cdot 3^5$ and thus $|W| < 3$, a contradiction. In the second case, a Sylow 5-subgroup P of G must centralize $\mathbf{Z}(K/F)$ and so $|\mathrm{Syl}_5(G)| = |K/Z : \mathbf{C}_{K/Z}(P)| \leq 2^4 \cdot 3^4$. By (10.6), $|W|^{9t} \leq 2^8 \cdot 3^8$. Since $3 \mid e$, $3 \mid |W| - 1$ and hence $|W| = 4$, $t = 1$ and $|V| = |W|^e = 4^9$. Now $P \leq \mathbf{C}_G(Z)$ and $\mathbf{C}_V(P)$ is a Z-submodule of V_Z. Thus $|\mathbf{C}_V(P)| = 4^j$ for some j. By Step 12, $|\mathbf{C}_V(P)| \leq |V|^{1/2} = 4^{9/2}$ and $j < 5$. Since $V = \mathbf{C}_V(P) \oplus [V, P]$, 5 must divide $4^{9-j} - 1$, and hence $|\mathbf{C}_V(P)| \leq 4^3$. By Step 9, $4^9 = |V| \leq |\mathrm{Syl}_5(G)| \cdot |\mathbf{C}_V(P)| \leq 2^4 \cdot 3^4 \cdot 4^3$. This contradiction completes the step.

Step 14. $p = 2$.

Proof. By Step 13, we may assume that $p = 3$. It then follows from Step 12 that

$$|\text{Syl}_3(G)| \geq |W|^{te/2}. \tag{10.8}$$

Since by Step 11 $P \leq \mathbf{C}_G(Z)$, $|\text{Syl}_p(G)| \leq |K/F| \cdot |F/Z| = |K/F| \cdot e^2$. By (10.8) and Step 10 (c), (d), we have

$$e^{12} \geq 4|W|^e, \qquad \text{and} \tag{10.9}$$

$$e^{12} \geq 2^{10} \cdot |W|^e \quad \text{if} \quad f_1 = 2. \tag{10.10}$$

Since $|W| \geq 3$, it follows from (10.9) that $e < 48$.

Since $p = 3$, $3 \nmid e$. Now $e = f_1 \ldots f_m$ and no f_i is 4, 8, 16, or 32 by Step 7 (b), (c), (d). By Step 6 (c), we also have that $f_i \neq 2$ for $i > 1$. Also $e \geq 5$, by Step 8 (a). Assume $f_1 = 2$ so that $m \geq 2$ and $f_2 = q_2^{n_2}$ with $q_2 \geq 5$. Since $2 \mid |F|$, it follows that $|W| \neq 8$, and $q_2 \mid |W| - 1$ implies $|W| \geq 11$. As $e \geq 10$, (10.10) gives a contradiction. Hence $f_i = q_i^{n_i}$ for a prime $q_i \geq 5$ ($i = 1, \ldots, m$). Then $|W| \geq 8$ and (10.9) implies that $e < 16$. Thus $e = f_1 = q_1$ is 5, 7, 11 or 13. Now G/F is an irreducible subgroup of $GL(2, q_1)$, $(G/F)' = K/F \neq 1$ and G/K is a 3-group. Thus part (c) of Theorem 2.11 applies, and since $\mathbf{O}^{3'}(G) = G$, $\mathbf{Z}(G/F)$ has index $2^2 \cdot 3$ in G/F. Hence $|\text{Syl}_3(G)| \leq 2^2 \cdot e^2$, and inequality (10.8) yields $2^2 \cdot e^2 \geq |W|^{e/2} \geq 8^{e/2}$. Consequently, $e < 5$. This contradiction completes the proof of this step.

Step 15. Conclusion.

Proof. It remains to consider $p = 2$. Since $P \leq \mathbf{C}_G(Z)$, it follows from Step 10 that $|\text{Syl}_2(G)| \leq |K/F| \cdot |F/Z| \leq e^5/2$. By Step 12,

$$|\text{Syl}_2(G)| \geq |W|^{te/3}, \quad \text{and so} \tag{10.11}$$

$$e^{15} \geq 8|W|^e. \tag{10.12}$$

Since $p = 2$, we have by Lemma 9.2 that $\text{char}(V) = 2$.

Applying (10.12), we get the following upper bounds for e, given $|W|$.

$|W|$	4	8	16	32	64	≥ 128
upper bound for e	64	32	16	16	6	4

Since $|Z| \mid |W| - 1$, indeed $|W| \geq 4$. Each prime divisor of e divides $|Z|$ and also divides $|W| - 1$. Since $e \geq 5$, the cases where $|W| \geq 32$ yield contradictions. Now $e = f_1 f_2 \cdots f_m \geq 5$ and no f_i is 3 by Step 7 (b). When $|W| = 16$, it follows that $e = f_1 = 5$ or $e = f_1 = 3^2$. When $|W| = 8$, e must be a power of 7 and so $e = f_1 = 7$. When $|W| = 4$, then $e = f_1$ is 3^2 or 3^3. Since $e = f_1$ in all cases, F/Z is a faithful irreducible G/F-module. Applying Step 7 and Theorem 2.11 to the action of G/F on F/Z, we are limited to the following possibilities:

$|W|$	$e = f_1$	$|K/F|$
16	5	3
16	3^2	5
8	7	3 or 3^2
4	3^3	at most $3^2 \cdot 13$
4	3^2	5.

By (10.11), $|W|^{e/3} \leq |\mathrm{Syl}_2(G)| \leq |K/F| \cdot e^2$. This yields a contradiction except in the following two cases: $|W| = 4$, $e = 3^2$, and $|W| = 8$, $e = 7$. When $e = 7$, it follows from Step 3 (e) that $G/F \leq Sp(2,7)$ and so $|K/F| = 3$. In both the remaining cases, $|K/F|$ is prime. Since by Step 4, G/K is elementary abelian and $\mathbf{O}_2(G/F) = 1$, it follows that G/F is a Frobenius group of order $2|K/F|$. Now G/F acts faithfully and irreducibly on F/Z and so $|\mathbf{C}_{F/Z}(P)| = |F/Z|^{1/2} = e$ (see Lemma 0.34). Therefore, by (10.11), $|W|^{e/3} \leq |\mathrm{Syl}_2(G)| \leq |K/F| \cdot e$, which is a contradiction in both cases. The proof of the theorem is now complete. $\square$

We next delete the hypothesis that $\mathbf{O}^{p'}(G) = G$ in the above theorem.

10.5 Theorem. *Assume that V is a finite faithful and pseudo-primitive G-module for a solvable group G. Suppose that $p \mid |G|$ but $p \nmid |G : \mathbf{C}_G(v)|$ for all $v \in V$ (p a fixed prime). Then V is an irreducible G-module and one of the following occurs:*

(i) $\mathbf{O}^{p',p}(G)$ *is a cyclic p'-group and $G \leq \Gamma(V)$;*

(ii) $|V| = 3^2$, $p = 3$ *and G is isomorphic to $SL(2,3)$ or $GL(2,3)$; or*

(iii) $|V| = 2^6$, $p = 2 = |G : \mathbf{F}(G)|$, $\mathbf{F}(G)$ *is extra-special of order 3^3 and exponent 3, $\mathbf{Z}(\mathbf{F}(G)) = \mathbf{Z}(G)$ and $\mathbf{O}^{2'}(G) = G$.*

Proof. Let $H = \mathbf{O}^{p'}(G)$ and $K = \mathbf{O}^p(H)$. The action of H on V satisfies the hypotheses of Theorem 10.4; in particular, V_H is irreducible. Since each $v \in V$ is centralized by some $P_v \in \mathrm{Syl}_p(H)$ and $H = KP_v$, even V_K is irreducible. If K is cyclic, then $G \leq \Gamma(V)$ by Theorem 2.1. If secondly $|V| = 3^2$, $p = 3$ and $H = SL(2,3)$, then $G = H$ or $G = GL(2,3)$, as $SL(2,3)$ has index 2 in $GL(2,3)$.

By Theorem 10.4, we may assume that $K = \mathbf{F}(H)$ is non-abelian of order 3^3 and exponent 3, $|H/K| = 2 = p$, $\mathbf{Z}(K) = \mathbf{Z}(H)$ and $|V| = 2^6$. In this case however Example 10.3 (c) shows that $G = H$. $\qquad\square$

Corollary 10.6. *Suppose that V is a finite faithful and pseudo-primitive G-module for a solvable group G. Assume that $\pi \neq \emptyset$ is a set of prime divisors of $|G|$, and that $\mathbf{C}_G(v)$ contains a Hall-π-subgroup of G for all $v \in V$. Then V is an irreducible G-module and one of the following occurs:*

(i) *$G \leq \Gamma(V)$, $G/\mathbf{F}(G)$ is cyclic and $\mathbf{F}(G)$ is a π'-group;*

(ii) *$|V| = 3^2$, $\pi = \{3\}$ and G is isomorphic to $SL(2,3)$ or $GL(2,3)$;*

(iii) *$|V| = 2^6$, $\pi = \{2\}$, $\mathbf{F}(G)$ is extra-special of order 3^3 and exponent 3, $|G/\mathbf{F}(G)| = 2$, $\mathbf{Z}(\mathbf{F}(G)) = \mathbf{Z}(G)$ and $\mathbf{O}^{2'}(G) = G$.*

Proof. Choose some $p \in \pi$ and apply Corollary 10.5. If $G \leq \Gamma(V)$, then $G/\mathbf{F}(G) \leq \Gamma(V)/\Gamma_0(V)$ is cyclic. Finally, the hypotheses imply that $\mathbf{O}_\pi(G) \leq \mathbf{C}_G(V) = 1$ and so $\mathbf{F}(G)$ is a π'-group. $\qquad\square$

The presentation of sections 9 and 10 is based on Wolf [Wo 3], Gluck & Wolf [GW 1] and Manz & Wolf [MW 1].

§11   Arithmetically Large Orbits

Suppose G is a solvable irreducible subgroup of $GL(V)$ where V is a finite vector space of order q^n, q a prime power. The intent of this section is to show that G has a large orbit $\{v^G\}$ on V in the sense that $|\{v^G\}|$ is divisible by many prime divisors of $|G|$. Of course, exceptions occur. Most notably if $G = \Gamma(V)$, then the orbit sizes are 1 and $q^n - 1$ while $|G| = n(q^n - 1)$.

Choose $H \leq G$ such that $V = W^G$ for a primitive irreducible H-module W. If $H/\mathbf{C}_H(W) \not\leq \Gamma(W)$, then we show there exists $v \in V$ with $|\{v^G\}|$ divisible by each prime divisor p of $|G|$ with $p \geq 5$. A large part of the proof is devoted to the case when $V = W$ is primitive. The proof here is similar to and uses results of Section 10. To pass from the primitive to the imprimitive case, we use Corollary 5.7 (a) to Gluck's permutation theorem. This corollary states that given a solvable permutation group G on a set Ω, there exists a subset $\Delta \subseteq \Omega$ such that $\mathrm{stab}_G(\Delta)$ is a $\{2,3\}$-group. Since this corollary cannot be improved (to delete 2 or 3), this explains, in part, why we only consider prime divisors p of $|G|$ with $p \geq 5$. These small primes pose other difficulties too.

For $H \leq G$, we let $\pi_0(G : H)$ be the set of those primes $p \geq 5$ that divide $|G : H|$. Likewise, we define $\pi_0(G)$. Our first lemma examines small linear groups as in Section 2.

11.1 Lemma. *Suppose that G is a solvable, irreducible subgroup of* $GL(n, q)$, q *prime.*

(a) *If $q^n = 2^4$, then $\pi_0(G) = \varnothing$ or $G \lesssim \Gamma(2^4)$;*

(b) *If $q^n = 2^6$, then $\pi_0(G) = \varnothing$ or G is isomorphic to a subgroup of* $\Gamma(2^3) \,\mathrm{wr}\, Z_2$ *or* $\Gamma(2^6)$;

(c) *If $q^n = 2^8$, then $\pi_0(G) = \varnothing$ or G is isomorphic to a subgroup of* $\Gamma(2^4) \,\mathrm{wr}\, Z_2$ *or* $\Gamma(2^8)$;

(d) *If $q^n = 2^{10}$, then G is isomorphic to a subgroup of $S_3 \,\mathrm{wr}\, F_{20}$,* $\Gamma(2^5) \,\mathrm{wr}\, Z_2$ *or* $\Gamma(2^{10})$;

(e) *If $q^n = 3^4$, then $\pi_0(G) \subseteq \{5\}$; and*

(f) *If $q^n = 3^6$, then $|\pi_0(G)| \leq 2$ and G has a normal Sylow p-subgroup for all $p \in \pi_0(G)$.*

Proof. Parts (a), (b) and (d) are immediate from Corollary 2.15.

We first observe that if $K \leq GL(p^m, p)$ is irreducible, quasi-primitive and solvable (with p prime), then $K \leq \Gamma(p^{p^m})$. Since $\mathbf{O}_p(K) = 1$, Corollary 2.5 implies that $\mathbf{F}(K)$ is abelian. Then Corollary 2.3 yields that $K \leq \Gamma(p^{p^m})$.

(c) If $q^n = 2^8$, then the last paragraph implies that $G \leq \Gamma(2^8)$, that $G \cong S_3$ wr S_4, or $G \leq H$ wr Z_2 for a solvable irreducible $H \leq GL(4, 2)$. Part (c) now follows from part (a).

(e, f) Suppose now that q^n is 3^4 or 3^6 and that G is not a $\{2, 3\}$-group. First assume that G is not quasi-primitive. Then $G \leq H$ wr S for an irreducible linear group H and solvable primitive permutation group $S \leq S_m$, with $m \mid n$. Since n is 4 or 6, $m \leq 4$. Thus S is a $\{2, 3\}$-group and hence H is not. Thus H is an irreducible subgroup of $GL(3, 3)$, $n = 6$ and $G \lesssim H$ wr Z_2. Since H is not a $\{2, 3\}$-group, H is quasi-primitive. By the next to last paragraph, $H \leq \Gamma(3^3)$. Conclusion (f) holds because $\pi_0(G) \subseteq \{13\}$ and G has a normal Sylow 13-subgroup.

We thus assume the corresponding G-module V is quasi-primitive, $|V| = 3^4$ or 3^6. We apply Corollary 1.10 and adopt its notation. Since we may assume that $G \nleq \Gamma(V)$, then $e := |F : T|^{1/2} > 1$ by Corollary 2.3. Since $\mathbf{O}_3(G) = 1$ and $e \mid n$, e is 2 or 4. Let W be an irreducible U-submodule of V. Then $\dim(W) \mid n/e$ (by Corollary 2.6) and $|U| \mid |W| - 1$.

First suppose that $e = 4$. Since $e \mid n$, we have that $n = 4$, $\dim(W) = 1$ and $|U| = 2 = |T|$. In this case, $A = G$ and F/T is faithful G/F-module of order 2^4. Furthermore, F/T is an irreducible G/F-module or the direct product of two irreducible modules of order 2^2. By (a), $\pi_0(G) = \pi_0(G/F) \subseteq \{5\}$. Conclusion (e) holds.

Finally assume that $e = 2$. Now F/T is a completely reducible A/F-module of order 2^2. Thus $\pi_0(A/T) = \pi_0(A/F) = \varnothing$. Now $\dim(W) \mid n/2$ and so $|W|$ is 3, 3^2, or 3^3. Then $|U|$ divides 8 or 26. Since $A = \mathbf{C}_G(Z)$ and U is cyclic, $\pi_0(G/T) = \pi_0(G/A) = \varnothing$ and $\pi_0(G) = \pi_0(T) = \pi_0(U) \subseteq \{13\}$. Furthermore, $\pi_0(G) = \{13\}$ is possible only when $|W| = 3^3$ and $n = 6$.

Conclusions (e) and (f) hold. $\square$

11.2 Proposition. *Let G be solvable. Then the number $|\mathrm{Syl}(G)|$ of distinct Sylow subgroups of G (for all primes) is at most $|G|$.*

Proof. By induction on $|G|$. We note that equality holds when $|G| \leq 2$. We may choose a maximal normal subgroup M of G and set $q = |G/M|$, a prime. By the inductive hypothesis, $|\mathrm{Syl}(M)| \leq M$. If $P \in \mathrm{Syl}_p(G)$ for $p \neq q$, then $P \in \mathrm{Syl}_p(M)$, and so the number of Sylow subgroups of G for all primes other than q is at most $|M|$. But $|\mathrm{Syl}_q(G)| \leq |G|/q = |M|$. Hence $|\mathrm{Syl}(G)| \leq 2|M| \leq |G|$. $\square$

11.3 Theorem. *Let V be a finite faithful quasi-primitive G-module for a solvable group G. Assume that each $v \in V$ is centralized by a non-trivial Sylow p-subgroup of G for some prime $p \geq 5$ (dependent on v). Then $G \subseteq \Gamma(V)$.*

Proof. We let π be the set of prime divisors $p \geq 5$ of $|G|$ for which $\mathbf{C}_V(P) \neq 0$ for $P \in \mathrm{Syl}_p(G)$. The hypotheses imply that each $v \in V$ is centralized by a Sylow p-subgroup for some $p \in \pi$. Thus $\pi \neq \varnothing$. By Theorem 10.5, we may assume that $|\pi| \geq 2$.

Let $F = \mathbf{F}(G)$. For $1 \neq Q \in \mathrm{Syl}(F)$, $\mathbf{C}_V(Q) = \{0\}$ by the irreducibility of V. Thus F is a π'-group. In particular, $\pi \subseteq \pi_0(G/F)$ and hence $|\pi_0(G/F)| \geq 2$.

Since V is quasi-primitive, Corollary 1.10 applies and we adopt the notation there. Set $e^2 = |F : T|$. By Corollary 2.3, we assume that $e > 1$. We set $C/F = \mathbf{C}_{G/F}(F/T)$ and observe by Corollary 1.10 (vii) that $G' \leq A$ and $A \cap C = F$. Thus $C/F \leq \mathbf{Z}(G/F)$.

Step 1. Let $D/U \in \mathrm{Hall}_\pi(C/U)$. Then $D \trianglelefteq G$, D/U is abelian, and D/U is G-isomorphic to a Hall π-subgroup of C/F. Furthermore, $U = \mathbf{C}_D(Z)$.

Proof. Let $D_0/F \in \mathrm{Hall}_\pi(C/F)$. Because $C/F \leq \mathbf{Z}(G/F)$, note that $D_0 \trianglelefteq G$ and D_0/F is abelian. Since no $p \in \pi$ divides $|F|$, $D_0 = FD$ and $F \cap D = U$. Since D centralizes F/T and T/U, we have that $D_0/U = F/U \times D/U$ and D/U char D_0/U. Thus $D \trianglelefteq G$ and $D/U \cong D_0/F$ is abelian. Now $U \leq D \cap A = D \cap C \cap A = D \cap F = U$ and thus $\mathbf{C}_D(Z) = A \cap D = U$.

Step 2. Let W be an irreducible U-submodule of V. Then

(a) $|V| = |W|^{te}$ for an integer t;

(b) $|U| \mid |W| - 1$.

Proof. By hypothesis, $|V|$ is finite. Part (a) is just Corollary 2.6. Part (b) is immediate because V_U is homogeneous and U is cyclic.

Step 3. Let $p \in \pi$ and $P \in \mathrm{Syl}_p(G)$. Then

(a) $|\mathbf{C}_V(P)| \leq |V|^{1/2}$;

(b) If $1 \neq P_1 \leq P \cap D$, then $|\mathbf{C}_V(P_1)| \leq |V|^{1/5}$ and $p \mid t \cdot \dim(W)$.

Proof. Let $1 \neq P_0 \leq P$ with $|P_0| = p$. Recall that $p \nmid |F|$. First suppose that $p \mid |D|$ and assume without loss of generality that $P_0 \leq D$. Since $U = \mathbf{C}_D(U)$ by Step 1 and $p \nmid |U|$, we may choose $1 \neq Y \leq Z$ with YP_0 a Frobenius group. Note $\mathbf{C}_V(Y) = 0$ because $Y \trianglelefteq G$. Then $\dim(V) = p \cdot \dim(\mathbf{C}_V(P))$ by Lemma 0.34. Since $\dim(V) = te \dim(W)$ and $p \nmid |F|$, in fact $p | t \cdot \dim(W)$. Parts (a) and (b) follow when $p \mid |D|$.

Choose $Q \in \mathrm{Syl}_q(F)$ such that $P_0 \nleq \mathbf{C}_G(Q)$. Apply Lemma 7.2 or Lemma 0.34 to conclude $|\mathbf{C}_V(P)| \leq |V|^{1/2}$ (recall $p \geq 5$). This step follows.

Step 4.

(a) $\displaystyle\sum_{p \in \pi} \sum_{P \in \mathrm{Syl}_p(G)} |\mathbf{C}_V(P)| \geq |V|$;

(b) $\displaystyle|G| \geq \sum_{p \in \pi} |\mathrm{Syl}_p(G)| \geq |V|^{1/2}$; and

(c) Some $p \in \pi$ does not divide $|D|$. In particular, $|G/C|$ is divisible by p.

Proof. (a, b) Since each $v \in V$ is centralized by some Sylow p-subgroup for some $p \in \pi$, $\bigcup_{p \in \pi} \bigcup_{P \in \mathrm{Syl}_p(G)} \mathbf{C}_V(P) = V$. Thus $\sum_{p \in \pi} \sum_{P \in \mathrm{Syl}_p(G)} |\mathbf{C}_V(P)| \geq |V|$. Applying Step 3 (a) and Proposition 11.2,

$$|V| \leq \sum_{p \in \pi} \sum_{P \in \mathrm{Syl}_p(G)} |V|^{1/2} \leq \sum_{p \in \pi} |\mathrm{Syl}_p(G)| \, |V|^{1/2} \leq |G| \, |V|^{1/2}.$$

We have proven (a) and (b).

(c) Assume each $p \in \pi$ divides $|D|$. Since $D \trianglelefteq G$, the intersection of D with a Sylow subgroup of G is a Sylow subgroup of D. Thus each $v \in V$ is centralized by a non-trivial Sylow p-subgroup of D for some $p \in \pi$. As above, $|V| \leq \sum_{p \in \pi} \sum_{P \in \mathrm{Syl}_p(D)} |\mathbf{C}_V(P)|$. By Step 3 (b), $|\mathbf{C}_V(P)| \leq |V|^{1/5}$ for $P \in \mathrm{Syl}_p(D)$, $p \in \pi$. Hence $\sum_{p \in \pi} |\mathrm{Syl}_p(D)| \geq |V|^{4/5}$. For $p \in \pi$, $|\mathrm{Syl}_p(D)| \leq |U|$ because D/U is abelian. By Step 1, D/U acts faithfully on the cyclic group $Z \leq U$ and hence

$$|\pi| \leq \log_2(|D/U|) \leq \log_2(|U|).$$

Thus

$$|V|^{4/5} \leq \sum_{p \in \pi} |\mathrm{Syl}_p(D)| \leq |\pi| \, |U| \leq \log_2(|U|) \, |U|.$$

By Step 2,

$$|V| = |W|^{te} \geq |W|^e \geq |U|^e.$$

Thus

$$|U|^{4e/5} \leq |U| \log_2(|U|) \leq |U|^2 \text{ and } e < 3.$$

Since $e > 1$, $e = 2$ and $|U|^{3/5} \leq \log_2 |U|$. Then $|U| < 2$, a contradiction. So some $p \in \pi$ does not divide $|D|$.

Step 5.

(a) $e^{13} > 4|W|^{te-4}$.

(b) $e \leq 32$.

(c) No prime larger than 3 divides e.

Proof. By Corollary 3.7, $|G| \leq e^{13/2}|U|^2/2$. By Steps 4 (b) and 2 (b),

$$|W|^2 e^{13/2}/2 > |U|^2 e^{13/2}/2 \geq |G| \geq |V|^{1/2} = |W|^{te/2}$$

and

$$e^{13} > 4|W|^{te-4} \geq 4|W|^{e-4}. \tag{11.1}$$

Since $|W| \geq 3$, (11.1) yields $e < 64$. If $|W|$ is 3, 4 or 5, then $|U|$ and hence e are both powers of 2 or both are powers of 3. In this case, $e \leq 32$. If, however, $|W| \geq 7$, then (11.1) implies that $e \leq 32$. This proves (a) and (b).

We now assume that $s|e$ for a prime $s \geq 5$. First suppose that $s > 7$. Then $|W| \geq 23$ and (11.1) implies that $e < 17$, whence $e = 11$ or 13. Suppose that $e = 11$. By (11.1), $|W|$ is 23 or 67. Since $Z \leq U$ and $|U| \mid |W| - 1$, $|Z| \mid 66$. But $C/F \lesssim \text{Aut}(Z)$ and so C/F is a $\{2, 5\}$-group. Now $G/C \leq GL(2, 11)$ and so G/C is a $\{2, 3, 5, 11\}$-group. Every prime in π divides $|G/C|$ or $|C/F|$ and is at least 5. But no prime in π divides $|F|$ or e, in particular. Thus $\pi \subseteq \{5\}$, a contradiction because $|\pi| \geq 2$ (see the first paragraph of the theorem's proof). So $e \neq 11$. Should $e = 13$, then $|W| = 27$ and we similarly derive the contradiction $\pi \subseteq \{7\}$. Hence $s \leq 7$.

Next consider $s = 7$. Should $|W| \geq 29$, then (11.1) implies $e < 14$, whence $e = 7$. If $|W| < 29$ then $|W| = 8$ and part (b) implies that $e = 7$. Then $G/C \leq GL(2, 7)$ is a $\{2, 3, 7\}$-group. Since $7|e$, no prime in π divides $|G/C|$, contradicting Step 4 (c). Hence s must be 5 and $|W| \geq 11$. Now (11.1) implies that $e < 20$. Now $F/T = \prod_{i=1}^{\ell} (F_i/T)$ is a completely reducible and faithful G/C-module with each $|F_i/T| =: e_i^2$ (see Corollary 1.10). Thus $e = e_1 \cdots e_\ell$ is 5, 5 · 2, or 5 · 3. Since $GL(2, r)$ is a $\{2, 3, 5\}$-group whenever $r \in \{2, 3, 5\}$, G/C is a $\{2, 3, 5\}$-group. Since $5|e$, we have that G/C is a π'-group, again contradicting Step 4 (c). This step is complete.

Step 6. Writing $e = e_1 \cdots e_\ell$ as in the last paragraph, we may assume that $e_1 \geq 2^3$. Furthermore, if $e_1 = 2^3$ then $e_2 = 2^2$ and $e = 2^3 \cdot 2^2$.

Proof. Now $F/T = F_1/T \times \cdots \times F_\ell/T$ where each F_i/T is an irreducible G/C-module of order e_i^2. Set $C_i = \mathbf{C}_G(F_i/T)$ so that $C = \cap_{i=1}^{\ell} C_i$. We have by Step 4 (c) that at least one prime $p \in \pi$ divides $|G/C|$. We may assume that $p \mid |G/C_1|$. Thus e_1 is not 2 or 3. Since $e \leq 32$ and e is divisible by no primes larger than 3 (see Step 5 (b, c)), we may assume for this step that e_1 is 2^2 or 2^3.

First suppose that e_1 is 2^2 and every $e_i \leq 4$. By Lemma 11.1 (a), each G/C_i is a $\{2,3,5\}$-group with a normal (possibly trivial) Sylow 5-subgroup. Since $\bigcap C_i = C$ and $C/F \leq \mathbf{Z}(G/F)$, it follows that G/C is a $\{2,3,5\}$-group and G/F has a normal Sylow 5-subgroup. Each $q > 5$ in π must divide $|C/F|$ and hence $|D/U|$. Also for $q > 5$ in π and $Q \in \mathrm{Syl}_q(D)$, we have $1 \neq Q \in \mathrm{Syl}_q(G)$. Let $P \in \mathrm{Syl}_5(G)$. Each $v \in V$ is centralized by some conjugate of P or by a Sylow q-subgroup of D with $q > 5$, $q \in \pi$. Thus

$$|\mathrm{Syl}_5(G)|\,|\mathbf{C}_V(P)| + \sum_{5 < q \in \pi} \sum_{Q \in \mathrm{Syl}_q(D)} |\mathbf{C}_V(Q)| \geq |V|.$$

Since $FP \trianglelefteq G$, $|\mathrm{Syl}_5(G)| \leq |F/T|\,|U| = e^2|U|$. Applying Step 3, Proposition 11.2 and Step 1,

$$\sum_{5 < q \in \pi} \sum_{Q \in \mathrm{Syl}_q(D)} |\mathbf{C}_V(Q)| \leq |D|\,|V|^{1/5} \leq |U|^2|V|^{1/5}.$$

Now $|\mathbf{C}_V(P)| \leq |V|^{1/2}$ by Step 3 and $|U| < |W|$ by Step 2. Thus

$$|V| \leq e^2|W|\,|V|^{1/2} + |W|^2|V|^{1/5}.$$

Now $|V|^{3/10} \geq |W|^{3e/10} \geq |W|^{12/10} \geq |W|$ and so $|W||V|^{1/2} \geq |W|^2|V|^{1/5}$. Hence $2e^2|W||V|^{1/2} \geq |V|$. Since $|V| \geq |W|^e$, we have $4e^4 \geq |W|^{e-2}$. Since $e \geq 4$, $|W| < 32$. Since Z is cyclic and $|Z| \mid |W| - 1$, $\mathrm{Aut}(Z)$ and D/U are easily checked to be $\{2,3,5\}$-groups. Hence 5 is the only prime in π, a contradiction. Hence e_1 is not 2^2.

For this step, we next assume that $e_1 = 2^3$. Since $e \leq 32$, we may assume that $e_i \leq 3$ for $i > 1$ and so G/C_i is a $\{2,3\}$-group ($i > 1$). Since $p \mid |G/C_1|$, it follows from Lemma 11.1 (b) that $p = 7$ is the unique $p \in \pi$ dividing $|G/C_1|$ and that G/C_1 has a normal Sylow 7-subgroup. As above G/C and G/F have normal Sylow 7-subgroups. As in the previous paragraph

$4e^4 \geq |W|^{e-2}$. Since $e \geq 8$, $|W| < 6$. Then Aut (Z) and D/U are $\{2,3\}$-groups and $|\pi| = 1$, a contradiction. This step is complete.

Step 7. $D = U$.

Proof. By Step 5, we have that

$$e^{13} > |W|^{te-4} \cdot 4. \qquad (11.2)$$

If $t \geq 5$, then $e^{13} \geq |W|^{5e-4} \cdot 4 \geq 4 \cdot 3^{5e-4}$ and $e < 4$, contradicting Step 6. So $t < 5$. We may assume that $D > U$ and choose q a prime dividing $|D/U|$. By Step 3, $q | \dim(W)$ and note $q \geq 5$.

Because $e \geq 8$, (11.2) implies that $|W| \leq 2^{37/4} \leq 620$. Since $q \geq 5$ divides $\dim(W)$, we must have that $|W|$ is 2^5, 2^7 or 3^5. In the first two cases, $|W| - 1$ is a prime and so $|U|$ is 31 or 127, whence 31 or 127 divides e, contradicting Step 5 (b, c). Thus $|W| = 3^5$. By (11.2), $e < 16$. Since each prime divisor of e divides $|W| - 1$, e is a power of 2 by Step 5 (c). Applying Step 6, we get $e = 2^3 \cdot 2^2$, a contradiction. Thus $D = U$.

Step 8. $e = e_1 = 2^5$.

Proof. First we show $e_1 \neq 3^2$. If $e_1 = 3^2$, observe that $e_i \leq 3$ for $i > 1$ (Step 5 (b)). Then G/C_1 and G/C are $\{2,3,5\}$-groups by Lemma 11.1. Since $D = U$, $\pi \subseteq \{5\}$, a contradiction. Thus $e_1 \neq 3^2$.

By Steps 5 (b) and 6, we have that $e = e_1 \cdots e_\ell$ is $2^3 \cdot 2^2$, 2^4, $2^4 \cdot 2$, 3^3 or 2^5. For now, exclude the last case. By Lemma 11.1 and Step 7, $|\pi| \leq 2$ and each Sylow p-subgroup of G/C for $p \in \pi$ is normal in G/C. Since $C/F \leq \mathbf{Z}(G/F)$, $FP/F \trianglelefteq G/F$ whenever $P \in \mathrm{Syl}_p(G)$, $p \in \pi$. Since $e \geq 16$, $|W| < 23$ by Step 5. Because $|U| \mid |W| - 1$ and $|T/U| \leq 2$, Aut(U) is a $\{2,3\}$-group and $P \leq \mathbf{C}_G(T)$. Thus $|\mathrm{Syl}_p(G)| \leq |F/T| = e^2$. Hence, by Steps 4 and 2,

$$3^{e/2} \leq |W|^{e/2} \leq |V|^{1/2} \leq \sum_{p \in \pi} |\mathrm{Syl}_p(G)| \leq |\pi| e^2 \leq 2e^2.$$

Since $e \geq 16$, this is a contradiction, completing this step.

Step 9. Conclusion.

Proof. We have that $e = e_1 = 2^5$ and char$(W) \neq 2$. By Step 5, $|W| = 3$. Thus $U = T$ has order 2 and $A = G$. Furthermore $\pi \subseteq \pi_0(G/F)$ has at least 2 elements. Since F/U is an irreducible faithful G/F-module, $\pi \subseteq \{11, 31, 5\}$ by Lemma 11.1. In fact, $G/F \lesssim \Gamma(2^5)$ wr Z_2 or $G/F \lesssim \Gamma(2^{10})$. Observe that G/F has normal Sylow p-subgroups for $p = 11$ and 31, and $|G/F|$ divides $31^2 \cdot 5^2 \cdot 2$ or $33 \cdot 31 \cdot 5 \cdot 2$. Since $U \leq \mathbf{Z}(G)$,

$$|\mathrm{Syl}_{11}(G)| \leq 2^{10}, \quad |\mathrm{Syl}_{31}(G)| \leq 2^{10},$$

and

$$|\mathrm{Syl}_5(G)| \leq 2^{21}.$$

By Steps 4 and 2,

$$3^{16} \leq |W|^{e/2} \leq |V|^{1/2} \leq \sum_{p \in \pi} |\mathrm{Syl}_p(G)| \leq 2^{10} + 2^{10} + 2^{21} \leq 2^{22}.$$

This contradiction completes the proof of the theorem.   □

One should note again that in the action of $GL(2,3)$ on its natural module W, each $w \in W$ is centralized by a Sylow 3-subgroup, but $GL(2,3) \nleq \Gamma(3^2)$.

We next proceed to imprimitive modules. If V is an irreducible, faithful G-module, then $V = W^G$ for a primitive module W of a subgroup $H \leq G$. If $H/\mathbf{C}_H(W) \nleq \Gamma(W)$ (we assume V finite and G solvable), we show that there exists some $v \in V$ with $\pi_0(G : \mathbf{C}_G(v)) = \pi_0(G)$. This uses the above Theorem 11.3 together with Corollary 5.7 of Gluck's permutation theorem 5.6.

11.4 Theorem. *Suppose that V is a finite faithful irreducible G-module and that $V = W^G$ for an (irreducible) primitive module W of H for some*

$H \leq G$ (possibly $H = G$). Assume that $H/\mathbf{C}_H(W) \not\leq \Gamma(W)$, but G is solvable. Then there exists $v \in V$ such that $\pi_0(G : \mathbf{C}_G(v)) = \pi_0(G)$.

Proof. Since $V = W^G$, we may write $V = X_1 \oplus \cdots \oplus X_m$ for subspaces X_i of V that are transitively permuted by G with $W = X_1$. For $H \leq J \leq G$, W^J is irreducible and hence $H = \mathbf{N}_G(W) = \mathbf{N}_G(X_1)$. Since $H/\mathbf{C}_H(W) \not\leq \Gamma(W)$, Theorem 11.3 implies there exists $0 \neq x_1 \in W$ such that

$$\pi_0(H/\mathbf{C}_H(x_1)) = \pi_0(H/\mathbf{C}_H(W)).$$

Let $H_i = \mathbf{N}_G(X_i)$, a conjugate of H. But the $H_i/\mathbf{C}_{H_i}(X_i)$ are isomorphic as linear groups and so there exist, by Theorem 11.3, $0 \neq x_j \in X_j$ such that

$$\pi_0(H_j/\mathbf{C}_{H_j}(x_j)) = \pi_0(H_j/\mathbf{C}_{H_j}(X_j)) = \pi_0(H/\mathbf{C}_H(W))$$

for each j.

Next let $C = \bigcap_{i=1}^m H_i = \bigcap_{i=1}^m \mathrm{stab}_G(X_i)$, so that G/C faithfully and transitively permutes $\{X_1, \ldots, X_m\}$. By Corollary 5.7, we may now assume without any loss of generality that there exists $1 \leq \ell \leq m$ such that $\mathrm{stab}_{G/C}\{X_1, \ldots, X_\ell\}$ is a $\{2,3\}$-group, Let $x = x_1 + \cdots + x_\ell$ and suppose that $Q \in \mathrm{Syl}_q(G)$ for a prime $q \geq 5$ and $Q < \mathbf{C}_G(x)$. Then Q must stabilize $\{X_1, \ldots, X_\ell\}$ and hence $Q \leq C = \bigcap_{i=1}^m H_i$. Now Q must centralize $x_1, x_2, \ldots, x_\ell$. Since $q \nmid |H_i : \mathbf{C}_{H_i}(x_i)|$, the first paragraph implies that $q \nmid |H_i/\mathbf{C}_{H_i}(X_i)|$ for $i = 1, \ldots, \ell$. Since $C/\mathbf{C}_C(X_i) \cong C/\mathbf{C}_C(X_j)$ for $j = 1, \ldots, \ell, \ldots, m$ and $\bigcap_{i=1}^m \mathbf{C}_C(X_i) = 1$, C is a q'-group. Since $Q \leq C$, $Q = 1$. Hence $\pi_0(G) = \pi_0(G : \mathbf{C}_G(x))$. $\square$

11.5 Corollary. *Suppose V is a faithful finite module for a solvable group G. Suppose $V = V_1 \oplus \cdots \oplus V_n$ for irreducible G-modules V_i (we allow V to have "mixed characteristic"). For each i, write $V_i = W_i^G$ for a primitive module W_i of a subgroup $H_i \leq G$. Assume $H_i/\mathbf{C}_{H_i}(W_i)$ is not isomorphic to a subgroup of $\Gamma(W_i)$ for all i. Then there exists $v \in V$ such that $\pi_0(G : \mathbf{C}_G(v)) = \pi_0(G)$.*

Proof. Applying Theorem 11.4, there exists, for each i, $v_i \in V_i$ such that $\pi_0(G : \mathbf{C}_G(v_i)) = \pi_0(G/\mathbf{C}_G(V_i))$. Set $v = v_1 + \cdots + v_n$. If $p > 3$ and $P \in \mathrm{Syl}_p(G)$ centralizes v, then P centralizes each v_i and so $P \leq \mathbf{C}_G(V_i)$ for all i. This implies $P = 1$ and $p \nmid |G|$. So $\pi_0(G : \mathbf{C}_G(v)) = \pi_0(G)$. $\qquad\qquad\square$

Chapter IV
PRIME POWER DIVISORS OF CHARACTER DEGREES

§12   Characters of p'-degree and Brauer's
Height-Zero Conjecture

Suppose $N \unlhd G$, $\theta \in \mathrm{Irr}\,(G)$, and $\chi(1)/\theta(1)$ is a p'-number for all irreducible constituents χ of θ^G. The bulk of work in this section will be aimed at proving that G/N has an abelian Sylow p-subgroup, provided G/N is solvable. With little extra work, we see that p can be replaced by a set of primes. As a consequence of this and Fong reduction (Lemma 0.25 and Theorem 0.28), we then prove Brauer's height-zero conjecture for solvable G. Namely, if B is a p-block of a solvable group, then all the ordinary characters in B have height zero if and only if the defect group for B is abelian. The contents of this section are [Wo 3, GW 1], and while the arguments are essentially the same, some improvements and refinements should improve the reading thereof. Brauer's height-zero conjecture was extended to p-solvable G in [GW 2], with the help of the classification of simple groups.

In the key Theorem 12.9 of this section, we have $N \unlhd G$, $\theta \in \mathrm{Irr}\,(N)$ and $\chi(1)/\theta(1)$ a p'-number for all $\chi \in \mathrm{Irr}\,(G|\theta)$. The aim is to show that G/N has abelian Sylow p-subgroup, at least when G/N is solvable. In a minimal counterexample, there exists an abelian chief factor M/N of G such that each $\lambda \in \mathrm{Irr}\,(M/N)$ is invariant under some Sylow p-subgroup of G/M. Consequently, the results of Sections 9 and 10 play an important role.

12.1 Proposition. *Suppose that H acts on an abelian group A. Then*

(a) *H acts faithfully on A if and only if H acts faithfully on $\mathrm{Irr}\,(A)$.*

(b) *H acts irreducibly on A if and only if H acts irreducibly on $\mathrm{Irr}\,(A)$.*

Proof. Note that the actions of H on A and $A^* = \text{Irr}\,(A)$ satisfy $\lambda^h(a^h) = \lambda(a)$ for $a \in A$, $h \in H$, and $\lambda \in A^*$. Likewise H acts on A^{**} and $b^h(\lambda^h) = b(\lambda)$ for $b \in A^{**}$, $\lambda \in A^*$. There is a natural isomorphism from A to A^{**} given by $a \mapsto a^{**}$ where $a^{**}(\lambda) = \lambda(a)$ (see [Hu, V, 6.4]). Now

$$(a^h)^{**}(\lambda^h) = \lambda^h(a^h) = \lambda(a) = a^{**}(\lambda) = (a^{**})^h(\lambda^h)$$

for $a \in A$, $h \in H$, $\lambda \in A^*$. This natural isomorphism is hence an H-isomorphism. Thus H acts faithfully (irreducibly) on A if and only if H acts faithfully (irreducibly, resp.) on A^{**}.

If $h \in H$ centralizes a group B, h acts trivially on $\text{Irr}\,(B)$. Thus

$$\mathbf{C}_H(A) \leq \mathbf{C}_H(A^*) \leq \mathbf{C}_H(A^{**}) = \mathbf{C}_H(A).$$

This proves (a). If $1 < D < A$ is H-invariant then $1 < (A/D)^* < A^*$ is H-invariant. Applying this twice,

A^{**} irreducible $\implies A^*$ irreducible $\implies A$ irreducible $\implies A^{**}$ irreducible.

This proves (b). $\square$

Alternatively, to prove 12.1 (a), one can use Brauer's permutation lemma [Is, 6.32]. We employ this in the next lemma, which is related to Proposition 12.1 (a).

12.2 Lemma. *Assume that S acts on G with $(|S|, |G|) = 1$. If S fixes every irreducible character of G, then S centralizes G.*

Proof. With no loss of generality, we may assume that $S \neq 1$ is a cyclic p-group for some prime p. Since S is cyclic, Brauer's permutation lemma [Is, 6.32] implies that S fixes each conjugacy class $\mathcal{C}$ of G. Since $p \nmid |\mathcal{C}|$ and S is a p-group, $\mathcal{C} \cap \mathbf{C}_G(S) \neq \varnothing$. Since this is valid for each conjugacy class of G, we have that $G = \bigcup_{g \in G} \mathbf{C}_G(S)^g$. Since G is finite, $G = \mathbf{C}_G(S)$. $\square$

12.3 Proposition. *Let $N \trianglelefteq G$, $\theta \in \mathrm{Irr}\,(N)$ and $\chi \in \mathrm{Irr}\,(G|\theta)$. The following are equivalent:*

(i) $\chi_N = e\theta$ *with* $e^2 = |G : N|$;

(ii) $I_G(\theta) = G$ *and* χ *vanishes off* N; *and*

(iii) $I_G(\theta) = G$ *and* χ *is the unique irreducible constituent of* θ^G.

Proof. This is Exercise 6.3 of [Is]. □

In Proposition 12.3 above, we say that χ and θ are *fully ramified* with respect to G/N. Before proceeding with the main result of this section, we need some information on fully ramified sections.

12.4 Lemma. *Suppose that G/N is abelian, $\varphi \in \mathrm{Irr}\,(N)$, and $\chi \in \mathrm{Irr}\,(G|\varphi)$ is fully ramified with respect to G/N. Suppose that S acts on G fixing N, φ, and (hence) χ. Assume that $(|S|, |G/N|) = 1$. Set $C/N = \mathbf{C}_{G/N}(S)$ and $D/N = [G/N, S]$. Then*

(a) χ *is fully ramified with respect to both G/C and G/D; and*

(b) φ *is fully ramified with respect to both C/N and D/N.*

Proof. Since G/N is abelian, $C \trianglelefteq G$. Because χ is the unique irreducible constituent of φ^G, the irreducible constituents of φ^C and χ_C coincide. By Lemma 0.17 (b, f), χ_C has a unique irreducible S-invariant constituent, whereas every irreducible constituent of φ^C is S-invariant. Thus φ^C has a unique irreducible constituent β and $I_G(\beta) = G = I_G(\varphi)$. By Proposition 12.3, χ and φ are fully ramified with respect to G/C and C/N (respectively). A similar argument shows that χ and φ are fully ramified with respect to G/D and D/N (respectively). □

12.5 Lemma. *Suppose that $N \trianglelefteq G$ with G/N abelian. Assume that $\varphi \in \mathrm{Irr}\,(N)$ is fully ramified with respect to G/N. Then*

(a) $G/N \cong A \times A$ *for an abelian group A; and*

(b) *If G/N is a p-group, if Q is a q-group acting on G fixing N and φ, if $q \neq p$ and $Q/\mathbf{C}_Q(G/N)$ is abelian, then* $\operatorname{rank}(Q/\mathbf{C}_Q(G/N)) \leq \operatorname{rank}(G/N)/2$.

Proof. (a) $\Longrightarrow$ (b). By induction on $|Q/Q_0|$, where $Q_0 = \mathbf{C}_Q(G/N)$. By (a), we may assume that $\operatorname{rank}(G/N) \geq 2$ and thus we may assume that Q/Q_0 is not cyclic. We may choose $Q_0 < Q_1 < Q$ such that

$$\operatorname{rank}(Q/Q_0) = \operatorname{rank}(Q/Q_1) + \operatorname{rank}(Q_1/Q_0).$$

By Fitting's Lemma 0.6, $G/N = C/N \times D/N$ where $C/N = \mathbf{C}_{G/N}(Q_1)$ and $D/N = [G/N, Q_1]$. Now Q/Q_1 acts faithfully on C/N and Q_1/Q_0 acts faithfully on D/N. Applying Lemma 12.4 and the inductive hypothesis,

$$\operatorname{rank}(Q/Q_0) = \operatorname{rank}(Q/Q_1) + \operatorname{rank}(Q_1/Q_0)$$
$$\leq \operatorname{rank}(C/N)/2 + \operatorname{rank}(D/N)/2 = \operatorname{rank}(G/N)/2.$$

Hence (a) implies (b).

We next prove (a) by induction on $|G/N|$. By [Is, Theorem 11.28], we may assume that φ is linear and faithful. Since φ is also G-invariant, $N \leq \mathbf{Z}(G)$. If χ is the unique irreducible constituent of φ^G, then χ vanishes off N by Proposition 12.4 and so $N = \mathbf{Z}(G)$. Note that N is cyclic and $G' \leq N$.

Choose $x \in G$ such that $o(Nx)$ is maximal. Set $D = \langle N, x \rangle$ and $C = \mathbf{C}_G(D) = \mathbf{C}_G(x) \geq D$. Then G/C acts faithfully on D, while centralizing both D/N and N. By Lemma 1.5 and its proof, we have that $|G/C| \mid |D/N|$ and $G/C \lesssim \operatorname{Hom}(D/N, N)$. Since D/N and N are cyclic, G/C is also cyclic. Let $\lambda \in \operatorname{Irr}(D|\varphi)$ and $\gamma \in \operatorname{Irr}(C|\lambda)$. Note $\chi \in \operatorname{Irr}(G|\gamma)$. Since D is abelian and centralized by C, it follows that λ is linear and $\gamma_D = \gamma(1)\lambda$. By [Is, Lemma 2.29], $\gamma(1) \leq |C : D|^{1/2}$. Thus, as $|G/C| \mid |D/N|$,

$$|G/N|^{1/2} = \chi(1) = (\chi(1)/\gamma(1))\gamma(1)$$
$$\leq |G : C| |C : D|^{1/2}$$
$$\leq (|G/C| |D/N|)^{1/2} |C/D|^{1/2}$$
$$\leq |G/N|^{1/2}.$$

Hence $|G/C| = |D/N|$. We may choose $y \in G$ with $G/C = \langle Cy \rangle$ and $o(Cy) = o(Nx)$. By choice of x, it follows that $o(Ny) = o(Nx)$ and $G/N = C/N \oplus \langle Ny \rangle$. Also $\langle Nx, Ny \rangle = \langle Nx \rangle \oplus \langle Ny \rangle$. If y centralizes x^i ($i \in \mathbb{Z}$) then $\mathbf{C}(x^i) \supseteq \langle C, y \rangle = G$ and $x^i \in \mathbf{Z}(G) = N$. Setting $U = \langle N, x, y \rangle$, it easily follows that $N = \mathbf{Z}(U)$.

Let $V = \mathbf{C}_G(U)$ so that $G/N = U/N \times V/N$ by Corollary 1.7. Let $\eta \in \mathrm{Irr}\,(V|\varphi)$. Then $\chi \in \mathrm{Irr}\,(G|\eta)$ and $I_G(\eta) = G$ because $U \leq \mathbf{C}_G(V)$. Since χ vanishes off N and η is the unique irreducible constituent of χ_V, we see that η vanishes on $V \setminus N$. Certainly φ is V-invariant and so Proposition 12.3 implies that φ is fully ramified with respect to V/N. By the inductive hypothesis, $V/N \cong B \oplus B$ for an abelian group B. Now $G/N \cong A \oplus A$ for an abelian group A, because $G/N \cong V/N \oplus \langle Nx \rangle \oplus \langle Ny \rangle$ and $o(Nx) = o(Ny)$. $\qquad\qquad\square$

If $\chi \in \mathrm{Irr}\,(G)$ and $\chi_N \in \mathrm{Irr}\,(N)$ with $N \trianglelefteq G$, then Gallagher's Theorem 0.9 tells us $\beta \mapsto \beta\chi$ is a bijection from $\mathrm{Irr}\,(G/N)$ onto $\mathrm{Irr}\,(G|\chi_N)$. Lemma 0.10 strengthens this. Propositions 0.11 and 0.12 and Theorem 0.13 give sufficient conditions for a G-invariant $\varphi \in \mathrm{Irr}\,(N)$ to extend to G. These results will be used repeatedly throughout Section 12, often without reference.

12.6 Lemma. *Suppose that G/N is abelian and $\varphi \in \mathrm{Irr}\,(N)$ is G-invariant. Then there exists a unique $N \leq M \leq G$ such that every $\gamma \in \mathrm{Irr}\,(M|\varphi)$ extends φ and is fully ramified with respect to G/M.*

Proof. We first show uniqueness. If K also satisfies the conclusion, then each $\chi \in \mathrm{Irr}\,(G|\varphi)$ vanishes off $M \cap K$. But $\chi_M = f \cdot \gamma$ for an integer f and extension γ of φ. Thus γ vanishes off $M \cap K$, yet $\gamma_{M \cap K}$ is irreducible. Thus $M \leq K$ and by symmetry $M = K$. This establishes uniqueness.

We now choose $M \leq G$ maximal such that φ has a G-invariant extension $\sigma \in \mathrm{Irr}\,(M)$. Since $\mathrm{Irr}\,(M|\varphi) = \{\lambda\sigma \mid \lambda \in \mathrm{Irr}\,(M/N)\}$ and since G/N is abelian, every $\gamma \in \mathrm{Irr}\,(M|\varphi)$ is a G-invariant extension of φ. Thus we need just show that σ is fully ramified with respect to G/M. Let $\chi \in \mathrm{Irr}\,(G|\sigma)$

and let $M \leq U \leq G$ with U/M cyclic. By Proposition 12.3, it suffices to show that χ vanishes on $U \setminus M$.

Let $\beta \in \mathrm{Irr}(U|\sigma)$ with $[\chi_U, \beta] \neq 0$. Since U/M is cyclic, β extends σ. For $g \in G$, $\beta^g = \alpha_g \beta$ for a unique $\alpha_g \in \mathrm{Irr}(U/M)$. Since G/M is abelian, $\alpha_{gh} = \alpha_g \alpha_h$ for $g, h \in G$. Thus there is a subgroup $A \leq \mathrm{Irr}(U/M)$ such that $\{\alpha\beta \mid \alpha \in \mathrm{Irr}(A)\}$ is the set of G-conjugates of β. By [Hu, V, 6.4], $A = \mathrm{Irr}(U/K)$ for some subgroup K with $M \leq K \leq U$. Now β_K is G-invariant and hence $K = M$ by the choice of M. Set $\rho = \sum_{\lambda \in \mathrm{Irr}(U/M)} \lambda$ (the regular character of U/M), so that $\chi_U = f\rho\beta$ for an integer f. Since ρ vanishes off the identity (see [Is, Lemma 2.10]), χ vanishes on $U \setminus N$. $\square$

12.7 Proposition. *Assume that $G/N \cong Q_8$ and $\lambda \in \mathrm{Irr}(N)$ is G-invariant. Then λ extends to G.*

Proof. By [Is, Theorem 11.28], we may assume that λ is linear and faithful, whence $N \leq \mathbf{Z}(G)$. Set $Z/N = \mathbf{Z}(G/N)$. For $x \in G \setminus Z$, $x^2 \in Z \setminus N$, and $Z \leq \mathbf{C}_G(x)$. Thus $Z = \mathbf{Z}(G)$. Pick $y \in G$ with $[x, y] \neq 1$. By usual commutator identities, $1 = [x, y^2] = [x, y][x, y]^y = [x, y]^2$. But $[x, y] \notin N$ and so Z splits over N. Note $[x, y] = [nx, my]$ for $n, m \in N$ and so $Z = N \times U$ where $U = \langle [x, y] \rangle = G'$ has order 2. Thus λ extends to $\lambda^* \in \mathrm{Irr}(Z)$ with $U \leq \ker \lambda^*$. Now λ^* extends to G because G/U is abelian. $\square$

In some ways, our main Theorem 12.9 generalizes the following result of Ito.

12.8 Lemma. *If G is p-solvable and $p \nmid \chi(1)$ for all $\chi \in \mathrm{Irr}(G)$, then G has a normal abelian Sylow-p-subgroup.*

Proof. This is immediate from [Is, Theorem 12.33]. We also give a proof of this and more in Theorem 13.1. $\square$

12.9 Theorem. *Suppose that $N \trianglelefteq G$, $\theta \in \mathrm{Irr}(N)$ and $\chi(1)/\theta(1)$ is a π'-number for all $\chi \in \mathrm{Irr}(G|\theta)$ and a set π of primes. Assume that G/N is solvable. Then G/N has an abelian Hall π-subgroup. In particular, G/N has π-length at most one.*

Proof. By induction on $|G/N|$. The hypotheses imply that $|G : I_G(\theta)|$ is π' and so $I_G(\theta)/N$ contains a Hall π-subgroup of G/N. We may assume that $G = I_G(\theta)$.

Let $M/N = \mathbf{O}_\pi(G/N)$. If $\varphi \in \mathrm{Irr}(M|\theta)$, then $\varphi(1)/\theta(1) \in \pi'$ and $\varphi(1)/\theta(1) \mid |M/N|$. Thus each $\varphi \in \mathrm{Irr}(M|\theta)$ extends θ. In particular, if $\lambda \in \mathrm{Irr}(M/N)$, then $\lambda\varphi \in \mathrm{Irr}(M|\theta)$ and $\lambda\varphi$ extends θ. Thus λ is linear and M/N is abelian.

Suppose that $N \vartriangleleft N_1 \vartriangleleft G$ and $\theta_1 \in \mathrm{Irr}(N_1|\theta)$. Then $\theta_1(1)/\theta(1)$ is π' and $\chi(1)/\theta_1(1)$ is π' for all $\chi \in \mathrm{Irr}(G|\theta_1)$. The inductive hypothesis implies that G/N_1 and N_1/N have abelian Hall π-subgroups. In particular, we may assume that $\mathbf{O}^{\pi'}(G/N) = G/N$ and $\mathbf{O}_{\pi'}(G/N) = 1$. Since $M/N = \mathbf{O}_\pi(G/N)$ is abelian and $\mathbf{O}_{\pi'}(G/N) = 1$, we have that $M/N = \mathbf{C}_{G/N}(M/N)$ by Lemma 0.19.

We may assume $M < G$ and choose a maximal normal subgroup K of G with $M \leq K$. Since $\mathbf{O}^{\pi'}(G/N) = G/N$ and $M/N = \mathbf{O}_\pi(G/N)$, we have that $|G/K| = p$ for a prime $p \in \pi$ and that $M < K$. Since K/N has an abelian Hall π-subgroup and $M/N = \mathbf{C}_{G/N}(M/N)$, K/M is a π'-group and $(G/M)' = K/M$.

We claim that M/N is a chief factor of G. If not, then whenever $N < J < M$ with $J \trianglelefteq G$, G/J has an abelian Hall π-subgroup. Thus $M/J \leq \mathbf{Z}(\mathbf{O}^{\pi'}(G/J)) = \mathbf{Z}(G/J)$. In particular, K/M centralizes M/J whenever M/J is a chief factor of G with $N \leq J$. Since $|M/N|$ is π and $|K/M|$ is π', this implies $M/N \leq \mathbf{Z}(K/N)$, a contradiction since $M/N = \mathbf{C}_{G/N}(M/N)$ and $M < K$. Thus $M/N = \mathbf{C}_{G/N}(M/N)$ is a chief factor of G and a faithful irreducible G/M-module. Since $K/M = (G/M)'$, in fact $K/N = (G/N)'$.

We have now established:

Step 1. (a) M/N is an elementary abelian q-group for a prime $q \in \pi$.

(b) M/N is a faithful irreducible G/M-module.

(c) K/M is a non-trivial π'-group.

(d) $K/N = (G/N)'$ is the unique maximal normal subgroup of G/N.

(e) $|G/K| = p$ for a prime $p \in \pi$.

Step 2. There is a G-invariant extension $\theta^* \in \mathrm{Irr}\,(M)$ of θ.

Proof. Since $(|K/M|, |M/N|) = 1$ and $I_K(\theta) = K$, Lemma 0.17 (d) yields the existence of a K-invariant $\theta^* \in \mathrm{Irr}\,(M|\theta)$. Since $\theta^*(1)/\theta(1)$ is a π'-number, and M/N is a π-group, θ^* extends θ. The hypotheses imply that $I_G(\theta^*)$ contains a Hall π-subgroup of G. But $K \leq I_G(\theta^*)$ and $|G/K| = p \in \pi$. Thus θ^* is G-invariant.

Step 3. Let $V = \mathrm{Irr}\,(M/N)$. Then V is a faithful irreducible G/M-module. Furthermore each $\lambda \in V$ is centralized by a Sylow p-subgroup of G/M.

Proof. By Step 1 (b) and Proposition 12.1, V is a faithful irreducible G/M-module. Now $\lambda \mapsto \lambda\theta^*$ defines a bijection from V onto $\mathrm{Irr}\,(M|\theta)$. In particular, $I_G(\lambda\theta^*) = I_G(\lambda)$ for all $\lambda \in V$. The hypotheses imply that $p \nmid |G : I_G(\lambda\theta^*)|$ and hence $p \nmid |G : I_G(\lambda)|$ for all $\lambda \in V$.

Step 4. (a) V is not a quasi-primitive G/M-module.

(b) If $M \leq A \lhd K$ with $A \unlhd G$ and K/A a cyclic t-group for some prime t, then $t \mid |A/M|$.

Proof. First we prove (b). Assume that $t \nmid |A/M|$. Also $p \nmid |A/M|$. By Proposition 0.17 (d), there exists $\mu \in \mathrm{Irr}\,(A|\theta^*)$ that is G-invariant. Each Sylow subgroup of G/M is cyclic, and so there exists $\mu^* \in \mathrm{Irr}\,(G|\mu)$ extending μ. Then $\sigma\mu^* \in \mathrm{Irr}\,(G|\theta^*)$ for all $\sigma \in \mathrm{Irr}\,(G/A)$. Thus $p \nmid \sigma(1)$ for all $\sigma \in \mathrm{Irr}\,(G/A)$. By Ito's Theorem 12.8, G/A has a normal Sylow p-subgroup, a contradiction as $(G/A)' = K/A \neq 1$. This proves (b).

Now $(G/M)' = K/M$. If V is quasi-primitive, then Theorem 10.4 implies that K/M is cyclic or that $K/M \cong Q_8$ and $|G/K| = 3$. By (b), $K/M \neq 1$ is not cyclic. Now, by Proposition 12.7, θ^* extends to K and hence to G. Again this implies that each $\sigma \in \mathrm{Irr}\,(G/M)$ has $3'$-degree, contradicting Lemma 12.8. This proves (a).

We have that the action of G/M on V satisfies the hypotheses of Theorem 9.3 (proof below). We lift the notation of that theorem to G/M.

Step 5. We have normal subgroups $M \leq C \leq L \leq K \leq G$ such that

   (i) $V_C = V_1 \oplus \cdots \oplus V_n$ for homogeneous components V_i of V_C, with $|V_i| = q^m$ (q prime as in Step 1).

   (ii) $G/C \cong D_6$, D_{10}, or $A\Gamma(2^3)$, $p = 2, 2$, or 3 (respectively), $n = 3, 5$, or 8 (resp.), and G/C faithfully and primitively permutes the V_i.

   (iii) L/C is the unique minimal normal subgroup of G/C, $|L/C| = n$, and L/C transitively permutes the V_i.

   (iv) $C/\mathbf{C}_C(V_i)$ transitively permutes $V_i^{\#}$, for each i.

   (v) $q^m \neq 3^2$ nor 3^4.

Proof. Choose $M \leq C \lhd G$ maximal such that V_C is not homogeneous by Step 4 (a). Since $(K/M) = (G/M)'$ has index p in G/M, we see that $M \leq C \leq K \leq G$ and $p \mid |G/C|$. Writing $V_C = V_1 \oplus \cdots \oplus V_n$ for homogeneous components V_i of V_C and $n > 1$, Proposition 0.2 (i) applies and G/C faithfully and primitively permutes $\{V_1, \ldots, V_n\}$. By Step 3, each $\lambda \in V$ is centralized by a Sylow p-subgroup. Theorem 9.3 applies to the action of G/M on V. Conclusions (i), (ii), (iii) and (iv) follow from Theorem 9.3. Note that $K = L$ when $p = 2$, and that $|K/L| = 7$ when $p = 3$.

Suppose that $q^m = 3^2$ or 3^4. By Lemma 9.2, $q = 3$ and so $G/C \cong A\Gamma(2^3)$. Since $7 \nmid |GL(m,3)|$, we have that $7 \nmid |C/\mathbf{C}_C(V_i)|$. Since $\bigcap \mathbf{C}_C(V_i) = M$, it follows that 7 does not divide $|C/M|$ or $|L/M|$. This contradicts Step 4 (b), as G/L is a Frobenius group of order 21. This contradiction proves (v).

Step 6. Set $C_i = \mathbf{C}_C(V_i)$, $F_i/C_i = \mathbf{F}(C/C_i)$ and $F = \bigcap_{i=1}^n F_i$. Then

(i) $\bigcap C_i = M$.

(ii) C/F_i and F_i/C_i are cyclic.

(iii) $F/M = \mathbf{F}(C/M)$. Also F/M and C/F are abelian of rank at most n.

(iv) There exists a prime $r \mid |F/M|$ and $R \in \mathrm{Syl}_r(F/M)$ such that $r \nmid |C/F|$ and $F/M = \mathbf{C}_{C/M}(R/M)$.

(v) If $C > F$, then $F = \mathbf{C}_G(R/M)$ and $r \nmid |L/F|$.

Proof. Part (i) is immediate because V is a faithful G/M and C/M-module. Since $q^m \neq 3^2$ or 3^4 (Step 5 (v)), parts (ii), (iii) and (iv) follow from Corollary 9.7 and Lemma 9.8 (in case $q^m = 2^6$, $2 \in \pi$ and $|C/M|$ must be odd).

For (v), we assume that $C > F$. By Lemma 9.10 (b), $\mathbf{C}_G(C/F) = C$. By (iv), $F = \mathbf{C}_C(R/M)$ and so $\mathbf{C}_G(R/M) \cap C = F$. Thus $\mathbf{C}_G(R/M)$ centralizes C/F and so $\mathbf{C}_G(R/M) \leq C$, i.e. $\mathbf{C}_G(R/M) \leq \mathbf{C}_C(R/M) = F$. So we need just prove $r \nmid |L/F|$.

But we know $r \nmid |C/F|$. Furthermore $|L/C|$ is 3, 5, or 2^3 and $p = 2, 2,$ or 3 (respectively). Without loss of generality, r is 3, 5, or 2 (resp.). Since r is a Zsigmondy prime divisor of $q^m - 1$, we must have that m is 1, 2, or 4 (not necessarily respectively). But $\exp(C/F)|m$ by Lemma 9.8 (b) and $C/F \neq 1$. Hence m is 2 or 4 and C/F is a 2-group. Since $m > 1$, the Zsigmondy prime r is 3 or 5 and thus $p = 2$. This is a contradiction as C/F is a π'-group.

Step 7. If $M \leq T \leq F$ with $T \trianglelefteq G$ and $\mathbf{C}_{G/M}(T/M) \not\leq C/M$, then T/M is cyclic and $T/M \leq \mathbf{Z}(K/M)$.

Proof. That T/M is cyclic follows from Step 5 (v) and Corollary 9.9. Since $K/M = (G/M)'$, $T/M \leq \mathbf{Z}(K/M)$.

Step 8. Suppose that $M \leq T \leq F$ with $T \trianglelefteq G$ and T/M an s-group for a prime s, $s \nmid |L/F|$. Assume there is a G-invariant extension $\delta \in \mathrm{Irr}\,(T)$ of θ^*. Then $s = 7 = |K/L|$, $p = 3$, $T/M \leq \mathbf{Z}(G/M)$ and T/M is cyclic.

Proof. Now $\alpha \mapsto \alpha\delta$ defines a bijection from $\mathrm{Irr}\,(T/M)$ onto $\mathrm{Irr}\,(T|\theta^*)$. Since δ is G-invariant, $I_G(\alpha) = I_G(\alpha\delta)$ for each $\alpha \in \mathrm{Irr}\,(T/M)$. Thus the hypotheses of the theorem imply that $p \nmid |G : I_G(\alpha)|$ for all $\alpha \in \mathrm{Irr}\,(T/M)$.

First suppose that $T/M \leq \mathbf{Z}(L/M)$. By Step 7, T/M is cyclic and central in K/M. Since $p \nmid |G : I_G(\alpha)|$ for all $\alpha \in \mathrm{Irr}\,(T/M)$, we see that G/K acts on T/M fixing all $\alpha \in \mathrm{Irr}\,(T/M)$. Thus $T/M \leq \mathbf{Z}(G/M)$. Let $T/M \leq S/M \in \mathrm{Syl}_s(F/M)$. Because $s \nmid |L/F|$, Fitting's Lemma 0.6 implies that $S/M = S_1/M \times U/M$ where $U/M = [S, L]\,M/M$ and $S_1/M = \mathbf{C}_{S/M}(L/F) \geq T/M$. By Step 7, $S_1/M \cong S/U$ is cyclic. Now $1 \neq T/M \leq \mathbf{Z}(G/M)$. If $s \nmid |G/S|$, then $S_1/M \leq \mathbf{Z}(G/M)$ and $S/U \leq \mathbf{Z}(G/U)$. Also $\mathbf{O}^s(G/M) < G/M$, contradicting Step 1 (d). Hence $s \mid |G/F|$, but $s \nmid |L/F|$. So $s \mid |G/K|$. Because $p \nmid |K/M|$, certainly $p \neq s$ and so $s \mid |K/L|$. Since $K = L$ when $p = 2$, the only possibility is $p = 3$ and $s = 7 = |K/L|$. The conclusion of this step is satisfied when $T/M \leq \mathbf{Z}(L/M)$.

It suffices to show that $T/M \leq \mathbf{Z}(L/M)$. Assume not. Since we have $(|L/F|, |T/M|) = 1$, we may find a chief factor T_0/M of G with $T_0/M \nleq \mathbf{Z}(L/M)$. Since δ restricted to T_0 is a G-invariant extension of θ^*, we may assume without loss of generality that $T = T_0$, i.e. T/M is a chief factor of G. Let $B = \mathbf{C}_G(T/M)$. Since B does not contain L, we have that $K/M = (G/M)' \nleq B/M$ and T/M is not cyclic. By Step 7, it follows that $F \leq B \leq C$.

Let $X = \mathrm{Irr}\,(T/M)$. Then X is an irreducible faithful G/B-module by Proposition 12.1. As in the first paragraph of this step, each $\chi \in X$ is centralized by a Sylow p-subgroup of G/B. If $p = 2$, then $s = 2$ by Lemma 9.2, contradicting Step 1 as $p \nmid |K/M|$. Thus $p = 3$, $n = 8$, and $G/C \cong A\Gamma(2^3)$. Then G/C and G/B are not metacyclic, whence Theorem 10.4 implies X is not a quasi-primitive G/B-module. Since $\mathbf{O}^{p'}(G/B) = G/B$,

Theorem 9.3 and Step 6 yield that

$$\text{rank}\,(X) = \dim(X) \geq 8 \geq n \geq \text{rank}\,(S/M) \geq \text{rank}\,(T/M).$$

Hence $\text{rank}\,(X) = 8 = \text{rank}\,(T/M)$ and $|T/M| = s^8$.

Let $T_i = \mathbf{C}_T(V_i) = T \cap C_i$. Since $T/T_i \lesssim F_i/C_i$, it follows from Step 6 that T/T_i is cyclic and acts fixed-point-freely on V_i. Since $\text{rank}\,(T/M) = 8$, we have that

$$T > T_1 > T_1 \cap T_2 > \cdots > \bigcap_{i=1}^{8} T_i = M \qquad (12.1)$$

is a properly descending chain with each factor group cyclic of order s.

Now $V = V_1 \oplus \cdots \oplus V_8$ and we choose $\delta \in V$ with $\delta = (\delta_1, \ldots, \delta_5, 1, 1, 1)$ with $\delta_1, \ldots, \delta_5$ non-principal. Now $3 \nmid |G : I_G(\delta)|$ by Step 3 and $CI_G(\delta)/C$ must stabilize $\{V_1, \ldots, V_5\}$. Thus $I_G(\delta)/I_C(\delta) \cong CI_G(\delta)/C$ has order 3. Since $s \nmid |C/S|$ and $s \neq p = 3$, we have $|I_G(\delta)/I_S(\delta)|$ is not divisible by s. By Lemma 0.17 (d), there exists $\delta^* \in \text{Irr}\,(I_S(\delta)|\delta)$ that is invariant in $I_G(\delta)$. Now $\delta \in V = \text{Irr}\,(M/N)$ and $s \nmid |M/N|$. Thus δ extends to $I_S(\delta)$. Since $I_S(\delta)/M \leq S/M$ is abelian, indeed δ^* extends δ.

Let $I = I_T(\delta) \leq I_S(\delta)$. Since $T/T_i \lesssim F_i/C_i$ acts fixed-point-freely on V_i, $I = T_1 \cap \cdots \cap T_5$. By the hypothesis of this step, $\gamma \in \text{Irr}\,(T)$ is a G-invariant extension of $\theta^* \in \text{Irr}\,(M)$ (which in turn is a G-invariant extension of $\theta \in \text{Irr}\,(N)$). In particular γ_I and $\delta_I^* \gamma_I$ are $I_G(\delta)$-invariant extensions of θ^* and $\delta\theta^*$ (resp.). Since $I = T_1 \cap \cdots \cap T_5$, it follows from (12.1) that $|I/M| = s^3$, $I \cap T_j$ ($j = 6, 7, 8$) are distinct subgroups of I and $|I : I \cap T_j| = s$ ($j = 6, 7, 8$). Let $\epsilon \in \text{Irr}\,(I/I \cap T_6)$ be faithful. Since ϵ and δ^* are linear, $\epsilon\delta_I^*\gamma_I$ extends $\delta\theta^* \in \text{Irr}\,(M|\theta)$. Since $I = I_T(\delta) = T \cap I_G(\delta) \trianglelefteq I_G(\delta) = I_G(\delta\theta^*)$, there exists $P \in \text{Syl}_3(I_G(\delta))$ such that $\epsilon\delta_I^*\gamma_I$ is P-invariant. Since P fixes γ_I which extends θ^*, it follows (see Lemma 0.9) that P fixes $\epsilon\delta_I^*$. But δ^* is invariant in $I_G(\delta)$ and δ_M^* is irreducible. Since $P \leq I_G(\delta)$, indeed ϵ is P-invariant. In particular, P fixes $\ker(\epsilon) = I \cap T_6$. But $P \in \text{Syl}_3(I_G(\delta))$, and $I_G(\delta)C/C$ has order 3 and stabilizes $\{V_1, \ldots, V_5\}$. Then $PC/C = I_G(\delta)C/C$ transitively permutes $\{V_6, V_7, V_8\}$ and thus also transitively permutes $\{T_6, T_7, T_8\}$. But P fixes T_6, whence $T_6 = T_7 = T_8$. This contradiction completes Step 8.

Step 9. Suppose that W/M is a chief factor of G and an s-group for a prime s, $s \nmid |L/F|$. Assume there is an L-invariant extension $\gamma \in \mathrm{Irr}\,(W|\theta^*)$. Then $|W/M| = s = 7$, $p = 3$, and $W/M \leq \mathbf{Z}(K/M)$.

Proof. Now $\alpha \mapsto \alpha\gamma$ is a bijection from $\mathrm{Irr}\,(W/M)$ onto $\mathrm{Irr}\,(W|\theta^*)$. For $g \in G$, $\gamma^g = \alpha_g\gamma$ for a unique $\alpha_g \in \mathrm{Irr}\,(W/M)$ and

$$L = L^g = I_L(\gamma^g) = I_L(\alpha_g\gamma) = I_L(\alpha_g).$$

By Step 8, we may assume that γ is not G-invariant and hence $L = I_L(\alpha)$ for some $1 \neq \alpha \in \mathrm{Irr}\,(W/M)$. Since W/M is a chief factor of G, $W/M \leq \mathbf{Z}(L/M)$. By Step 7, W/M is cyclic of prime order s and $W/M \leq \mathbf{Z}(K/M)$.

By Lemma 0.17 (d), there exists $\gamma_1 \in \mathrm{Irr}\,(W|\theta^*)$ that is invariant in a Hall s'-subgroup H/M of G/M. Now $\gamma_1 = \delta\gamma$ for a linear $\delta \in \mathrm{Irr}\,(W/M)$. Since $W/M \leq \mathbf{Z}(K/M)$, certainly δ and $\gamma_1 = \delta\gamma$ are L-invariant. Thus it is no loss to assume that γ is LH-invariant. By Step 8 again, we may assume γ is not G-invariant. Thus $s \mid |G/L|$. Since $s \neq p = |G : K|$ and since $K = L$ when $p = 2$, indeed $p = 3$ and $s = |K : L| = 7$.

Step 10. Assume that $S/M \in \mathrm{Syl}_s(F/M)$ for a prime s, $s \nmid |L/F|$. Then one of the following holds:

(i) θ^* is fully ramified with respect to S/M, or

(ii) $p = 3$, $s = 7$, and S/M is cyclic.

Proof. By Lemma 12.6, there exists $M \leq H \leq S$ such that each $\sigma \in \mathrm{Irr}\,(H|\theta^*)$ extends θ^* and is fully ramified with respect to S/H. Also $H \trianglelefteq G$ because θ^* is G-invariant. We may assume $H > M$. By Lemma 0.17 (d), there exists an L-invariant $\lambda \in \mathrm{Irr}\,(S|\theta^*)$. Now λ is fully ramified with respect to S/H and the unique irreducible constituent σ of λ_H is an L-invariant extension of θ^* to H. If $M \leq W \leq H$ and $W \trianglelefteq G$, then σ_W is an L-invariant extension of θ^* to W. Since $H > M$, Step 9 implies that $p = 3$ and $s = 7$.

Since H/M is abelian and central in F/M and since $7 \nmid |L/F|$, Fitting's

Lemma 0.6 implies that $H/M = A/M \times A_0/M$ for A, $A_0 \trianglelefteq G$ and $A/M = \mathbf{C}_{H/M}(L/F)$. Suppose that $A_0 > M$ and choose $M < W \leq A_0$ with W/M a chief factor of G. Since σ_W is an L-invariant extension of θ^*, Step 9 implies that W/M is cyclic and $W/M \leq \mathbf{Z}(K/M)$. Then W/M is centralized by L/F, a contradiction. Thus $A_0 = M$ and H/M is centralized by L/F.

We again apply Fitting's Lemma 0.6 to write $S/M = D/M \times E/M$ with D, $E \trianglelefteq G$ and $D/M = \mathbf{C}_{S/M}(L/F) \geq H/M$. By Step 7, D/M is cyclic. By the first paragraph of this step, we have $\sigma \in \mathrm{Irr}\,(H)$ and $\lambda \in \mathrm{Irr}\,(S|\sigma)$ that are fully ramified with respect to S/H. By Lemma 12.4, σ is fully ramified with respect to D/H. Since D/H is cyclic, in fact $D = H$ by Lemma 12.5. Because $H > M$, $\mathrm{rank}\,(E/M) = \mathrm{rank}\,(S/M) - 1 \leq 7$ using Step 6 (iii). Now $E/M \cong S/H$ has even rank by Lemma 12.5, and so $2 \leq \mathrm{rank}\,(E/M) \leq 6$ or $E = M$. By Step 7, $E = M$ or $\mathbf{C}_{G/M}(E/M) \leq C/M$. By Lemma 9.12, $E = M$. Thus $S = D = H$ and S/M is cyclic, completing this step.

Step 11. If $C > F$, then $p = 3$ and C/F is a 2-group.

Proof. By Step 6 (iv, v), there exist a prime r and Sylow r-subgroup R/M of F/M such that $F/M = \mathbf{C}_{G/M}(R/M)$ and $r \nmid |L/R|$. Since $(G/M)' = K/M > F/M$, we see that R/M is not cyclic. By Step 10, θ^* is fully ramified with respect to R/M. Thus $\mathrm{rank}\,(R/M)$ is even, and by Step 6 $\mathrm{rank}\,(R/M) \leq n$. If $p = 2$, then Corollary 9.11 yields that $\mathrm{rank}\,(R/M) = n$ because $C > F$. For $p = 2$, n is odd, a contradiction. Hence $p = 3$ and $n = 8$.

Pick $1 \neq X/F \in \mathrm{Syl}(C/F)$ for some prime. By Lemma 12.5 (b), $\mathrm{rank}(X/F) \leq \mathrm{rank}\,(R/M)/2 \leq 4$. By Lemma 9.10 (c), X/F is a 2-group and thus so is C/F.

Step 12. $p = 2$.

Proof. Suppose then that $p = 3$. Let $S/M \in \mathrm{Syl}_7(F/M)$ and $T/M \in \mathrm{Syl}_7(G/M)$. By the last paragraph, C/F and hence L/F are 2-groups. In particular, $|T : S| = 7 = |K : L|$.

We claim that no $\mu \in \text{Irr}(S|\theta)$ is T-invariant. For if μ were T-invariant, then Lemma 0.17 (d) guarantees the existence of $\eta \in \text{Irr}(L|\mu)$ such that η is invariant in $K = LT$. Since $\eta \in \text{Irr}(L|\theta)$ and $3 \nmid \chi(1)$ for all $\chi \in \text{Irr}(G|\theta)$, indeed η is G-invariant. But G/L is non-abelian of order 21 and so η extends to $\eta^* \in \text{Irr}(G|\theta)$. But also there exists $\sigma \in \text{Irr}(G/L)$ with $\sigma(1) = 3$ and $\sigma\eta^* \in \text{Irr}(G|\theta)$, contradicting the hypotheses. Hence the claim holds.

If θ^* is fully ramified with respect to S/M, the unique irreducible constituent of $(\theta^*)^S$ is G-invariant, contradicting the last paragraph. By Step 10, S/M is cyclic. Since $K/M = (G/M)'$, $S/M \leq \mathbf{Z}(K/M)$. In particular, T/M is abelian. If T/M is cyclic, then θ^* extends to $\zeta \in \text{Irr}(T)$. Then $\zeta_S \in \text{Irr}(S|\theta)$ is T-invariant, contradicting the last paragraph. So T/M is abelian but not cyclic.

Since $V = \text{Irr}(M/N)$ is a faithful irreducible G/M-module and S/M is a cyclic normal subgroup of G/M, $\mathbf{C}_{S/M}(v) = 1$ for all $v \in V^{\#}$. Because T/M is abelian and not cyclic, it is not the case that $\mathbf{C}_{T/M}(v) = 1$ for all $0 \neq v \in V$. So we may choose $\lambda \in \text{Irr}(M/N)$ such that $I_T(\lambda)/M$ has order s and $I_S(\lambda)/M = 1$. Now $I_T(\lambda\theta^*) = I_T(\lambda)$ and $(\lambda\theta^*)^S \in \text{Irr}(S|\theta)$, invariant in $SI_T(\lambda) = T$. This is a contradiction to the second paragraph of this step. Hence $p \neq 3$.

Step 13. Let $t \in G$ with Mt an involution in G/M. Then t fixes exactly one V_j. Furthermore the centralizer in V_j of t has order $2^{m/2}$ or 2^m.

Proof. Now $G/C \cong D_6$ or D_{10} permutes $V_1, \ldots, V_n$ with $n = 3$ or 5 (respectively). An involution in G/C fixes exactly one V_j and permutes the others in pairs. Now $\mathbf{N}_G(V_j)/C$ has order 2. Since $F = C$ by Step 11, C/C_j is cyclic of odd order. Since each $v \in V_j$ is centralized by an involution of $\mathbf{N}_G(V_j)$ and since we may assume that $t \in \mathbf{N}_G(V_j)$ does not centralize V_j, we have that $\mathbf{O}_2(\mathbf{N}_G(V_j)/C_j) = 1$ and $\mathbf{N}_G(V_j)/C_j$ acts faithfully on V_j. Since $\mathbf{N}_G(V_j)/C_j$ has a cyclic normal subgroup of odd order and index 2, there is a Frobenius group $H/C_j \leq \mathbf{N}_G(V_j)/C_j$ with Frobenius complement of order 2. Without loss of generality, $t \in H$ and Lemma 0.34 implies that the centralizer of t in V_j has order $|V_j|^{1/2} = 2^{m/2}$.

Step 14. Conclusion.

Since $p = 2$, $F = C$ by Step 11. We have that $L = K$ has index 2 in G, and L/C regularly and transitively permutes $\{V_1, \ldots, V_n\}$ ($n = 3$ or 5). Let $X = \{(v_1, v_2, \ldots, v_n) \mid$ all v_i are non-zero$\}$. Since $F = C$, the cyclic group C/C_i acts fixed-point-freely on V_i. Thus $I_C(\alpha) = \bigcap C_i = M$ for all $\alpha \in X$ and $I_G(\alpha)/M \lesssim G/C \cong D_{2n}$. Since $2 \mid |I_G(\alpha)/M|$, $I_C(\alpha)/M \cong D_{2n}$ or has order two. In either case, $\alpha\theta^*$ extends to $I_G(\alpha) = I_G(\alpha\theta^*)$. Since each $\gamma \in \mathrm{Irr}\,(I_G(\alpha\theta^*) \mid \alpha\theta^*)$ must have odd degree, each irreducible character of $I_G(\alpha)/M$ must also have odd degree because $\alpha\theta^*$ extends to $I_G(\alpha)$. Thus $|I_G(\alpha)/M| = 2$ for all $\alpha \in X$.

Now let $Y = \{(v_1, \ldots, v_n) \mid$ exactly one $v_i = 0\}$. Let $\beta \in Y$ so that $CI_G(\beta)/C$ has order 2. Without loss of generality, $\beta = (0, v_2, \ldots, v_n)$ with $v_i \neq 0$ for $i > 1$. Then $CI_G(\beta) = \mathbf{N}_G(V_1)$ and so $I_G(\beta)/I_C(\beta) \cong \mathbf{N}_G(V_1)/C$ has order 2. Since C/C_i acts fixed-point-freely on V_i, $I_C(\beta) = C_2 \cap \cdots \cap C_n$, and $I_C(\beta)/M$ is isomorphic to a subgroup of the cyclic group C/C_1. Thus all Sylow subgroups of $I_G(\beta)/M$ are cyclic. Thus $\beta\theta^*$ extends to an irreducible character of $I_G(\beta) = I_G(\beta\theta^*)$. Since each $\eta \in \mathrm{Irr}\,(G \mid \beta\theta^*)$ has odd degree, every $\delta \in \mathrm{Irr}\,(I_G(\beta)/M)$ has odd degree. Since $|I_G(\beta)/C_2 \cap \cdots \cap C_n| = 2$ and $C_2 \cap \cdots \cap C_n/M$ is cyclic of odd order, in fact $I_G(\beta)/M$ is cyclic. In particular, β is fixed by a unique Sylow 2-subgroup of G/M, and so is each element of Y.

Now $I_G(\beta)/M \lesssim \mathbf{N}_G(V_1)/C_1$ in its action on V_1. If Mt is the involution of $I_G(\beta)/M$, then the centralizer in V_1 of t is $C_2 \cap \cdots \cap C_n$-invariant. Hence t centralizes V_1 or $C_2 \cap \cdots \cap C_n$ does not act irreducibly on V_1.

Assume for this paragraph that $C_2 \cap \cdots \cap C_n = M$. Consequently, the intersection of any $(n-1)$ distinct C_i is M. Set $Z = \{(v_1, \ldots, v_n) \mid$ exactly two v_i are zero}. Let $\gamma \in Z$, say $\gamma = (0, 0, v_3, \ldots, v_n)$. Then $I_C(\gamma)/M = C_3 \cap \cdots \cap C_n/M \lesssim C/C_2$ and $I_C(\gamma)/M$ is cyclic. We argue as in the second paragraph of this step to show that $I_G(\gamma)/M$ contains a unique involution. If $C_2 \cap \cdots \cap C_n = M$, then each element of Z is fixed by a unique involution of G/M.

Now Mt is an involution of G/M stabilizing V_1 and permuting the other V_i in pairs. Let $2^l = |\mathbf{C}_{V_1}(t)|$, so that $l = m$ or $l = m/2$ by Step 13, recalling $|V_1| = 2^m$. Observe that t centralizes $(2^l - 1)(2^m - 1)^{(n-1)/2}$ elements of X, $(2^m - 1)^{(n-1)/2}$ elements of Y and $(2^l - 1)(2^m - 1)^{(n-3)/2}(n-1)/2$ elements of Z (of course n is 3 or 5). A Sylow 2-subgroup of G/M has order 2 and all involutions of G/M are conjugate. The first and second paragraphs of this step show that each element of $X \cup Y$ is fixed by a unique involution of G/M. Thus

$$(2^l - 1)(2^m - 1)^{(n-1)/2}|\mathrm{Syl}_2(G/M)| = |X| = (2^m - 1)^n \qquad (12.2)$$

and

$$(2^m - 1)^{(n-1)/2}|\mathrm{Syl}_2(G/M)| = |Y| = n \cdot (2^m - 1)^{(n-1)}. \qquad (12.3)$$

Now (12.2) and (12.3) yield that $2^m - 1 = n(2^l - 1)$ and $l \neq m$. Thus $l = m/2$ by Step 13 and $n = 2^{m/2} + 1$. By (12.2), we now have

$$n = 3, \ |V_1| = 2^2, \ l = 1, \ |C/C_i| = 3, \ \text{and} \ |\mathrm{Syl}_2(G)| = 9$$

or

$$n = 5, \ |V_1| = 2^4, \ l = 2, \ |C/C_i| = 15, \ \text{and} \ |\mathrm{Syl}_2(G)| = 3^2 \cdot 5^3.$$

If, in addition, $C_2 \cap \cdots \cap C_n = M$, the last paragraph implies that each element of Z is centralized by a unique involution, whence

$$(2^l - 1)(2^m - 1)^{(n-3)/2}(n-1)|\mathrm{Syl}_2(G/M)| = 2|Z| = n(n-1)(2^m - 1)^{n-2}$$

and $(2^l - 1)|\mathrm{Syl}_2(G/M)| = n(2^m - 1)^{(n-1)/2}$. For $n = 5$, this is a contradiction. Hence $C_2 \cap \cdots \cap C_n > M$ when $n = 5$.

Suppose that $n = 5$. Now $1 \neq (C_2 \cap \cdots \cap C_5)/M \lesssim C/C_1$ in its action on V_1. Since Mt does not centralize V_1, the third paragraph of this step implies that $(C_2 \cap \cdots \cap C_5)/M$ does not act irreducibly on V_1. Since $|C/C_1| = 15$ and $|V_1| = 2^4$, it follows that $|(C_2 \cap \cdots \cap C_5)/M| = 3$. By conjugation, $|H/M| = 3$ whenever H is the intersection of any four distinct C_i. But a Sylow 3-subgroup T/M of C/M is elementary abelian of order at most 3^5 (see Step 6 (iii)). Thus $|T/M| = 3^5$. Since $T/M \in \mathrm{Syl}_3(G/M)$, θ^* is fully ramified with respect to T/M by Step 10. Hence $\mathrm{rank}\,(T/M)$ is even by Lemma 12.5. By this contradiction, $n = 3$.

Since $|C/C_i| = 3 = |K/C|$, we have that K/M is a 3-group of order at most 3^4. By Step 4, K/M is not cyclic. If $J/M = (K/M)'$, then K/J is not cyclic. Because $\mathbf{O}^{2'}(G/M) = G/M$ and $|\mathrm{Syl}_2(G/M)| = 3^2$, it follows that G/K acts fixed-point-freely on K/J and $|K/J| = 3^2$. Furthermore J/M must be centralized by a Sylow 2-subgroup of G/M. Since $\mathbf{O}^{2'}(G/M) = G/M$, in fact $J/M = \mathbf{Z}(G/M)$. Since $\exp(C/M) = 3$, Step 7 yields that $|J/M| = 3$. Now $J/M \leq C/M$ must act on V_1 non-trivially. In particular, $\mathbf{C}_G(V_1)J/\mathbf{C}_G(V_1)$ is isomorphic to J/M and is a central subgroup of $\mathbf{N}_G(V_1)/\mathbf{C}_G(V_1)$. Since $\mathbf{N}_G(V_1)/\mathbf{C}_G(V_1) \lesssim S_3$, in fact $\mathbf{N}_G(V_1)/\mathbf{C}_G(V_1)$ has order 3. Thus Mt centralizes V_1. This final contradiction completes the proof of the theorem. $\qquad \square$

The motivation for Theorem 12.9 is the next theorem, Brauer's height-zero conjecture for solvable groups. Recall that this conjecture states that a defect group D of a p-block B is abelian if and only if every $\chi \in \mathrm{Irr}(B)$ has height zero. If D is abelian and G is p-solvable, then Fong [Fo 1] proved that every $\chi \in \mathrm{Irr}(B)$ has height zero via "Fong reduction". Indeed the proof below in this direction works just as well for p-solvable. For the converse direction the proof is Theorem 12.9 and Fong reduction (below). Fong proved the converse direction for the principal block B (via Fong reduction and Ito's Theorem 12.8) or when p is the largest prime divisor of G [Fo 2]. We mention that Theorem 12.10 extends to p-solvable G (see [GW 2]).

12.10 Theorem. *Let D be a defect of a p-block B of a solvable group G.*

Then every $\chi \in \mathrm{Irr}\,(B)$ *has height zero if and only if* D *is abelian.*

Proof. By induction on $|G : \mathbf{O}_{p'}(G)|$. Let $K = \mathbf{O}_{p'}(G)$ and choose $\varphi \in \mathrm{Irr}\,(K)$ such that B covers $\{\varphi\}$. Let $I = I_G(\varphi)$. By Lemma 0.25, there exists a block b of I such that character induction is a height-preserving bijection from $\mathrm{Irr}\,(b)$ onto $\mathrm{Irr}\,(B)$. Furthermore b and B share a common defect group $D_1 \cong D$.

Now every $\psi \in \mathrm{Irr}\,(b)$ has height zero if and only if every $\chi \in \mathrm{Irr}\,(B)$ has height zero. If $I < G$, the inductive hypothesis implies that every $\psi \in \mathrm{Irr}\,(b)$ has height zero if and only if D is abelian. The result then follows easily. Hence we assume that $G = I_G(\varphi)$.

By Theorem 0.28, $\mathrm{Irr}\,(B) = \mathrm{Irr}\,(G|\varphi)$, and the defect groups of B are Sylow p-subgroups of G. If each $\chi \in \mathrm{Irr}\,(B)$ has height zero, then $p \nmid \chi(1)$ for all $\chi \in \mathrm{Irr}\,(G|\varphi)$. By Theorem 12.9, G has abelian Sylow p-subgroups, i.e. the defect groups of B are abelian. To prove the converse, now assume G has abelian Sylow p-subgroups. We show $p \nmid \beta(1)$ for all $\beta \in \mathrm{Irr}\,(G|\varphi)$.

Let $M/K = \mathbf{O}_{p'p}(G/K)$. By Lemma 0.19, $M/K \geq \mathbf{C}_{G/K}(M/K)$. Since $D \in \mathrm{Syl}_p(G)$ is abelian, KD/K centralizes M/K. Thus $M = KD$. Since φ is M-invariant, φ extends to $\hat{\varphi} \in \mathrm{Irr}\,(M)$ by Theorem 0.13. Since $D \cong M/K$ is abelian, every $\eta \in \mathrm{Irr}\,(M|\varphi)$ extends φ. In particular $p \nmid \eta(1)$ for all $\eta \in \mathrm{Irr}\,(M|\varphi)$. Since G/M is a p'-group, $p \nmid \chi(1)$ for all $\chi \in \mathrm{Irr}\,(G|\varphi)$. $\qquad\square$

One can ask a number of questions related to the theorems in this section. For example, if $\theta \in \mathrm{Irr}\,(N)$ with $N \trianglelefteq G$ and $p^{e+1} \nmid \chi(1)/\theta(1)$ for all $\chi \in \mathrm{Irr}\,(G|\theta)$, can we give a bound for $\mathrm{dl}(P)$ where $P \in \mathrm{Syl}_p(G/N)$? The answer is yes, namely $2e + 1$. We refer the reader to Corollary 14.7 (a) for a proof. Consequently one can bound the derived length of a defect group of a block B of a p-solvable group in terms the maximum height in $\mathrm{Irr}\,(B)$. By the way, there is no generalization of the converse of Brauer's height conjecture, since it is easy to find p-groups of derived length two that have arbitrarily large degrees of irreducible characters.

Similar questions about Brauer characters may also be asked. But another twist may be added, since we can discuss the situation where $q \nmid \varphi(1)$ for all φ in some subset of $\mathrm{IBr}_p(G)$ in the two cases, $q = p$ and $q \neq p$. In Section 13, we discuss and prove analogues of Ito's Theorem for both $p = q$ and $p \neq q$. In Section 14, we show there is no analogue of Theorem 12.9 for Brauer characters when $p = q$ (i.e. $\{q\} = \pi$), but do give a result when $p \neq q$ and some generalizations.

§13 Brauer Characters of q'-degree and Ito's Theorem

In this section, we investigate the set of p-Brauer characters $\mathrm{IBr}_p(G)$. If G is a p'-group, we shall freely use the existence of the natural bijection $\mathrm{Irr}(G) \to \mathrm{IBr}_p(G)$ (see Lemma 0.31). For example, if A is an abelian p'-group and H acts on A, then H acts faithfully (irreducibly, resp.) on A if and only if H acts faithfully (irreducibly, resp.) on $\mathrm{IBr}_p(A)$ by Proposition 12.1.

13.1 Theorem (Ito). *Let p be a prime and $P \in \mathrm{Syl}_p(G)$.*

(a) *If $P \trianglelefteq G$ and $P' = 1$, then $p \nmid \chi(1)$ for all $\chi \in \mathrm{Irr}(G)$.*

(b) *If $P \trianglelefteq G$, then $p \nmid \beta(1)$ for all $\beta \in \mathrm{IBr}_p(G)$.*

(c) *If G is p-solvable, then the converses of* (a) *and* (b) *also hold.*

Proof. (a) By [Is, 6.15], $\chi(1) \mid |G/P|$ and the conclusion holds.

(b) Recall that $\mathbf{O}_p(G)$ is contained in the kernel of each p-modular irreducible representation (see Proposition 0.20). Thus β can be considered as an element of $\mathrm{IBr}_p(G/P)$ and hence in $\mathrm{Irr}(G/P)$. Therefore, (b) follows from (a).

(c) We first show in both cases that $P \trianglelefteq G$. Let M be a minimal normal subgroup of G. By induction on $|G|$, G/M has a normal Sylow p-subgroup $N/M = MP/M$. We may assume that $N = G$ and also that $p \nmid |M|$, since otherwise G would have a normal Sylow p-subgroup. The hypotheses imply that $I_G(\varphi) = G$ for all $\varphi \in \text{Irr}(M)$, i.e. P acts trivially on $\text{Irr}(M)$. By Lemma 12.2, $P \leq \mathbf{C}_G(M)$ and thus $P \trianglelefteq G$. For the converse of (a), note that Clifford's Theorem implies that $p \nmid \psi(1)$ for all $\psi \in \text{Irr}(P)$. Since $\psi(1) \mid |P|$, we have that ψ is linear and P is abelian. $\square$

Part (c) above is not the full truth, namely the hypothesis "p-solvable" is superfluous, as Michler [Mi 1,2] and Okuyama [Ok 1] showed. This depends on the classification of finite simple groups and we comment on the proof at the end of this section.

Theorem 13.1 provokes the following question: *What can be said about G if there exists a prime $q \neq p$ such that $q \nmid \beta(1)$ for all $\beta \in \text{IBr}_p(G)$?* For solvable G, we show that the q-length $l_q(G)$ of G is at most 2 and $l_p(\mathbf{O}^{q'}(G)) \leq 3$. These results rely on Sections 9 and 10.

13.2 Example. Let p and r be distinct primes, and let q be a prime and $b \in \mathbb{N}$ such that $q \nmid r^{qb} - 1$. We consider $H \trianglelefteq A\Gamma(r^{qb})$, where the field automorphisms in H only consist of a group of order q; hence $|H| = r^{qb} \cdot q \cdot (r^{qb} - 1)$. Since $|H/H'| = q(r^b - 1)$, we may take $G \trianglelefteq H$ of order $|G| = r^{qb} \cdot q \cdot (r^{qb} - 1)/(r^b - 1)$. Now $A\Gamma(r^{qb})$ transitively permutes the non-principal characters of $\text{IBr}_p(A\Gamma(r^{qb}))$. The inertia group in $A\Gamma(r^{qb})$ of $1 \neq \lambda \in \text{IBr}_p(A(r^{qb}))$ has index $r^{qb} - 1$. Since $A(r^{qb}) \leq G \trianglelefteq H \trianglelefteq A\Gamma(r^{qb})$, $q \nmid |G : I_G(\lambda)|$. Assume now in addition that $(r^{qb} - 1)/(r^b - 1) = p^a$ for some $a \in \mathbb{N}$. Then all $\beta \in \text{IBr}_p(G)$ have degree 1 or p^a; in particular, $q \nmid \beta(1)$ for all β.

Observe that $q = r$ is not forbidden. In this case, G has q-length 2. Small examples are $A\Gamma(2^2) \cong S_4$ with $p = 3$ and $q = 2 = r$, and $A\Gamma(2^3)$ with $p = 7$, $q = 3$ and $r = 2$.

As in Section 12, the primes $q = 2$ and 3 require some extra considerations.

13.3 Lemma. *Let p be an odd prime and assume that $2 \nmid \beta(1)$ for all $\beta \in \text{IBr}_p(G)$. If $\mathbf{O}^{2'}(G)$ is solvable, then $\mathbf{O}^{2'}(G)$ is a $\{2, p\}$-group.*

Proof. We argue via induction on $|G|$, and may thus assume that $G = \mathbf{O}^{2'}(G)$. Let M be a minimal normal subgroup of G. Then G/M is a $\{2, p\}$-group and $(|M|, 2p) = 1$. We may also assume that M is the unique minimal normal subgroup of G. Therefore $M = \mathbf{C}_G(M)$, because $(|M|, |G/M|) = 1$. We consider the faithful and irreducible action of G/M on $\mathrm{IBr}_p(M)$. By our hypotheses, $2 \nmid |G : I_G(\lambda)|$ for all $\lambda \in \mathrm{IBr}_p(M)$ and Lemma 9.2 yields $\mathrm{char}(M) = 2$, a contradiction. $\square$

Next is an immediate consequence of Theorem 9.3.

13.4 Lemma. *Suppose that $G = \mathbf{O}^{2'}(G)$ is a $\{2, p\}$-group for an odd prime p. Assume that G acts faithfully and irreducibly on a finite vector space V in such a way that $2 \nmid |G : \mathbf{C}_G(v)|$ for all $v \in V$. If there exists $C \leq G$ such that C is maximal with respect to $C \trianglelefteq G$ and V_C non-homogeneous, then*

(i) *$G/C \cong D_6$ and $p = 3$;*

(ii) *$V_C = V_1 \oplus V_2 \oplus V_3$ for homogeneous components V_i that are faithfully permuted by G/C;*

(iii) *$|V_i| = 2^2$ and C acts transitively on $V_i^{\#}$; and*

(iv) *There is a non-zero vector $x \in V$ such that $\mathbf{C}_G(v)$ has a normal 2-complement.*

Proof. Since $\mathbf{O}^2(G/C) < G/C$, it follows from Proposition 0.2 and Theorem 9.3 that $G/C \cong D_{2n}$ for $n = 3$ or 5, and that $V_C = V_1 \oplus \cdots \oplus V_n$ for homogeneous components V_i that are faithfully and primitively permuted by G/C. Also, C acts transitively on $V_i^{\#}$. By Lemma 9.2, $\mathrm{char}(V) = 2$, say $|V_i| = 2^a$ for some $a \geq 2$. Since C is a $\{2, p\}$-group, we conclude that $2^a - 1 = p^b$ for some integer b. As the $\{2, p\}$-group $G/C \cong D_{2n}$ transitively permutes the V_i, it follows that $n = p = 3$ and therefore $3 = p = 2^a - 1$ (cf. Proposition 3.1). Thus $|V_i| = 2^2$ and (i)–(iii) have been proved.

To establish (iv), let $x = (x_1, 0, 0) \in V$ with $x_1 \in V_1^{\#}$. Clearly $2 =$

$|\mathbf{C}_G(x)C/C| = |\mathbf{C}_G(x)/\mathbf{C}_C(x)|$. As

$$\mathbf{C}_C(x) \lesssim \prod_{i=1}^{3} C/\mathbf{C}_C(V_i) \lesssim S_3 \times S_3 \times S_3,$$

$\mathbf{C}_G(x)$ has a normal 2-complement.   □

13.5 Proposition. *Let M be a minimal normal subgroup of a solvable group G, let $D = \mathbf{C}_G(M)$ and p be an odd prime. Suppose that $\mathbf{O}^{2'}(G) = G$, $p \nmid |M|$ and $2 \nmid \beta(1)$ for all $\beta \in \mathrm{IBr}_p(G)$. Then $l_2(G/D) \leq 1$ and $l_p(G/D) \leq 1$.*

Proof. We set $V = \mathrm{IBr}_p(M)$ and may clearly assume that $D < G$ and $|G/D|$ is even. By our hypotheses, we have that $2 \nmid |G : \mathbf{C}_G(v)|$ for all $v \in V$, and that V is a faithful and irreducible G/D-module. If V is quasi-primitive, it is immediate from Theorem 10.4 that the assertions hold.

We may thus assume that V is not quasi-primitive, and by Lemma 13.3, G is a $\{2, p\}$-group. Applying Lemma 13.4, we obtain that $p = 3$, $\mathrm{char}(V) = 2$ and there exists $\lambda \in V$ such that $\mathbf{C}_G(\lambda)/D$ has a normal 2-complement L/D. Set $I = \mathbf{C}_G(\lambda)$ and choose $\sigma \in \mathrm{IBr}_3(I|\lambda)$. Since $\sigma^G \in \mathrm{IBr}_3(G)$, our hypotheses imply that $2 \nmid \sigma^G(1) = \sigma(1) \cdot |G : I|$. Therefore I/L is isomorphic to a Sylow 2-subgroup of G/D and σ_L is irreducible. Thus $\sigma\mu \in \mathrm{IBr}_3(I|\lambda)$ for all $\mu \in \mathrm{IBr}_3(I/L)$ (see Lemma 0.9). Now $\sigma\mu$ in the role of σ yields that $2 \nmid (\sigma\mu)(1)$ and $\mu(1) = 1$, because I/L is a 2-group. Consequently, I/L and the Sylow 2-subgroups of G/D are abelian. By Lemma 0.19, $l_2(G/D) \leq 1$.

It remains to show that $l_3(G/D) \leq 1$. By Lemma 13.4, there exists $C \trianglelefteq G$ such that $G/C \cong S_3$ and $C/D \lesssim S_3 \times S_3 \times S_3$. Consequently, there exists a series $D \leq B \leq C \leq G$ of normal subgroups of G such that B/D is a 3-group and C/B a 2-group. If $\mathbf{O}_3(G/D) = B/D$, then G/B has a normal Sylow 2-subgroup, because $l_2(G/D) = 1$. This contradicts $G/C \cong S_3$. Therefore, G/D has a normal Sylow 3-subgroup and the proof is complete.   □

After these preparations for the prime $q = 2$, we turn to the other "critical" prime $q = 3$.

13.6 Lemma. *Let p be a prime distinct from 3. Assume that $G = \mathbf{O}^{3'}(G)$ is a solvable group that acts faithfully and irreducibly on a finite vector space V. Suppose that $3 \nmid |G : \mathbf{C}_G(v)|$ for all $v \in V$, and that $3 \nmid \beta(1)$ for all $\beta \in \mathrm{IBr}_p(G)$. If V is not quasi-primitive, then the following assertions hold:*

  (i) *$p = 7$ and $l_7(G) \leq 2$;*

  (ii) *$l_3(G) = 1$; and*

  (iii) *There exists $x \in V^{\#}$ such that $2 \mid |G : \mathbf{C}_G(x)|$.*

Proof. Choose C maximal with respect to $C \trianglelefteq G$ and V_C non-homogeneous. Since $3 \mid |G/C|$, but $3 \nmid |G : \mathbf{C}_G(v)|$ for all $v \in V$, it follows from Proposition 0.2 and Theorem 9.3 that $G/C \cong A\Gamma(2^3)$ and that $V_C = V_1 \oplus \cdots \oplus V_8$ for homogeneous components V_i which are primitively and faithfully permuted by G/C. Furthermore, C acts transitively on each $V_i^{\#}$. As $A\Gamma(2^3)$ has a factor group isomorphic to the Frobenius group F_{21} of order $3 \cdot 7$, the hypothesis about Brauer characters applied to F_{21} implies that $p = 7$. It also follows from Huppert's Theorem 6.8 that either C is metabelian or $7 \nmid |C|$. Consequently, $l_7(C) \leq 1$ and $l_7(G) \leq 2$, proving (i).

Remember that $G/C \cong A\Gamma(2^3)$ has a unique chief series $C < L < K < G$, where $|L/C| = 2^3$, $|K/L| = 7$ and $|G/K| = 3$. Also note that $\mathbf{O}_3(G) \leq \mathbf{C}_G(V)$ and thus $\mathbf{O}_3(G) = 1$. If $7 \nmid |C|$, then the condition on Brauer characters implies that every $\rho \in \mathrm{Irr}(C)$ has $3'$-degree. Since $\mathbf{O}_3(C) = \mathbf{O}_3(G) = 1$, Theorem 13.1 yields that $3 \nmid |C|$. Thus, to prove (ii), we may assume that $7 \mid |C|$. By the first paragraph, C is metabelian. Set $X = \mathbf{O}_{3'}(C)$. Then C/X is an abelian 3-group. We show that K/C centralizes C/X. If not, then there is a chief factor C/D of G such that $K/C \not\leq \mathbf{C}_{G/C}(C/D)$, because $3 \nmid |K/C|$. Set $W = \mathrm{IBr}_p(C/D)$. Then W is an irreducible G-module, and the hypothesis about Brauer characters implies that $3 \nmid |G : \mathbf{C}_G(\lambda)|$ for all $\lambda \in W$. Suppose that G/C acts faithfully on

W. Then W_L is homogeneous (by Theorem 9.3) and thus L/C is cyclic. This contradiction shows that $\mathbf{C}_G(W) = L$. Now $G/L \cong F_{21}$ and so each $\lambda \in W^{\#}$ is fixed by exactly one of seven Sylow 3-subgroups of G/L. Let $R \in \mathrm{Syl}_3(G/L)$, and set $|W| = 3^a$ and $|\mathbf{C}_W(R)| = 3^b$. Counting yields

$$3^a - 1 = |W| - 1 = |\mathrm{Syl}_3(G/L)| \cdot (|\mathbf{C}_W(R)| - 1) = 7 \cdot (3^b - 1).$$

In particular, $b \mid a$ and we obtain $7 = (3^b)^{(a/b)-1} + \cdots + 3^b + 1$, a contradiction. Thus $K/C \leq \mathbf{C}_{G/C}(C/X)$. So G has a normal 3-complement and assertion (ii) follows.

To prove (iii), consider $x = (x_1, 0, \ldots 0) \in V$ with $x_1 \in V_1^{\#}$. Since L/C regularly permutes the V_i, certainly $2 \mid |G : \mathbf{C}_G(x)|$ holds. $\square$

Before we reach the main results of this section, we observe the following fact.

13.7 Lemma. *Suppose that $q \nmid \beta(1)$ for all $\beta \in \mathrm{IBr}_p(G)$, where $q \neq p$. If $M \trianglelefteq N \trianglelefteq G$ such that N/M is a q-group, then N/M is abelian.*

Proof. By Clifford's Theorem and since $p \nmid |N/M|$, we have that $q \nmid \varphi(1)$ for all $\varphi \in \mathrm{Irr}(N/M)$. But $\varphi(1) \mid |N/M|$ for all such φ. So $\varphi(1) = 1$ and N/M is abelian. $\square$

13.8 Theorem. *Let p, q be distinct primes. Assume that $\mathbf{O}^{q'}(G)$ is solvable and that $q \nmid \beta(1)$ for all $\beta \in \mathrm{IBr}_p(G)$. Then*

(1) *In each q-series of G, the q-factors are abelian; and*

(2) *$l_q(G/\mathbf{O}_{p,q}(G)) \leq 1$.*

In particular, a Sylow q-subgroup of G is metabelian.

Proof. Since assertion (1) immediately follows from Lemma 13.7, it remains to prove assertion (2). As $\mathbf{O}_{p,q}(\mathbf{O}^{q'}(G)) \leq \mathbf{O}_{p,q}(G)$, we may assume that $G = \mathbf{O}^{q'}(G)$. We may also assume that $\mathbf{O}_p(G) = 1$. It thus suffices to show that $l_q(G/\mathbf{F}(G)) \leq 1$. By Gaschütz's Theorem 1.12, $\mathbf{F}(G)/\Phi(G) = V_1 \oplus \cdots \oplus V_n$ is a faithful completely reducible $G/\mathbf{F}(G)$-module with irreducible constituents V_i. Set $C_i = \mathbf{C}_G(V_i)$ and $W_i = \mathrm{IBr}_p(V_i)$. Since $\mathrm{char}\,(V_i) \neq p$ for all i, the W_i are faithful and irreducible G/C_i-modules. Our hypothesis about the degrees of Brauer characters implies that $q \nmid |G : \mathbf{C}_G(\lambda_i)|$ for all $\lambda_i \in W_i$, $i = 1, \ldots, n$.

If W_j is quasi-primitive, then Theorem 10.4 immediately yields that $l_q(G/C_j) \leq 1$. If however W_j is not quasi-primitive, then $q = 2$ or 3, by Theorem 9.3. In the first case, it follows from Proposition 13.5 (with V_j in the role of M) and in the second case from Lemma 13.6 that $l_q(G/C_j) \leq 1$. Since $\bigcap_i C_i = \mathbf{F}(G)$, we have shown that $l_q(G/\mathbf{F}(G)) \leq 1$, as desired.   □

We next bound the p-length of $\mathbf{O}^{q'}(G)$ under the hypotheses of Theorem 13.8.

13.9 Theorem. *Let p, q be distinct primes, let $N = \mathbf{O}^{q'}(G)$ be solvable and assume that $q \nmid \beta(1)$ for all $\beta \in \mathrm{IBr}_p(G)$. Then*

(1) $l_p(N/\mathbf{O}_p(N)) \leq 1$, *or*

(2) $l_p(N/\mathbf{O}_p(N)) = 2$ *and* $(p,q) = (7,3)$. *In this case, there exists* $\rho \in \mathrm{IBr}_7(G)$ *with* $\rho(1)$ *even.*

Proof. We may assume that $G = N$ and $\mathbf{O}_p(G) = 1$. We again write $\mathbf{F}(G)/\Phi(G) = V_1 \oplus \cdots \oplus V_n$ with irreducible G-modules V_i, set $C_i = \mathbf{C}_G(V_i)$ and $W_i = \mathrm{IBr}_p(V_i)$, and observe that $q \nmid |G : \mathbf{C}_G(\lambda_i)|$ for all $\lambda_i \in W_i$, $i = 1, \ldots, n$.

If W_j is quasi-primitive, then $l_p(G/C_j) \leq 1$ by Theorem 10.4. Otherwise, Theorem 9.3 yields $q = 2$ or 3. It follows from Proposition 13.5 and Lemma 13.6 that $l_p(G/C_j) \leq 1$ in the first case, and that $p = 7$ and $l_7(G/C_j) \leq 2$ in the second case. Since $\bigcap_i C_i = \mathbf{F}(G)$ and $p \nmid |\mathbf{F}(G)|$, we obtain $l_p(G) \leq 1$, or $l_p(G) \leq 2$ and $(p,q) = (7,3)$, as required.

We still have to prove the existence of the character ρ in the exceptional case (2). By the previous paragraph, we may assume that W_1 is not quasi-primitive. Applying Lemma 13.6 (iii), there then exists $\lambda \in W_1$ such that $2 \mid |G : I_G(\lambda)|$. For ρ we may thus choose any element of $\mathrm{IBr}_7(G|\lambda)$.   □

13.10 Corollary. *Assume that G is solvable and $\beta(1)$ is a p-power for all $\beta \in \mathrm{IBr}_p(G)$. Then $\mathbf{O}^p(G)/\mathbf{O}_p(\mathbf{O}^p(G))$ has p-length at most 1.*

Proof. Without loss of generality, $\mathbf{O}_p(G) = 1$. Set $K = \prod_{q \neq p} \mathbf{O}^{q'}(G) = \mathbf{O}^p(G)$. It follows from Theorem 13.9 that $l_p(\mathbf{O}^{q'}(G)) \leq 1$ for each q (the exceptional case with $(p,q) = (7,3)$ is ruled out by the existence of $\rho \in \mathrm{Irr}(\mathbf{O}^{q'}(G))$ of even degree). Thus $l_p(K) \leq 1$.   □

Example 13.2 shows that the assertion of Theorem 13.8 is best possible. Under certain circumstances however we have a statement analogous to Theorem 13.1 (c), as we shall see next in Theorem 13.12.

13.11 Lemma. *Let p, q be distinct primes and assume that the solvable group $G = \mathbf{O}^{q'}(G) \neq 1$ acts faithfully and irreducibly on a finite vector space V. Suppose that $q \nmid |G : \mathbf{C}_G(v)|$ for all $v \in V$ and $q \nmid \beta(1)$ for all $\beta \in \mathrm{IBr}_p(G)$. Then $q \mid p - 1$ or $(p,q) = (2,3)$.*

Proof. Suppose that V is quasi-primitive. We may assume by Theorem 10.4 that $N := \mathbf{O}^q(G)$ is cyclic. Now $p \nmid |G/\mathbf{O}_p(N)|$ and thus each $\alpha \in \mathrm{Irr}(G/\mathbf{O}_p(N))$ has q'-degree. By Theorem 13.1, and the hypothesis that $\mathbf{O}^{q'}(G) = G$, it follows that $G/\mathbf{O}_p(N)$ is a q-group. Then N is a cyclic p-group. Since $\mathbf{O}_q(G) = 1$, we obtain $q \mid p - 1$.

Suppose secondly that V is not quasi-primitive. By Theorem 9.3, we have that $q = 2$, or $q = 3$ and G has a factor group isomorphic to $A\Gamma(2^3)$. If $q = 2$, then p is odd and the assertion trivially holds. Since $A\Gamma(2^3)$ has a factor group isomorphic to the Frobenius group of order $3 \cdot 7$, the hypothesis about Brauer characters forces $p = 7$ and again $q \mid p - 1$.   □

13.12 Theorem. *Let p, q be distinct primes, let $\mathbf{O}^{q'}(G)$ be solvable and assume that $q \nmid \beta(1)$ for all $\beta \in \mathrm{IBr}_p(G)$. If $q \nmid p-1$ and $(p,q) \neq (2,3)$, then $G/\mathbf{O}_p(G)$ has a normal abelian Sylow q-subgroup.*

Proof. We may again assume that $G = \mathbf{O}^{q'}(G)$ and $\mathbf{O}_p(G) = 1$. By Lemma 13.7 it is enough to show that $G = \mathbf{F}(G)$. To do so, we write $\mathbf{F}(G)/\Phi(G) = V_1 \oplus \cdots \oplus V_n$ with irreducible G-modules V_i, set $C_i = \mathbf{C}_G(V_i)$ and $W_i = \mathrm{IBr}_p(V_i)$, and observe that our hypotheses imply $q \nmid |G : \mathbf{C}_G(\lambda_i)|$ for all $\lambda_i \in W_i$, $i = 1, \ldots, n$. Since $q \nmid p-1$ and $(p,q) \neq (2,3)$, it follows from Lemma 13.11 that $G/C_j = 1$ for all $j = 1, \ldots, n$. Now $\bigcap_j C_j = \mathbf{F}(G)$ and we get $G/\mathbf{F}(G) = 1$, as required. □

13.13 Remarks. (a) (Michler, Okuyama) Let p be a prime and $P \in \mathrm{Syl}_p(G)$.

   (i) If $p \nmid \chi(1)$ for all $\chi \in \mathrm{Irr}(G)$, then $P \trianglelefteq G$ and $P' = 1$.

   (ii) If $p \nmid \beta(1)$ for all $\beta \in \mathrm{IBr}_p(G)$, then $P \trianglelefteq G$.

To prove (i) or (ii), it suffices to show that $P \trianglelefteq G$. To this end, one may repeat the arguments of Theorem 13.1 (c) to conclude that G has a minimal normal subgroup M, that M is non-solvable and $p \mid |M|$. In particular, $M = E \times \cdots \times E$ for a non-abelian simple group E such that $p \mid |E|$. One then wishes to obtain contradictions by showing the existence of $\chi \in \mathrm{Irr}(E)$ and $\mu \in \mathrm{IBr}_p(E)$ with $p \mid \chi(1)$ and $p \mid \mu(1)$. This can be done with the help of the classification of simple groups, although Okuyama [Ok 1] has given a direct proof that a simple non-abelian group has a Brauer character of even degree. For p odd, Michler [Mi 2] has shown that the simple group E has a block B that is not of maximal defect, i.e. a defect group is not a Sylow p-subgroup. Then the degree of every $\chi \in \mathrm{Irr}(B) \cup \mathrm{IBr}_p(B)$ is divisible by p (see Lemma 0.24).

(b) Does Theorem 13.8 also allow an extension to arbitrary finite groups G? Clearly, it makes no sense to speak about the q-length of G any longer. But we may ask the following question:

Let p, q be distinct primes and suppose that $q \nmid \beta(1)$ for all $\beta \in \mathrm{IBr}_p(G)$. Is Q metabelian for $Q \in \mathrm{Syl}_q(G)$?

If so, it would be best possible. The solvable group S_4 has a dihedral Sylow 2-subgroup and satisfies these hypotheses with $q = 2$, $p = 3$ (see Example 13.2). To see a non-solvable example, we mention that the p-Brauer character degrees of $\mathrm{PSL}(2,p)$ are $\{1, 3, 5, \ldots, p\}$, but a Sylow 2-subgroup of $\mathrm{PSL}\,(2,p)$ is dihedral (see [HB, VII, 3.10] and [Hu, II, 8.10 (b)]).

It was mentioned in §12 that Theorem 12.9 remains valid for p-solvable groups, depending on the classification of finite simple groups. One of the facts needed is the following.

13.14 Theorem. If $p \nmid |E|$ for a simple non-abelian group E, then $\mathrm{Out}\,(E)$ has a cyclic and central Sylow p-subgroup.

Proof. See [GW 2, Lemma 1.3]. □

Also Theorem 13.8 extends to p-solvable groups. As the arguments are much easier than those needed to extend Theorem 12.9, we present a proof.

13.15 Corollary. Let p, q be distinct primes, and assume that G is p-solvable. If $q \nmid \beta(1)$ for all $\beta \in \mathrm{IBr}_p(G)$, then $\mathbf{O}^{q'}(G)$ is solvable. In particular, the assertions of Theorems 13.8 and 13.9 apply.

Proof. We assume without loss of generality that $G = \mathbf{O}^{q'}(G)$ and $\mathbf{O}_p(G) = 1$. Choose a non-solvable chief factor M/N of G with $|M|$ as large as possible. Thus G/M is solvable. Since $p \nmid |M/N|$, the hypothesis about Brauer degrees implies that $q \nmid \chi(1)$ for all $\chi \in \mathrm{Irr}\,(M/N)$, and 13.13 (a) yields $q \nmid |M/N|$. Since $G = \mathbf{O}^{q'}(G)$, it follows that $M < G$, and by the maximality of M, G/M is isomorphic to subgroup of $\mathrm{Out}\,(M/N)$. Suppose at first that M/N is simple. Then Theorem 13.14 implies that G/M is a q-group. By our hypotheses, every irreducible Brauer and ordinary character of M/N is invariant in G. Hence $Q \in \mathrm{Syl}_q(G/N)$ centralizes M/N by Lemma 12.2. This contradicts $G = \mathbf{O}^{q'}(G)$.

We may now assume that M/N is not simple. Then G induces a non-trivial transitive permutation group on the simple components of M/N. We may write $M/N = S_1 \times \cdots \times S_n$ with $n > 1$ isomorphic non-solvable groups S_i that are primitively permuted by G. Let $C \trianglelefteq G$ be the kernel of this permutation action, and fix $1 \neq \alpha_i \in \mathrm{Irr}(S_i)$, $i = 1, \ldots, n$. For $\Delta \subseteq \{1, \ldots, n\}$, consider $\prod_{i \in \Delta} \alpha_i \in \mathrm{Irr}(M/N)$. Since this character must be invariant under some $Q \in \mathrm{Syl}_q(G)$, we conclude that $1 \neq QC/C \in \mathrm{Syl}_q(G/C)$ stabilizes Δ. As G/C is solvable, it follows from Corollary 5.8 that G/C is isomorphic to D_6, D_{10} or $A\Gamma(2^3)$, $n = 3$, 5 or 8 (resp.) and $q = 2$, 2 or 3 (resp.). We now choose $Q_0 \in \mathrm{Syl}_q(G)$ that stabilizes $\alpha_1 \cdots \alpha_q \in \mathrm{Irr}(M/N)$. Since Q_0 transitively permutes $\{S_1, \ldots, S_q\}$, it follows that $\alpha_1(1) = \cdots = \alpha_q(1)$. Consequently, S_1 has at most two distinct ordinary character degrees. This however forces S_1 to be solvable (see [Is, Theorem 12.5]), a contradiction. $\qquad \square$

Most of the material of this section appeared in [MW 1]. Corollary 13.10 has been proved by R. Gow [Go 2] for groups of odd order.

§14   The p-Part of Character Degrees

In the two previous sections, we were concerned with the situation where the degrees of ordinary or p-Brauer characters (resp.) of a solvable group G are coprime to a given prime q. We extend this question a little bit and consider the largest q-power which can occur as a factor in some character degree.

14.1 Definition. Let G be a group and q be a prime number.

(a) By $e_q(G)$ we denote the smallest non-negative integer e such that $q^{e+1} \nmid \chi(1)$ for all $\chi \in \mathrm{Irr}(G)$.

(b) For a prime p, analogously $\bar{e}_q(G)$ denotes the smallest non-negative integer f such that $q^{f+1} \nmid \beta(1)$ for all $\beta \in \mathrm{IBr}_p(G)$.

Note that $^-$ always refers to the given characteristic p. Also observe that for $N \trianglelefteq G$, $e_q(N) \leq e_q(G)$ and $e_q(G/N) \leq e_q(G)$ hold. The analogous statements hold for $\bar{e}_q$. We are interested in bounding invariants of a solvable group G in terms of $e_q(G)$ and $\bar{e}_q(G)$, respectively. Such invariants are the q-length $l_q(G)$, the q-rank $r_q(G)$ and the derived length $\mathrm{dl}\,(Q)$ for $Q \in \mathrm{Syl}_q(G)$. We remind the reader that the q-rank $r_q(G)$ is the maximum dimension of all q-chief factors of G.

Before we start, we state some useful relationships between the above invariants. The first fact is rather elementary (see [Hu, VI, 6.6 (c)]).

14.2 Lemma. *Let G be solvable and q be a prime. Then $l_q(G) \leq r_q(G)$.*

The second bound does not at all lie at the surface. For odd primes q, it is a consequence of Hall–Higman B (cf. [HB, IX, 5.4]). The case $q = 2$ however was obtained more recently by Bryukhanova [Br 1], improving an earlier result of Berger and Gross [BG].

14.3 Theorem. *Let G be solvable, q a prime and $Q \in \mathrm{Syl}_q(G)$. Then*

$$l_q(G) \leq \mathrm{dl}\,(Q).$$

Indeed, 14.2 and 14.3 also hold for q-solvable groups, as will some of the results of this section. However we shall restrict ourselves to solvable groups only. Also see Remarks 14.12. The next proposition is valid for all p-solvable G, as we prove later in Theorem 23.5, but it is not valid for arbitrary G even when $N = 1$.

14.4 Proposition. *Suppose that $\alpha \in \mathrm{IBr}_p(N)$, $N \trianglelefteq G$ and $\chi \in \mathrm{IBr}_p(G|\alpha)$. If G/N is solvable, then $\chi(1)/\alpha(1) \mid |G/N|$.*

Proof. Arguing by induction on $|G : N|$, we may assume that N is a maximal normal subgroup. By the solvability hypothesis, G/N is cyclic. By Clifford's Theorem 0.8 and the inductive hypothesis, we can assume α is G-invariant. By Proposition 0.11 and Lemma 0.9, $\chi = \lambda\mu$ where $\mu \in \mathrm{IBr}_p(G)$ extends α and $\lambda \in \mathrm{IBr}_p(G/N)$. Since G/N is cyclic, $\lambda(1) = 1$ and $\chi_N = \alpha$. $\square$

14.5 Lemma. *Let q be a prime and $Q \in \mathrm{Syl}_q(G)$. Assume that G is solvable and acts faithfully and completely reducibly on a $GF(q)$-vector space V. Suppose that $q \nmid |G : \mathbf{C}_G(v)|$ for all $v \in V$. Then*

(i) $\mathrm{dl}(Q) \le 2$; *and*

(ii) *if $q \ge 5$, then $\mathrm{dl}(Q) \le 1$.*

Proof. We may clearly assume that $G = \mathbf{O}^{q'}(G) \ne 1$. If $V = V_1 \oplus \cdots \oplus V_n$ with irreducible G-modules V_i, then $G \le \prod_i G/\mathbf{C}_G(V_i)$ and $Q_i \in \mathrm{Syl}_q(G/\mathbf{C}_G(V_i))$ satisfies (i) or (ii), respectively, by induction. We may thus assume that V is faithful and irreducible. If V is quasi-primitive, it follows from Theorem 10.5 that either $q^2 \nmid |G|$ or Q is cyclic. Therefore Q is abelian.

We may now assume that V is not quasi-primitive, and choose $C \trianglelefteq G$ maximal with respect to $V_C = V_1 \oplus \cdots \oplus V_n$ non-homogeneous, where the V_i denote the homogeneous components. By Theorem 9.3, we have that $q \le 3$, $q^2 \nmid |G/C|$ and $C/\mathbf{C}_C(V_i)$ acts transitively on $V_i^{\#}$. In particular, assertion (ii) holds. To establish assertion (i), it suffices to show that C has abelian Sylow q-subgroups. As $\mathbf{O}_q(C) = 1$, we may assume that $C'' \ne 1$ and thus that $C/\mathbf{C}_C(V_i) \not\le \Gamma(V_i)$. By Huppert's Theorem 6.8, it follows that $|V_i| = 3^2$ or 3^4 and $3^2 \nmid |C/\mathbf{C}_C(V_i)|$. Since $\bigcap_i \mathbf{C}_C(V_i) = 1$, a Sylow-3-subgroup of C is abelian also in this exceptional case. $\square$

The following is a modular analogue of Theorem 12.9, but with weaker assertions.

14.6 Theorem. Let $N \trianglelefteq G$, G/N solvable, q and p distinct primes, $Q/N \in \mathrm{Syl}_q(G/N)$ and $\alpha \in \mathrm{IBr}_p(N)$. Suppose that $q \nmid \beta(1)/\alpha(1)$ for all $\beta \in \mathrm{IBr}_p(G|\alpha)$. Then

(i) $\mathrm{dl}(Q/N) \leq 3$; and

(ii) if $q \geq 5$, then $\mathrm{dl}(Q/N) \leq 2$.

Proof. We argue by induction on $|G : N|$. Observe that if $N \leq K \trianglelefteq G$ and $\tau \in \mathrm{IBr}_p(K|\alpha)$, then $q \nmid \tau(1)/\alpha(1)$ and $q \nmid \gamma(1)/\tau(1)$ for all $\gamma \in \mathrm{IBr}_p(G|\tau)$. By Proposition 14.4, we may hence assume that $\mathbf{O}_{q'}(G/N) = 1$, but $q \mid |G/N|$. Let $I = I_G(\alpha)$. If $\tau \in \mathrm{IBr}_p(I|\alpha)$, then $\tau^G \in \mathrm{IBr}_p(G|\alpha)$ and $\tau^G(1)/\alpha(1) = |G : I| \cdot \tau(1)/\alpha(1)$. The hypothesis on character degrees implies that $q \nmid \tau(1)/\alpha(1)$ and that $Q \leq I$ (up to conjugacy). We may thus assume that α is invariant in G.

Set $M/N = \mathbf{O}_q(G/N) \neq 1$ and let $\sigma \in \mathrm{IBr}_p(M|\alpha)$. Then $q \nmid \sigma(1)/\alpha(1)$ and since $\sigma(1)/\alpha(1) \mid |M/N|$, we obtain that $\sigma_N = \alpha$. In particular, the map $\lambda \mapsto \sigma \cdot \lambda$ yields a bijection from $\mathrm{IBr}_p(M/N)$ onto $\mathrm{IBr}_p(M|\alpha)$ (see Lemma 0.9). It follows that all $\lambda \in \mathrm{IBr}_p(M/N)$ are linear and M/N is abelian, because $q \neq p$.

Let H/N be a Hall q'-subgroup of G/N. We apply Lemma 0.17 (d) to find $\varphi \in \mathrm{IBr}_p(M|\alpha)$ which is fixed by H. Hence $|G : I_G(\varphi)|$ is a q-power, and the hypothesis about character degrees implies that $I_G(\varphi) = G$. Lemma 0.9 and the hypotheses imply that $I_G(\varphi \cdot \lambda) = I_G(\lambda)$ has q'-index in G for all $\lambda \in \mathrm{IBr}_p(M/N)$.

As $\mathbf{O}_{q'}(G/N) = 1$, we have that $\mathbf{F}(G/N) = M/N$ is an abelian q-group. Let $N = N_0 < N_1 < \ldots N_m = M$ such that N_i/N_{i-1} is irreducible as G-module, and define $C_i = \mathbf{C}_G(N_i/N_{i-1}) \geq M$. Observe that $\bigcap_i C_i = M$ and that $V_i := \mathrm{IBr}_p(N_i/N_{i-1})$ is an irreducible and faithful G/C_i-module $(i = 1, \ldots, m)$. Since M/N is abelian, each $\beta_i \in V_i$ is the restriction of some character in $\mathrm{IBr}_p(M/N)$. By the previous paragraph, $q \nmid |G : I_G(\beta_i)|$. We apply Lemma 14.5 to the action of G/C_i on V_i. As $\bigcap_i C_i = M$, a Sylow q-

subgroup of G/M has derived length at most 2, and is even abelian provided that $q \geq 5$. Since $(M/N)' = 1$, the result follows. $\square$

Next is a consequence of both Theorem 12.9 and Theorem 14.6.

14.7 Corollary. *Let q be a prime, $N \trianglelefteq G$, let G/N be solvable and let $Q/N \in \mathrm{Syl}_q(G/N)$.*

    (a) *Suppose that $\varphi \in \mathrm{Irr}(N)$ and $q^{e+1} \nmid \chi(1)/\varphi(1)$ for all $\chi \in \mathrm{Irr}(G|\varphi)$. Then $\mathrm{dl}(Q/N) \leq 2e + 1$.*

    (b) *Suppose that $p \neq q$, $\alpha \in \mathrm{IBr}_p(N)$ and $q^{e+1} \nmid \beta(1)/\alpha(1)$ for all $\beta \in \mathrm{IBr}_p(G|\alpha)$. Then*

        (i) *$\mathrm{dl}(Q/N) \leq 4e + 3$; and*

        (ii) *if $q \geq 5$, then $\mathrm{dl}(Q/N) \leq 3e + 2$.*

Proof. (a) By Theorem 12.9, we may assume that $e \geq 1$ and therefore $\mathrm{dl}(Q/N) \geq 2$. Walking along an ascending chief series of the solvable group G, we find $N \leq K \trianglelefteq G$ such that $\mathrm{dl}(K \cap Q/N) = 2$. Again by Theorem 12.9, there exists $\tau \in \mathrm{Irr}(K|\alpha)$ such that $q \mid \tau(1)/\alpha(1)$. Therefore $q^e \nmid \chi(1)/\tau(1)$ for all $\chi \in \mathrm{Irr}(G|\tau)$, and induction on $|G : N|$ yields that $\mathrm{dl}(QK/K) \leq 2(e - 1) + 1$. Consequently

$$\mathrm{dl}(Q/N) \leq \mathrm{dl}(Q \cap K/N) + \mathrm{dl}(QK/K) \leq 2e + 1,$$

as desired.

    (b) The proof relies on exactly the same arguments, but using Theorem 14.6 in place of Theorem 12.9. $\square$

We now combine Theorem 14.3 and Corollary 14.7 for $N = 1$ to obtain the following result.

14.8 Corollary. *Let q be a prime, $Q \in \mathrm{Syl}_q(G)$ and G solvable. Then*

    (a) *$l_q(G) \leq \mathrm{dl}(Q) \leq 2 \cdot e_q(G) + 1$.*

    (b) *Let $p \neq q$. Then*

    (i) $l_q(G) \leq \mathrm{dl}(Q) \leq 4 \cdot \bar{e}_q(G) + 3$; and

    (ii) if $q \geq 5$, then $l_q(G) \leq \mathrm{dl}(Q) \leq 3 \cdot \bar{e}_q(G) + 2$.

If $\beta \in \mathrm{IBr}_p(G)$, then $\mathbf{O}_p(G) \leq \ker(\beta)$ and therefore $\bar{e}_p(G) = \bar{e}_p(G/\mathbf{O}_p(G))$. What remains is the question whether one can bound $l_p(G)$ and the derived length of $P/\mathbf{O}_p(G) \in \mathrm{Syl}_p(G/\mathbf{O}_p(G))$ in terms of $\bar{e}_p(G)$. Although such bounds turn out to exist, we proceed to show that they cannot be derived "locally" as done in Corollary 14.7.

14.9 Example. Let p be a prime. For each non-negative integer n there exists a solvable group G_n whose center Z_n is a cyclic p'-group, and a faithful $\lambda_n \in \mathrm{IBr}_p(Z_n)$ such that the following statements hold:

    (1) $\mathrm{IBr}_p(G_n | \lambda_n) = \{\chi_n\}$ and $p \nmid \chi_n(1)$;

    (2) $l_p(G_n/Z_n) = n$;

    (3) $\mathbf{O}_{p'}(G_n/Z_n) = 1$; and

    (4) $\mathbf{O}_p(G_n/Z_n)$ is abelian.

In particular it follows by Theorem 14.3 that the derived length of a Sylow p-subgroup of $(G_n/Z_n)/\mathbf{O}_p(G_n/Z_n)$ tends to infinity as $n \to \infty$.

Proof. We set $G_0 = 1$ and construct the groups G_n iteratively. Assume now that G_n has been found with the given properties. Let q be a prime with $2 \neq q \neq p$ and $q \nmid |G_n|$. For sufficiently large m, G_n/Z_n can be embedded into $GL(m, q)$. Since

$$A \mapsto \begin{pmatrix} A & 0 \\ 0 & (A^t)^{-1} \end{pmatrix}$$

embeds $GL(m, q)$ into $Sp(2m, q)$, G_n/Z_n may be embedded into $Sp(2m, q)$. Let Q be extra-special of order q^{2m+1} and exponent q. Then G_n/Z_n acts faithfully on Q and on $Q/\mathbf{Z}(Q)$, while centralizing $\mathbf{Z}(Q)$ (cf. [Hu, III, §13]).

Let H be the semi-direct product $Q \cdot G_n$ and $Z_{n+1} = \mathbf{Z}(H) = \mathbf{Z}(Q) \times Z_n$. Since $p \neq q$ and $q \nmid |G_n|$, the inductive hypothesis implies that Z_{n+1} is a cyclic p'-group. We fix a faithful $\lambda \in \mathrm{Irr}(\mathbf{Z}(Q))$ and let $\theta \in \mathrm{Irr}(Q)$ be the unique irreducible constituent of λ^Q. Let $\tau = \theta \times 1_{Z_n} \in \mathrm{Irr}(QZ_n)$. Now

$Z_n = \ker(\tau)$ and $(|H : QZ_n|, |QZ_n : \ker(\tau)|) = 1$. Since λ is H-invariant, so are θ and τ. By Theorem 0.13, τ extends to $\chi \in \mathrm{Irr}(G|\theta)$. Since $p \nmid |QZ_n|$, we have that $\tau \in \mathrm{IBr}_p(QZ_n)$. Now the restriction μ of χ to p-regular elements of H is a positive $\mathbb{Z}$-linear sum of irreducible Brauer characters of H, yet $\mu_{QZ_n} = \tau$. Thus $\mu \in \mathrm{IBr}_p(H)$ extending τ and $\theta \in \mathrm{IBr}_p(Q)$. By Lemma 0.9, $\alpha \mapsto \alpha\mu$ is a bijection from $\mathrm{IBr}_p(H/Q)$ onto $\mathrm{IBr}_p(H|\theta)$. We set $\lambda_{n+1} = \lambda \cdot \lambda_n$, a faithful irreducible character of the p'-group Z_{n+1}. Now $\theta \cdot \lambda_n \in \mathrm{Irr}(QZ_n) = \mathrm{IBr}_p(QZ_n)$ and $\theta \cdot \lambda_n$ is the unique (ordinary or Brauer) irreducible character of QZ_n lying over λ_{n+1}.

If $\eta \in \mathrm{IBr}_p(H|\theta \cdot \lambda_n)$, then $\eta \in \mathrm{IBr}_p(H|\theta)$ and consequently $\eta = \mu \cdot \alpha$ for a unique $\alpha \in \mathrm{IBr}_p(H/Q) = \mathrm{IBr}_p(H|1_Q)$. As $Z_n \le \ker(\tau) \le \ker(\mu)$, we even have that $\alpha \in \mathrm{IBr}_p(H|1_Q \cdot \lambda_n)$. Since $H/Q \cong G_n$, it follows from the inductive hypothesis that $\mathrm{IBr}_p(H|1_Q \cdot \lambda_n) = \{\beta\}$ for some β satisfying $p \nmid \beta(1)$. Therefore, $\mathrm{IBr}_p(H|\theta \cdot \lambda_n) = \{\eta = \mu \cdot \beta\}$ and $p \nmid q^m \cdot \beta(1) = \eta(1)$. Since $\theta \cdot \lambda_n$ is the unique irreducible constituent of $\lambda_{n+1}^{QZ_n} = (\lambda \cdot \lambda_n)^{Q \times Z_n}$, we also have $\mathrm{IBr}_p(H|\lambda_{n+1}) = \{\eta\}$.

Recall that G_n/Z_n acts faithfully on $Q/\mathbf{Z}(Q) \cong (Q \cdot Z_n)/Z_{n+1}$, and therefore a minimal normal subgroup of H/Z_{n+1} must be contained in $(Q \cdot Z_n)/Z_{n+1}$. We may thus choose a $GF(p)$-vector space V such that H/Z_{n+1} acts faithfully on V and $\mathbf{C}_V(Q) = 1$; in particular, $\mathbf{C}_V(H) = 1$ holds. We define G_{n+1} to be the semi-direct product $V \cdot H$. Observe that $\mathbf{Z}(G_{n+1}) = \mathbf{Z}(H) = Z_{n+1}$ is a cyclic p'-group, $\mathbf{O}_{p'}(G_{n+1}/Z_{n+1}) = 1$ and $\mathbf{O}_p(G_{n+1}/Z_{n+1}) = (V \cdot Z_{n+1})/Z_{n+1}$ is abelian. Applying the inductive hypothesis, we furthermore have that

$$l_p(G_{n+1}/Z_{n+1}) = 1 + l_p(H/Z_{n+1}) = 1 + l_p(G_n/Z_n) = 1 + n.$$

Since V is a p-group, $\sigma \mapsto \sigma_H$ defines a bijection from $\mathrm{IBr}_p(G_{n+1})$ onto $\mathrm{IBr}_p(H)$. Consequently, the last paragraph implies that $\mathrm{IBr}_p(G_{n+1}|\lambda_{n+1}) = \{\chi_{n+1}\}$ and $p \nmid \chi_{n+1}(1)$. $\square$

We next give estimates for the p-rank of $G/\mathbf{O}_p(G)$ in terms of $\bar{e}_p(G) = \bar{e}_p(G/\mathbf{O}_p(G))$. For that it is no loss to assume $\mathbf{O}_p(G) = 1$.

14.10 Lemma. *Let G be solvable and $\mathbf{O}_p(G) = 1$. Then*

(a) $r_p(G) \leq 2 \cdot \bar{e}_p(G)$; *and*

(b) $r_p(G) \leq \bar{e}_p(G)$ *provided that $|G|$ is odd or $p \notin \{2\} \cup \mathfrak{M}$.*

Proof. (a) Let M be a minimal normal subgroup of G and $N/M :=$ $\mathbf{O}_p(G/M)$. Since $\mathbf{O}_p(G) = 1$, M is a p'-group, and N/M acts faithfully on both M and $V := \mathrm{IBr}_p(M)$. Let $V_N = V_1 \oplus \cdots \oplus V_n$ with irreducible N-modules V_i and $C_i = \mathbf{C}_N(V_i)$. By Theorem 4.7, there exist λ_i, $\mu_i \in V_i$ such that $\mathbf{C}_N(\lambda_i) \cap \mathbf{C}_N(\mu_i) = C_i$ $(i = 1, \ldots, n)$. Setting $\lambda = \lambda_1 \cdots \lambda_n$ and $\mu = \mu_1 \cdots \mu_n$, we obtain $\mathbf{C}_N(\lambda) \cap \mathbf{C}_N(\mu) = \bigcap_i C_i = M$. Without loss of generality, we may thus assume that $|N/M| \mid |N : \mathbf{C}_N(\lambda)|^2$ and therefore $|N/M| \mid \varphi(1)^2$ for all $\varphi \in \mathrm{IBr}_p(N|\lambda)$. This implies that $t \leq 2 \cdot \bar{e}_p(G)$, where $p^t := |N/M|$. As $\mathbf{O}_p(G/N) = 1$, induction finally yields

$$r_p(G) \leq \max\{r_p(G/N), t\} \leq 2 \cdot \bar{e}_p(G),$$

as required.

(b) Under the hypotheses of (b), Theorem 4.4 yields the existence of $\nu_i \in V_i$ such that $\mathbf{C}_N(\nu_i) = C_i$ $(i = 1, \ldots, n)$. The result now follows along the same lines as in part (a). $\qquad\square$

14.11 Theorem. *Let G be solvable. Then*

(a) $l_p(G) \leq 2 \cdot \bar{e}_p(G) + 1$; *and*

(b) $l_p(G) \leq \bar{e}_p(G) + 1$ *provided that $|G|$ is odd or $p \notin \{2\} \cup \mathfrak{M}$.*

Proof. (a) Lemmas 14.2 and 14.10 yield

$$l_p(G) \leq l_p(G/\mathbf{O}_p(G)) + 1 \leq r_p(G/\mathbf{O}_p(G)) + 1 \leq 2 \cdot \bar{e}_p(G) + 1.$$

(b) Analogous. $\qquad\square$

14.12 Remarks. Let G be solvable and p be prime.

(a) There even exist logarithmic estimate for $l_p(G)$ in terms of $r_p(G)$, as first shown by Huppert [Hu 1]. The following improvement can be found in [Wo 5]:

(1) $l_p(G) \leq 2 + \log_s(r_p(G)/(p+1))$ where $s = p - 1 + 1/p$; and

(2) $l_p(G) \leq 1 + \log_p(r_p(G))$ if $p \notin \mathfrak{F}$.

(b) Combining (a) and Lemma 14.10 together, we also get the appropriate logarithmic estimates for $l_p(G)$ in terms of $\bar{e}_p(G)$.

(c) Observe that the proof of Lemma 14.10 works exactly the same way for $e_p(G)$ instead of $\bar{e}_p(G)$. Hence if $\mathbf{O}_p(G) = 1$, then

$$r_p(G) \leq 2 \cdot e_p(G).$$

(d) Therefore, (c) together with (a) yield logarithmic bounds for $l_p(G)$ in terms of $e_p(G)$. Note that these considerably improve the linear bounds obtained in Corollary 14.8 (a). We mention in this context that the assertion of Theorem 14.3 is best possible (cf. [HB, IX, 5.4]).

As announced, we next bound $\mathrm{dl}\,(P/\mathbf{O}_p(G))$ in terms of $\bar{e}_p(G)$, where $P \in \mathrm{Syl}_p(G)$. To do so, we take advantage of Theorem 14.11.

14.13 Lemma. Let G be p-nilpotent, $P \in \mathrm{Syl}_p(G)$ and $\mathbf{O}_p(G) = 1$. Then $\mathrm{dl}\,(P) \leq \bar{e}_p(G)$.

Proof. Set $N = \mathbf{O}_{p'}(G)$. As $\mathbf{O}_p(G) = 1$, it follows that $\mathbf{C}_P(N) = 1$. By Lemma 12.2, P faithfully permutes the elements of $\mathrm{Irr}\,(N)$ and $\mathrm{IBr}_p(N)$. Let $\Omega_1, \ldots, \Omega_n \subseteq \mathrm{IBr}_p(N)$ be the P-orbits of $\mathrm{IBr}_p(N)$, set $|\Omega_i| = p^{f_i}$ and assume without loss of generality that $f_1 \geq \cdots \geq f_n$. Then $P \leq S_{p^{f_1}} \times \cdots \times S_{p^{f_n}}$. As a Sylow p-subgroup of S_{p^f} has derived length f (see [Hu, III, 15.3]), we have that $\mathrm{dl}\,(P) \leq f_1$. On the other hand, if $\theta \in \Omega_1$, then $p^{f_1} \mid \varphi(1)$ for all $\varphi \in \mathrm{IBr}_p(G|\theta)$. Therefore

$$\mathrm{dl}\,(P) \leq f_1 \leq \bar{e}_p(G). \qquad \square$$

14.14 Theorem (Wang [Wa 1]). *Let G be solvable and $P \in \mathrm{Syl}_p(G)$. Then*

$$\mathrm{dl}\,(P/\mathbf{O}_p(G)) \leq l_p(G/\mathbf{O}_p(G)) \cdot \bar{e}_p(G).$$

Proof. We may assume that $\mathbf{O}_p(G) = 1$ and argue by induction on $l_p(G)$. Write $N = \mathbf{O}_{p'}(G)$ and $M = \mathbf{O}_{p',p}(G)$. Then M is p-nilpotent with $\mathbf{O}_p(M) = 1$, and Lemma 14.13 yields $\mathrm{dl}\,(M/N) \leq \bar{e}_p(M) \leq \bar{e}_p(G)$. By induction, we also have that

$$\mathrm{dl}\,(PM/M) \leq l_p(G/M) \cdot \bar{e}_p(G/M) \leq (l_p(G) - 1) \cdot \bar{e}_p(G).$$

Consequently,

$$\mathrm{dl}\,(P) \leq \mathrm{dl}\,(PM/M) + \mathrm{dl}\,(M/N) \leq l_p(G) \cdot \bar{e}_p(G),$$

as required. $\qquad\qquad\square$

Putting together Theorem 14.14 and Lemmas 14.2 and 14.10, we obtain the next corollary. This result can be improved somewhat if Remark 14.12 (a) is used in place of Lemma 14.2.

14.15 Corollary. *Let G be solvable and $P \in \mathrm{Syl}_p(G)$. Then*

  (a) $\mathrm{dl}\,(P/\mathbf{O}_p(G)) \leq 2 \cdot \bar{e}_p(G)^2$; *and*

  (b) $\mathrm{dl}\,(P/\mathbf{O}_p(G)) \leq \bar{e}_p(G)^2$ *provided that $|G|$ is odd or $p \notin \{2\} \cup \mathfrak{M}$.*

We finish this section with a result about $e_p(P)$ for $P \in \mathrm{Syl}_p(G)$. Note that $e_p(P)$ is the exponent of the largest character degree of P. The following actually is a consequence of Theorem 7.3.

14.16 Corollary (Espuelas [Es 1]). *Let G be solvable, p an odd prime and $P \in \mathrm{Syl}_p(G)$. If p^n is the p-part of $|G/\mathbf{O}_{p',p}(G)|$, then $e_p(P) \geq n$.*

Proof. It is clearly no loss to assume that $\mathbf{O}_{p'}(G) = 1$. Therefore, $\mathbf{F}(G) = \mathbf{O}_p(G)$, and $V := \mathbf{O}_p(G)/\Phi(G)$ is a faithful $G/\mathbf{O}_p(G)$-module of characteristic p. Also $\mathrm{Irr}\,(V)$ is a faithful $G/\mathbf{O}_p(G)$-module by Proposition 12.1, and since $p \neq 2$, there exists $\lambda \in \operatorname{Irr}(V)$ such that $\mathbf{C}_P(\lambda) = \mathbf{O}_p(G)$, by Theorem 7.3. Consequently, $p^n \mid \varphi(1)$ for all $\varphi \in \operatorname{Irr}(P|\lambda)$, and $e_p(P) \geq n$ follows. $\qquad\square$

For this section, the reader might also consult [MW 2] and [Wa 1]. We also note that Isaacs [Is 1] first derived Corollary 14.7 (a) in the case $N = 1$.

§15   McKay's Conjecture

Recall that $k(B) = |B \cap \operatorname{Irr}(G)|$ for a p-block B of G. We let $k_0(B) = |\{\chi \in \operatorname{Irr}(B)| \ \chi \text{ has height zero }\}|$ and $k_0(G) = |\{\chi \in \operatorname{Irr}(G) \mid p \nmid \chi(1)\}|$.

If $P \in \operatorname{Syl}_p(G)$, the McKay conjecture states that $k_0(G) = k_0(\mathbf{N}_G(P))$. Actually the original conjecture was only for $p = 2$ and G simple. The Alperin–McKay conjecture, a refinement of this conjecture, states that $k_0(B) = k_0(b)$ where b is the Brauer correspondent of B (i.e. b is a block of $\mathbf{N}_G(D)$ for a defect group D of B and $b = b^G$). Certainly, the Alperin–McKay conjecture implies the McKay conjecture. For p-solvable G, a slight strengthening of the McKay conjecture together with Fong reduction (see Chapter 0) implies the Alperin–McKay conjecture Theorem 15.12.

Isaacs [Is 1] first proved the McKay conjecture for groups of odd order. Wolf [Wo 1] extended this to solvable G, relying heavily on work of Dade [Da 1]. The work of Isaacs and Dade involves deep analysis of solvable groups with fully ramified sections (see Proposition 12.3). Dade [Da 2] announced a proof for p-solvable, albeit long and complicated. In [OW 2], Okuyama and Wajima gave a short proof for p-solvable G, even simplifying the proof for solvable G. The proof uses the Glauberman correspondence and a counting argument discussed in the next paragraphs. Unlike Dade's proof, no correspondence is given.

Suppose $N \trianglelefteq G$ and $\theta \in \operatorname{Irr}(N)$ is G-invariant. A result of Gallagher states that $k(G|\theta)$ equals the number of "θ-good" conjugacy classes of G/N (see the next two paragraphs). This appears as Exercise 11.10 of [Is]. An equivalent count, due to Schur [Sc 1], involves twisted group algebras. A modular version of Schur's result appears in [AOT]. Isaacs [Is 8] translated this to Brauer characters and showed this works even for "π-Brauer" characters. We will give a proof of this counting argument for Brauer characters. Of course, this leads to questions as to whether McKay's conjecture holds for Brauer characters. The answer is yes for solvable, even p-solvable G, but not arbitrary G. These questions have been studied extensively in [Wo 7] and are discussed below in Section 23.

Suppose $N \trianglelefteq G$ and $\varphi \in \mathrm{IBr}_p(N)$ is G-invariant. Let $g \in G$ and choose $\beta \in \mathrm{IBr}_p(\langle N, g \rangle)$ such that β extends φ (see Proposition 0.11). We say that g is φ-good if $\beta^x = \beta$ whenever $[x, g] \in N$, i.e. if β is invariant in C where $C/N = \mathbf{C}_{G/N}(g)$. If $\beta_1 \in \mathrm{IBr}_p(\langle N, g \rangle)$ also extends φ, then $\beta_1 = \lambda\beta$ for a unique linear $\lambda \in \mathrm{IBr}_p(\langle N, g \rangle/N)$ by Lemma 0.9. Since $\langle N, g \rangle/N$ is central in C/N, λ is C-invariant and $I_C(\beta_1) = I_C(\beta)$. Hence, the definition for g to be φ-good is independent of the choice of extension $\beta \in \mathrm{IBr}_p(\langle N, g \rangle)$ of φ.

It is clear from the definition that g is φ-good if and only if ng is φ-good whenever $n \in N$. For convenience, we also refer to $Ng \in G/N$ as being φ-good. Furthermore, for $y \in G$, it is easy to see that g is φ-good if and only if g^y is φ-good. Consequently, we will refer to φ-good conjugacy classes of G/N. Recall G^0 is the set of p-regular elements of G and that $\mathrm{cf}^0(G)$ is set of class functions on G^0.

15.1 Lemma. *Suppose that $N \trianglelefteq G$ and $\varphi \in \mathrm{IBr}_p(N)$ is G-invariant. Then there exists a right transversal T for N in G and $\sigma : G^0 \to \mathbb{C}$ such that $1 \in T$ and*

(i) *If Nt is p-regular, then t is p-regular;*

(ii) *$\sigma(t) = 1$ whenever $t \in T$ is p-regular;*

(iii) *If $\psi \in \mathrm{cf}^0(G|\varphi)$ and $g \in G$ is p-regular, then $\psi(g) = \sigma(g)\psi(t)$ where $t \in Ng \cap T$.*

Note: While σ is dependent upon T and φ, it is independent of the choice *of ψ. Of course, σ need not be a class function.*

Proof. Whenever Nx is a p-regular element of G/N, then Nx contains a p-regular element of G. Observe that in order to prove (iii), it suffices to prove that $\eta(g) = \sigma(g)\eta(t)$ whenever $\eta \in \mathrm{IBr}_p\left(\langle N, g\rangle | \varphi\right)$. Thus it involves no loss of generality to assume that $G/N = \langle Ng\rangle$ is a cyclic p'-group and g is p-regular. We then choose $t \in Ng \cap T$, define σ on $(Ng)^0$ and show that (i), (ii), and (iii) hold. For the coset N, we let $1 \in T$ and set $\sigma(n) = \varphi(n)/\varphi(1)$. Without loss of generality, $G > N$. Let $\theta \in \mathrm{Irr}(G)$ extend φ.

If θ vanishes on every p-regular element of Ng (this does not actually happen, as we shall see in the next corollary), we let t be any p-regular element of Ng and define $\sigma(g) = 1$. Certainly $\theta(g) = \sigma(g)\theta(t)$ in this case. Otherwise, we choose $t \in Ng$ so that t is p-regular and $\theta(t) \neq 0$. We let $\sigma(g) = \theta(g)/\theta(t)$. In all cases, $\theta(g) = \sigma(g)\theta(t)$.

Now let $\psi \in \mathrm{IBr}_p(G|\varphi)$. Since G/N is cyclic, $\psi = \beta\theta$ for a linear $\beta \in \mathrm{IBr}_p(G/N)$. Now $\psi(g) = \beta(g)\theta(g) = \beta(t)\sigma(g)\theta(t) = \sigma(g)\psi(t)$, as desired. This proves the lemma. $\quad\square$

Of course, the value $\sigma(g)$ is dependent upon the choice of $t \in Ng \cap T$. But it is not dependent upon the choice of the extension $\theta \in \mathrm{IBr}_p(G|\varphi)$. This is clear from the last paragraph.

15.2 Corollary. *Assume the notation of Lemma 15.1. Then*

(i) *For $n \in N$, $\sigma(n) = \varphi(n)/\varphi(1)$;*

(ii) *If $Ng = Nt$ with g and t p-regular and $t \in T$, then $\sigma(g) = \mu(g)/\mu(t)$ for every extension $\mu \in \mathrm{IBr}_p(\langle N, g\rangle)$ of φ. In particular $\mu(t) \neq 0$.*

Proof. Part (i) is immediate from Lemma 15.1 (iii) and the fact that $1 \in T$. By Lemma 15.1, we have that $\psi(x) = \sigma(x)\psi(s)$ whenever $x \in G$ is p-regular, $s \in T$, $x \in Ns$, and $\psi \in \mathrm{cf}^0(G|\varphi)$. To complete the proof, we may fix a p-regular element $t \in T$ and $\mu \in \mathrm{IBr}_p(\langle N, t\rangle | \varphi)$. It suffices to show $\mu(t) \neq 0$.

We may assume without loss of generality that $G = \langle N, t \rangle$, so that G/N is a cyclic p'-group. Assume that $\mu(t) = 0$. By Proposition 0.11 and Lemma 0.9, $\tau(t) = 0$ for all $\tau \in \mathrm{IBr}_p(G|\varphi)$. Hence $\eta(t) = 0$ for all $\eta \in \mathrm{cf}^0(G|\varphi)$. Now restriction $\eta \mapsto \eta_T$ defines a vector space homomorphism from $\mathrm{cf}^0(G|\varphi)$ into the vector space X of complex-valued functions on T. This is not onto, because $\eta(t) = 0$ for all $\eta \in \mathrm{cf}^0(G|\varphi)$. We do claim this map is 1–1. Assume that $\beta_T \equiv 0$ for some $\beta \in \mathrm{cf}^0(G|\varphi)$. If $x \in G$ is p-regular, then $\beta(x) = \sigma(x)\beta(s)$ for some $s \in T$ and $\beta(x) = 0$. Thus $\beta \equiv 0$ and the restriction map is 1–1. Hence $|G/N| = |\mathrm{IBr}_p(G|\varphi)| = \dim(\mathrm{cf}^0(G|\varphi)) < \dim(X) = |T| = |G/N|$. By this contradiction, $\mu(t) \neq 0$, as desired. $\square$

15.3 Theorem. *Let* G, N, φ, T *and* σ *be as in Lemma 15.1. Let* $\chi \in \mathrm{cf}^0(G)$. *The following are equivalent:*

(i) $\chi \in \mathrm{cf}^0(G|\varphi)$; *and*

(ii) *Whenever* $g \in Nt$ *for* $t \in T$ *and* g *is* p-regular, then $\chi(g) = \sigma(g)\chi(t)$.

Proof. By Lemma 15.1, (i) implies (ii). Assume (ii). Let $n \in N$ be p-regular. Since $1 \in T$, we have that $\chi(n) = \sigma(n)\chi(1)$. By Corollary 15.2, $\sigma(n) = \varphi(n)/\varphi(1)$ and so $\chi(n) = (\chi(1)/\varphi(1))\, \varphi(n)$. Hence $\chi_N = [\chi(1)/\varphi(1)]\varphi$, as desired. $\square$

15.4 Lemma. *Suppose that* $N \trianglelefteq G$ *and* $\varphi \in \mathrm{IBr}_p(N)$ *is* G-*invariant. If* $g \in G$ *is* p-regular *and* Ng *is not* φ-good, then $\psi(g) = 0$ *for all* $\psi \in \mathrm{cf}^0(G|\varphi)$.

Proof. Since Ng is not φ-good, there exists $\mu \in \mathrm{IBr}_p(\langle N, g \rangle|\varphi)$ that is not invariant in C, where $C/N = \mathbf{C}_{G/N}(Ng)$. Without loss of generality $G = C$. Set $Z = \langle N, g \rangle$ so that $Z/N \leq \mathbf{Z}(G/N)$. We may further assume that $\psi \in \mathrm{IBr}_p(G|\varphi)$. Since $\mu \in \mathrm{IBr}_p(Z)$ is not G-invariant and $Z/N \leq \mathbf{Z}(G/N)$, no extension of φ to Z is G-invariant. Thus we may assume that $\psi \in \mathrm{IBr}_p(G|\mu)$.

For $x \in G$, $\mu^x = \lambda_x \mu$ for a unique linear $\lambda_x \in \mathrm{IBr}_p(Z/N)$. Because $Z/N \leq \mathbf{Z}(G/N)$, $\lambda_{xy}\mu = \mu^{xy} = (\lambda_x\mu)^y = \lambda_x\lambda_y\mu$ and $\lambda_{xy} = \lambda_x\lambda_y$. Thus the

G-orbit of μ is $\{\lambda\mu \mid \lambda \in K\}$ for some subgroup $K \leq \mathrm{IBr}_p(Z/N)$. Because μ is not G-invariant, $K \neq 1$. Because K is a subgroup of the group of linear characters $\mathrm{IBr}_p(Z/N) = \mathrm{Irr}\,(Z/N)$, then $K = \mathrm{Irr}\,(Z/M)$ for some subgroup $N \leq M < Z$ (e.g. see [Hu, V, 6.4]). Thus ψ_Z is a multiple of $\rho_{Z/M}\mu$, where $\rho_{Z/M}$ is the regular character of Z/M. Because $N \leq M < Z = \langle N, g \rangle$ we have that $\psi(g) = \rho_{Z/M}(g)\mu(g) = 0$.   □

The second paragraph of the above proof repeats an argument in Lemma 12.6. That lemma could be used here at least for ordinary characters as it is possible to reduce to the case where G/N is abelian (i.e. $G = \langle N, g, x \rangle$ with $[x, g] \in N$ and $\mu^x \neq \mu$).

15.5 Proposition. *Assume the notation of Lemma 15.1. Let $g \in G$ and $t \in T$ be p-regular with $Ng = Nt$. Then*

(i) $\sigma(g^x) = \sigma(g)\sigma(t^x)$ *for $x \in G$.*

(ii) *If t is φ-good and if nt and mt are G-conjugate and p-regular with $n, m \in N$, then $\sigma(nt) = \sigma(mt)$.*

Proof. (i) Now $g^x, t^x \in Ns$ for a unique $s \in T$. Also s is p-regular. Let $\mu \in \mathrm{IBr}_p(\langle N, t \rangle)$ be an extension of φ. Then μ^x is an extension of φ to $\langle N, s \rangle = \langle N, t \rangle^x$. Applying Corollary 15.2 thrice,

$$\mu^x(g^x) = \mu(g) = \sigma(g)\mu(t),$$

$$\mu^x(t^x) = \sigma(t^x)\mu^x(s), \quad \text{and}$$

$$\mu^x(g^x) = \sigma(g^x)\mu^x(s).$$

Consequently, $\sigma(g^x)\mu^x(s) = \sigma(g)\mu(t) = \sigma(g)\mu^x(t^x) = \sigma(g)\sigma(t^x)\mu^x(s)$. Since $\mu^x(s) \neq 0$ (again Corollary 15.2), part (i) follows.

(ii) Choose $h \in G$ with $(nt)^h = mt$. Then h and t commute mod N. Because t is φ-good, μ must be $\langle h \rangle$-invariant. Now $\mu(nt) = \mu^h((nt)^h) = \mu^h(mt) = \mu(mt)$. Applying Corollary 15.2 (ii),

$$\sigma(nt) = \mu(nt)/\mu(t) = \mu(mt)/\mu(t) = \sigma(mt). \qquad \square$$

15.6 Theorem. Let $N \trianglelefteq G$ and $\varphi \in \mathrm{IBr}_p(N)$ be G-invariant. Then $|\mathrm{IBr}_p(G|\varphi)|$ equals the number of φ-good conjugacy classes of p-regular elements of G/N.

Proof. Choose a transversal T for N in G with T as in Lemma 15.1. Now choose a subset $S \subseteq T$ such that

(i) Each $s \in S$ is p-regular.

(ii) Each $s \in S$ is φ-good.

(iii) If Ng is p-regular and φ-good, then Ng is conjugate to (exactly) one Ns, $s \in S$.

Consequently $|S|$ equals the number of φ-good conjugacy classes of p-regular elements of G/N. Because $\mathrm{IBr}_p(G)$ is a basis for the $\mathbb{C}$-vector space $\mathrm{cf}^0(G)$, indeed $\mathrm{IBr}_p(G|\varphi)$ is a basis for $\mathrm{cf}^0(G|\varphi)$ and $|\mathrm{IBr}_p(G|\varphi)| = \dim(\mathrm{cf}^0(G|\varphi))$. If $V = \{f : S \to \mathbb{C}\}$, then $\dim(V) = |S|$. Consequently, it suffices to show that the restriction $\psi \mapsto \psi_S$ is an isomorphism from $\mathrm{IBr}_p(G|\varphi)$ onto V. Trivially it is a homomorphism.

Suppose that $\beta \in \mathrm{cf}^0(G|\varphi)$ and $\beta_S \equiv 0$. We wish to show that $\beta(g) = 0$ for all p-regular $g \in G$. By Lemma 15.4, we may assume that g is φ-good. By (iii) above, we may assume that $g = ns$ for some $s \in S$. By Theorem 15.3, $\beta(g) = \sigma(g)\beta(s) = 0$. Hence $\beta \equiv 0$ and the restriction map $\psi \mapsto \psi_S$ is one-to-one.

Now let $\alpha : S \to \mathbb{C}$. To complete the proof, we must show that there exists $\chi \in \mathrm{cf}^0(G|\varphi)$ with $\chi_S = \alpha$. We define χ as follows. Fix $x \in G$ p-regular. If x is not φ-good, we let $\chi(x) = 0$. If x is φ-good, there is a unique $s \in S$ such that x is conjugate to ns for some $n \in N$ (by (iii) above). In this case, we let $\chi(x) = \sigma(ns)\alpha(s)$ where σ is as in Lemma 15.1. If x is also conjugate to ms with $m \in N$, then Proposition 15.5 (ii) implies that $\sigma(ns) = \sigma(ms)$, because s is φ-good. Hence χ is well-defined. For $s \in S$, $\sigma(s) = 1$ by Lemma 15.1 (ii) and so $\chi(s) = 1 \cdot \alpha(s)$. Thus $\chi_S = \alpha$.

Suppose $y \in G$ is conjugate to x. Either both x and y are φ-good or both are not. In the latter case, $\chi(x) = 0 = \chi(y)$. In the former case, both x and y are conjugate to some ns with $n \in N$, $s \in S$. Then $\chi(x) = \chi(y)$ by definition of χ. So $\chi \in \mathrm{cf}^0(G)$. What needs to be shown is that $\chi \in \mathrm{cf}^0(G|\varphi)$.

Fix $g \in G$ p-regular and $t \in T$ with $g \in Nt$. By Theorem 15.3, it suffices to show that $\chi(g) = \sigma(g)\chi(t)$. If g is not φ-good, neither is t and $\chi(g) = 0 = \chi(t)$ in this case. We thus assume that g is φ-good and choose $x \in G$, $s \in S$ such that $g^x \in Ns$. Note that $t^x \in Ns$. By definition of χ, $\chi(g) = \sigma(g^x)\alpha(s)$ and $\chi(t) = \sigma(t^x)\alpha(s)$. By Proposition 15.5, $\sigma(g^x) = \sigma(g)\sigma(t^x)$. Thus $\chi(g) = \sigma(g)\sigma(t^x)\alpha(s) = \sigma(g)\chi(t)$, as desired. $\qquad\square$

Let $N \trianglelefteq G$ and $\theta \in \mathrm{Irr}\,(N)$. Then we let $k(G|\theta) = |\mathrm{Irr}\,(G|\theta)|$ and of course $k(G) = |\mathrm{Irr}\,(G)|$. Finally, we let $k_0(G|\theta) = |\{\chi \in \mathrm{Irr}\,(G|\theta) \mid p \nmid \chi(1)/\theta(1)\}|$.

15.7 Corollary. *Suppose that G/N is abelian and $\theta \in \mathrm{Irr}\,(N)$. Then θ extends to G if and only if $k(G|\theta) = |G/N|$.*

Proof. If $I_G(\theta) < G$, then θ does not extend to G and

$$k(G|\theta) = k(I_G(\theta) \mid \theta) \le k(I_G(\theta)/N) < |G/N|.$$

So we assume θ to be G-invariant. Now Lemma 12.6 shows there exists $N \le M \le G$ such that each $\tau \in \mathrm{Irr}\,(M|\theta)$ extends θ and is fully ramified with respect to G/M. Each $\chi \in \mathrm{Irr}\,(G|\theta)$ has degree $|G : M|^{1/2} \cdot \theta(1)$ and $k(G|\theta) = |M : N|$. Thus θ extends to G if and only if $M = G$, or equivalently $k(G|\theta) = |G/N|$. $\qquad\square$

15.8 Lemma. *Suppose that $M, K \trianglelefteq G$ with K and G/M p'-groups and M/K a p-group. Assume that G/M is abelian. Let $P \in \mathrm{Syl}_p(G)$ and set $C = \mathbf{C}_K(P)$. If $\theta \in \mathrm{Irr}\,(K)$ and $\beta = \theta\rho(K, P) \in \mathrm{Irr}(C)$ is the Glauberman correspondent of θ, then θ extends to G if and only if β extends to $\mathbf{N}_G(P)$.*

Proof. We argue by induction on $|G : K||P|$. If $P = 1$, then $C = K$ and $\beta = \theta$. Thus we may assume that $M > K$.

Set $H = \mathbf{N}_G(P)$. Observe that $M = KP$, $G = KH$ and $K \cap H = C$. Also $G/K \cong H/C$ and $M \cap H = C \times P$. Let $V \in \text{Hall}_{p'}(H)$. Then $C \leq V$ and $VK/K \in \text{Hall}_{p'}(G/K)$. Since $p \nmid |K|$, θ extends to M (see Theorem 0.13). By Proposition 0.12, θ extends to G if and only if θ extends to KV. Similarly, β extends to H if and only if β extends to V.

Let S/K be a minimal normal subgroup of G/K with $S \leq M$. Let $Q = S \cap P \cong S/K$ and $D = \mathbf{C}_K(Q)$. By Theorem 0.15, $\theta\rho(K,Q)\rho(D,P/Q) = \beta$. If $S < M$, we apply the inductive hypothesis twice to conclude that θ extends to SV if and only if $\theta\rho(K,Q)$ extends to DQV if and only if β extends to QV. By the last paragraph, θ extends to G if and only if β extends to H. We may thus assume $S = M$, i.e. M/K is a minimal normal subgroup of G/K. In particular, M/K is an elementary abelian p-group and an irreducible G/M-module.

Assume that $K = \mathbf{O}_{p'}(G)$. Then M/K is a faithful irreducible G/M-module. Since G/M is abelian, in fact G/M is cyclic by Lemma 0.5. Since $V/C \cong VK/K \cong G/M$, we have that θ extends to VK and β extends to V by Proposition 0.11. By the second paragraph, θ extends to G and β extends to H. We are done in this case. Letting $N = \mathbf{O}_{p'}(G)$, we thus assume that $N > K$.

Now $N/K = \mathbf{O}_{p'}(G/K)$ and is centralized by P. Also $N \cap H = \mathbf{O}_{p'}(H) = \mathbf{C}_H(P) = \mathbf{C}_N(P)$ and $N \cap H/C \cong N/K$. Because P centralizes N/K, every $\gamma \in \text{Irr}(N|\theta)$ is P-invariant by Lemma 0.17, and $\rho(N,P)$ maps $\text{Irr}(N|\theta)$ onto $\text{Irr}(N \cap H|\beta)$, by Lemma 0.16. Since $\rho(N,P)$ is 1–1, $|\text{Irr}(N|\theta)| = |\text{Irr}(N \cap H|\beta)|$. We may assume that θ extends to H or β extends to $N \cap H$, since the theorem is trivially true otherwise. Since $N/K \cong N \cap H/C \lesssim G/M$ and is abelian, it follows that

$$|N/K| = |\text{Irr}(N|\theta)| = |\text{Irr}(N \cap H \mid \beta)| = |N \cap H/C|.$$

Now Corollary 15.7 implies that both θ extends to N and β extends to $N \cap H$. Since N/K is abelian, we can apply the inductive hypothesis to

G/N to conclude that

$$\theta \text{ extends to } G \Longleftrightarrow \text{ some } \sigma \in \operatorname{Irr}(N|\theta) \text{ extends to } G$$

$$\Longleftrightarrow \text{ some } \tau \in \operatorname{Irr}(N \cap H|\beta) \text{ extends to } H$$

$$\Longleftrightarrow \beta \text{ extends to } H. \qquad \square$$

We now put the main ingredients, Theorem 15.6 and Lemma 15.8, together to get McKay's conjecture for p-solvable groups. While Theorem 15.9 (iv) is not the most general statement, it can be used with routine arguments to deduce Theorem 15.10 and it can be used with Fong reduction to prove the Alperin–McKay conjecture for p-solvable groups. The hypothesis in Theorem 15.9 that φ extends to P is met if $k_0(G|\varphi) \neq 0$ or $k_0(H|\varphi) \neq 0$.

15.9 Theorem. *Suppose that $L \leq K \leq M \leq G$ with $L, K, M \trianglelefteq G$ and $\varphi \in \operatorname{Irr}(L)$ is G-invariant. Assume K/L and G/M are p'-groups and M/K is a p-group. Let $P/L \in \operatorname{Syl}_p(G/L)$, $C/L = \mathbf{C}_{K/L}(P)$ and assume that φ extends to P. Then*

(i) *There is a bijection from $\{\theta \in \operatorname{Irr}(K|\varphi) \mid \theta$ is P-invariant $\}$ onto $\operatorname{Irr}(C|\varphi)$ given by $\theta \longleftrightarrow \beta$ if and only if $[\theta_C, \beta] \not\equiv 0 \pmod{p}$.*

(ii) *The map in (i) is preserved by conjugation by $H := \mathbf{N}_G(P)$.*

(iii) *Assume that $M \leq A \leq G$ with A/M abelian. If $\theta \longleftrightarrow \beta$ is as in part (i), then θ extends to A if and only if β extends to $H \cap A$.*

(iv) *$k_0(G|\theta) = k_0(H|\beta)$ whenever $\theta \in \operatorname{Irr}(K|\varphi)$ is P-invariant and $\theta \longleftrightarrow \beta \in \operatorname{Irr}(C)$.*

(v) *If, in addition, $p \nmid \varphi(1)o(\varphi)$, then also $p \nmid o(\theta)\theta(1)o(\beta)\beta(1)$. In particular θ and β have canonical extensions $\hat{\theta} \in \operatorname{Irr}(M)$ and $\hat{\beta} \in \operatorname{Irr}(CP)$. If $\lambda \in \operatorname{Irr}(M/K)$ is linear, then $k(G|\lambda\hat{\theta}) = k(H|\lambda_{CP}\hat{\beta})$.*

Proof. Since $H = \mathbf{N}_G(P)$ and $M = KP \trianglelefteq G$, the Frattini argument shows that $G = KH$ and $C = H \cap K$. Also $G/K \cong H/C$. If $\theta \in \mathrm{Irr}(K|\varphi)$ is P-invariant and $h \in H$, then θ^h is also P-invariant and $(\theta^h)_C = (\theta_C)^h$. Part (ii) follows from part (i).

(iii) $\implies$ (v) For convenience (which should become apparent further along in the proof), we next show that (iii) $\implies$ (v). Since $p \nmid \varphi(1)$, clearly, $p \nmid \theta(1)\beta(1)$ because K/L is a p'-group. Now $\theta_L = e\varphi$ for a p'-integer e and so $\det(\theta_L) = (\det\varphi)^e$ has p'-order. Since K/L is a p'-group, $o(\theta)$ is p'. Likewise, $p \nmid o(\beta)$. By Theorem 0.13, we let $\hat\theta$ and $\hat\beta$ be the canonical extensions of θ to M and β to CP.

We have $\lambda \in \mathrm{Irr}(M/K)$ is linear and we let $J = I_G(\lambda\hat\theta) \geq M$. Since $\lambda\hat\theta$ extends θ, we have that $I_G(\lambda\hat\theta) \leq I_G(\theta) = I_G(\hat\theta)$. So $J = I_G(\hat\theta) \cap I_G(\lambda)$. Similarly, $I_H(\lambda_{CP}\hat\beta) = I_H(\lambda_{CP}) \cap I_H(\hat\beta) = I_H(\lambda_{CP}) \cap I_H(\hat\theta)$. Hence $J \cap H = I_H(\lambda_{CP}\hat\beta)$. Also $M(J \cap H) = J$ and $J/M \cong J \cap H/CP$. We need to show that $k(J|\lambda\hat\theta) = k(J \cap H|\lambda_{CP}\hat\beta)$. To this end, it suffices to show that $j \in J \cap H$ is $\lambda_{CP}\hat\beta$-good if and only if j is a $\lambda\hat\theta$-good element of J. Note that j is $\lambda\hat\theta$-good if and only if $\lambda\hat\theta$ extends to $\langle M, j, x\rangle$ whenever $x \in J$ and $[j, x] \in M$. It suffices to show that whenever $M \leq A \leq J$ with A/M abelian, then $\lambda\hat\theta$ extends to A if and only if $\lambda_{CP}\hat\beta$ extends to $A \cap H$.

Suppose then $\lambda\hat\theta$ extends to A. In particular, θ extends to A and part (iii) implies that β extends to $\xi \in \mathrm{Irr}(H \cap A)$. Now $\xi_{CP} = \alpha\hat\beta$ for an $A \cap H$-invariant and linear $\alpha \in \mathrm{Irr}(CP/C)$. Since $\alpha, \lambda_{CP} \in \mathrm{Irr}(CP/C)$ are invariant in $A \cap H$ and since $(|A \cap H/CP|, |CP/C|) = 1$, both α and λ_{CP} extend to $A \cap H$. Say $\alpha_1, \lambda_1 \in \mathrm{Irr}(A \cap H)$ extend α and λ_{CP} (respectively). Then $\lambda_1\alpha_1^{-1}\xi \in \mathrm{Irr}(A \cap H)$ extending $\lambda_{CP}\hat\beta$. So we have shown that $\lambda_{CP}\hat\beta$ extends to $A \cap H$ whenever $\lambda\hat\theta$ extends. The proof of the converse is essentially identical. So (iii) implies (v).

(i), (iii), (iv). Since φ is G-invariant, it is no loss of generality to assume that φ is linear, via use of a character triple isomorphism (see [Is, Theorem 11.28]). Now $\varphi = \alpha\sigma$, for linear $\alpha, \sigma \in \mathrm{Irr}(L)$ with $o(\alpha)$ a p'-number and

$o(\sigma)$ a p-power. Since φ extends to P and φ is linear, σ also extends to P. By Proposition 0.12, σ extends to $\gamma \in \mathrm{Irr}\,(G)$. For $L \leq J \leq G$, the mapping $\tau \to \gamma_J^{-1}\,\tau$ is a 1–1 degree-preserving map $\mathrm{Irr}\,(J|\varphi)$ onto $\mathrm{Irr}\,(J|\alpha)$. Thus it is without loss of generality to assume that $\varphi = \alpha$, i.e. that φ is linear and $p \nmid o(\varphi)$. We may also assume that φ is faithful, so that now L and K are p'-groups. We now employ the Glauberman correspondence (Theorem 0.15) to prove (i) and Lemma 15.8 to prove (iii).

Since we now have that $p \nmid |K|$, we have canonical extensions $\hat\theta \in \mathrm{Irr}\,(M)$ of θ and $\hat\beta \in \mathrm{Irr}\,(CP)$ of $\hat\beta$. Repeating the argument of (iii) $\Longrightarrow$ (v), we have that $k(G|\lambda\hat\theta) = k(H|\lambda_{CP}\hat\beta)$ for all linear $\lambda \in \mathrm{Irr}\,(M/K)$. We may choose linear $\lambda_1, \ldots, \lambda_k \in \mathrm{Irr}\,(M/K)$ so that each linear character of M/K is H-conjugate to exactly one λ_i, $1 \leq i \leq k$. Since $G = KH$, we have that

$$\{\chi \in \mathrm{Irr}\,(G|\theta) \mid p \nmid \chi(1)\} = \mathrm{Irr}\,(G|\lambda_1\hat\theta) \,\dot{\cup} \cdots \dot{\cup}\, \mathrm{Irr}\,(G|\lambda_k\hat\theta).$$

Now restriction gives a bijection from $\mathrm{Irr}\,(M/K)$ onto $\mathrm{Irr}\,(CP/C)$ and each linear character of CP/C is H-conjugate to exactly one $(\lambda_i)_{CP}$, $1 \leq i \leq k$. So

$$\{\psi \in \mathrm{Irr}\,(H|\beta) \mid p \nmid \psi(1)\} = \mathrm{Irr}\,(H|(\lambda_1)_{CP}\hat\beta) \,\dot{\cup} \cdots \dot{\cup}\, \mathrm{Irr}\,(H|(\lambda_k)_{CP}\hat\beta).$$

Since $k(G|\lambda_i\hat\theta) = k(H|(\lambda_i)_{CP}\hat\beta)$ for each i, $k_0(G|\theta) = k_0(H|\beta)$. $\qquad\square$

15.10 Theorem. *Suppose that G/L is p-solvable and $\varphi \in \mathrm{Irr}\,(L)$ is P-invariant where $P/L \in \mathrm{Syl}_p(G/L)$. Set $H/L = \mathbf{N}_{G/L}(P/L)$. Then $k_0(G|\varphi) = k_0(H|\varphi)$.*

Proof. By induction on $|G : L|$. The result is trivially true if $P \trianglelefteq G$. Without loss of generality, $H < G$.

Let $I = I_G(\varphi)$. Then $P \leq H \cap I \leq I$. Since $|G : I||H : H \cap I|$ is a p'-number, the Clifford correspondence yields that $k_0(G|\varphi) = k_0(I|\varphi)$ and $k_0(H|\varphi) = k_0(H \cap I|\varphi)$. But $H \cap I/L = \mathbf{N}_{I/L}(P/L)$. If $I < G$, the inductive hypothesis implies that $k_0(I|\varphi) = k_0(H \cap I|\varphi)$. Then $k_0(G|\varphi) = k_0(H|\varphi)$, as desired. We thus assume that $I_G(\varphi) = G$.

We next let K/L be a chief factor of G and set $J = \mathbf{N}_G(KP/K)$. Suppose that θ, $\mu \in \mathrm{Irr}\,(K|\varphi)$ are P-invariant. We claim that θ and μ are G-conjugate if and only if they are J-conjugate. Indeed, assume that $\theta = \mu^g$ for some $g \in G$. Then P/L, $P^g/L \in \mathrm{Syl}_p(I_G(\theta)/L)$ and so $P^g = P^i$ for some $i \in I_G(\theta)$. Then $ig^{-1} \in \mathbf{N}_G(P) \le J$ and $\theta^{ig^{-1}} = \theta^{g^{-1}} = \mu$. The claim follows. Further note that if α, $\beta \in \mathrm{Irr}\,(K|\varphi)$ are J-conjugate, then α is P-invariant if and only if β is P-invariant.

Now we may choose $\theta_1, \ldots, \theta_t \in \mathrm{Irr}\,(K|\varphi)$ such that each θ_i is P-invariant and such that each P-invariant $\mu \in \mathrm{Irr}\,(K|\varphi)$ is J-conjugate to exactly one θ_j. Furthermore, we may assume there exists $0 \le k \le t$ such that $p \nmid \theta_j(1)/\varphi(1)$ if and only if $j \le k$. If $\chi \in \mathrm{Irr}\,(G|\varphi)$ and $p \nmid \chi(1)/\varphi(1)$, it follows from the last paragraph that $\chi \in \mathrm{Irr}\,(G|\theta_j)$ for a unique $j \le k$. Similarly, if $\psi \in \mathrm{Irr}\,(J|\varphi)$ and $p \nmid \psi(1)/\varphi(1)$, then $\psi \in \mathrm{Irr}\,(J|\theta_i)$ for a unique $i \le k$. The inductive hypothesis yields that $k_0(G|\theta_j) = k_0(J|\theta_j)$ for all $j \le k$. Hence $k_0(G|\varphi) = k_0(J|\varphi)$. If $J < G$, the inductive hypothesis implies that $k_0(J|\varphi) = k_0(H|\varphi)$ and hence $k_0(G|\varphi) = k_0(H|\varphi)$. We may thus assume $J = G$, i.e. G/K has a normal Sylow p-subgroup.

We may assume that $k_0(G|\varphi) \ne 0$ or $k_0(H|\varphi) \ne 0$. In either case, there exist $P \le N \le G$ and $\eta \in \mathrm{Irr}\,(N|\varphi)$ such that $p \nmid \eta(1)/\varphi(1)$. Since φ is G-invariant, there exists $\alpha \in \mathrm{Irr}\,(P|\varphi)$ with $[\eta_P, \alpha] \ne 0$ and $p \nmid \alpha(1)/\varphi(1)$. Since P/K is a p-group, φ extends to $\alpha \in \mathrm{Irr}(P)$.

If K/L is a p-group, then G/L has a normal Sylow p-subgroup and $H = G$, a contradiction. Thus K/L is a p'-group. In particular, $k = t$, i.e. $p \nmid \theta_i(1)/\varphi(1)$ for all i, $1 \le i \le t$. We have that each P-invariant $\mu \in \mathrm{Irr}\,(K|\varphi)$ is G-conjugate to exactly one θ_i $(1 \le i \le t)$. The Frattini argument shows that $G = KH$ and so each P-invariant $\mu \in \mathrm{Irr}\,(K|\varphi)$ is H-conjugate to exactly one θ_i. Let $C = K \cap H$ and note that $C/L = \mathbf{C}_{K/L}(P)$. By Theorem 15.9 (i), there exist $\beta_1, \ldots, \beta_t \in \mathrm{Irr}\,(C|\varphi)$ such that $[(\theta_i)_C, \beta_i] \not\equiv 0$

(mod p) and each $\tau \in \mathrm{Irr}(C|\varphi)$ is H-conjugate to exactly one β_i (note that Lemma 0.17 shows that every $\tau \in \mathrm{Irr}(C|\varphi)$ is P-invariant). If $\chi \in \mathrm{Irr}(G|\varphi)$ and $p \nmid \chi(1)$, then $\chi \in \mathrm{Irr}(G|\theta_j)$ for a unique j (see above). Thus $k_0(G|\varphi) = \sum_{i=1}^{t} k_0(G|\theta_i)$ and $k_0(H|\varphi) = \sum_{i=1}^{t} k_0(H|\beta_i)$. Since φ extends to P, then $k_0(G|\theta_i) = k_0(H|\beta_i)$ for all i by Theorem 15.9 (iv). Hence $k_0(G|\varphi) = k_0(H|\varphi)$. $\qquad\square$

Applying Theorem 15.10 with $L = 1$, we immediately get McKay's conjecture for p-solvable groups.

15.11 Corollary. *If G is p-solvable and $P \in \mathrm{Syl}_p(G)$, then $k_0(G) = k_0(\mathbf{N}_G(P))$.*

Next, we deduce the more refined Alperin–McKay conjecture for p-solvable G.

15.12 Theorem. *Let B be a p-block of a p-solvable group G. Let D be a defect group of B and let $b \in \mathrm{bl}(\mathbf{N}_G(D))$ be the Brauer correspondent of B. Then $k_0(B) = k_0(b)$.*

Proof. Argue by induction on $|G : \mathbf{O}_{p'}(G)|$. Let $K = \mathbf{O}_{p'}(G)$. We may choose $\varphi \in \mathrm{Irr}(K)$ covered by B so that $D \leq I := I_G(\varphi)$ (see Proposition 0.22 and Lemma 0.25). Applying Corollary 0.30 and Lemma 0.25, there exist blocks B_0 of I and b_0 of $I \cap \mathbf{N}_G(D)$ such that $k_0(B) = k_0(B_0)$, $k_0(b) = k_0(b_0)$, and b_0 is the Brauer correspondent of B_0. If $I < G$, the inductive hypothesis implies that $k_0(B_0) = k_0(b_0)$. Then $k_0(B) = k_0(b)$, as desired. Thus we assume that φ is G-invariant.

By Theorem 0.28, $D \in \mathrm{Syl}_p(G)$ and B is the unique p-block covering $\{\varphi\}$. Let $\mu = \varphi\rho(K, D) \in \mathrm{Irr}(\mathbf{C}_K(D))$ be the Glauberman correspondent of φ. By Theorem 0.29, b is the unique p-block of $\mathbf{N}_G(D)$ covering $\{\mu\}$. Thus $k_0(B) = k_0(G|\varphi)$ and $k_0(b) = k_0(\mathbf{N}_G(D)|\mu)$, as $D \in \mathrm{Syl}_p(G)$. By Theorem 15.10, $k_0(G|\varphi) = k_0(K\mathbf{N}_G(D)|\varphi)$. By Theorem 15.9 (iv), $k_0(K\mathbf{N}_G(D)|\varphi) =$

$k_0(\mathbf{N}_G(D)|\mu)$. Hence $k_0(B) = k_0(b)$. $\qquad\qquad\qquad\qquad\qquad\square$

We remark that both the McKay conjecture and Alperin–McKay conjecture remain open for arbitrary G. They have been verified for certain families of groups. But unlike some conjectures, there is no known method to reduce these questions to simple groups.

We let $l(G) = |\mathrm{IBr}_p(G)|$ and $l_0(G) = |\{\varphi \in \mathrm{IBr}_p(G) \mid p \nmid \varphi(1)\}|$. In light of the above results, one might ask whether $l_0(G) = l_0(\mathbf{N}_G(P))$ when $P \in \mathrm{Syl}_p(G)$. While this is not true for arbitrary G, it is true for p-solvable G and we give a proof below in Section 23 and we will discuss there a number of related questions.

Chapter V
COMPLEXITY OF CHARACTER DEGREES

§16   Derived Length and the Number of Character Degrees

We let $\mathrm{cd}(G) = \{\chi(1) | \chi \in \mathrm{Irr}(G)\}$. I. M. Isaacs proved that if $|\mathrm{cd}(G)| \leq 3$, then G is solvable and $\mathrm{dl}(G) \leq |\mathrm{cd}(G)|$ (see [Is, 12.6 and 12.15]). Since $|\mathrm{cd}(A_5)| = 4$, we cannot improve the first conclusion, but it has been conjectured by G. Seitz that $\mathrm{dl}(G) \leq |\mathrm{cd}(G)|$ for all solvable groups G. Isaacs gave the first general bound, namely $\mathrm{dl}(G) \leq 3 \cdot |\mathrm{cd}(G)|$ (or $2 \cdot |\mathrm{cd}(G)|$ if $|G|$ is odd). These are proved in Theorem 16.5 below. Lemma 16.4 is important here and further analysis allows us to present Gluck's improvement to $\mathrm{dl}(G) \leq 2 \cdot |\mathrm{cd}(G)|$ in Theorem 16.8. Using Theorem 8.4, we give Berger's proof of Seitz's conjecture for groups of odd order. The key result here is Theorem 16.6, which does not hold for arbitrary solvable groups.

The first proposition if quite important to this section. For $\chi \in \mathrm{Irr}(G)$, we let $D(\chi) = \bigcap\{\ker(\psi) \mid \psi \in \mathrm{Irr}(G) \text{ and } \psi(1) < \chi(1)\}$. Should χ be linear, then $D(\chi) = G$.

16.1 Proposition. *Let $\chi \in \mathrm{Irr}(G)$ and write $\chi = \theta^G$ for some $H \leq G$ and $\theta \in \mathrm{Irr}(H)$. Then $D(\chi) \leq D(\theta) \leq H$.*

Proof. Note that when χ is linear, then $\chi = \theta$ and $H = G = D(\chi)$. If $\psi \in \mathrm{Irr}(H)$ and $\psi(1) < \theta(1)$, then $\psi^G(1) < \theta^G(1) = \chi(1)$ and every irreducible constituent of ψ^G has degree less than $\chi(1)$. Thus $D(\chi) \leq \ker(\psi^G) \leq \ker(\psi) \leq H$. Hence $D(\chi) \leq D(\theta)$, except possibly when θ is linear and $H < G$. But in this case, observe that $1_H{}^G(1) = \theta^G(1) = \chi(1)$ and $1_H{}^G$ reduces. Thus $D(\chi) \leq \ker(1_H{}^G) \leq H = D(\theta)$. $\square$

We introduce a little more notation. For solvable groups G, we let $1 = f_1 < f_2 < \cdots < f_l$ be the l distinct character degrees of G. We let $D_i(G) = \bigcap\{\ker(\chi) \mid \chi \in \mathrm{Irr}\,(G)$ and $\chi(1) \leq f_i\}$. Thus $D_0(G) = G$, $D_1(G) = G'$ and $D(\chi) = D_{i-1}(G)$ should $\chi(1) = f_i$.

A group G is called an M-group if each $\chi \in \mathrm{Irr}\,(G)$ is induced from a linear character of a subgroup of G. Taketa proved that M-groups are solvable, in fact $\mathrm{dl}\,(G) \leq |\mathrm{cd}\,(G)|$ (see [Is, 5.12 and 5.13]). The strategy of this section is not dissimilar to, although more complicated than, the proof of Taketa's Theorem (as given in [Is]). To show that $\mathrm{dl}\,(G) \leq |\mathrm{cd}\,(G)|$ for an M-group G, it suffices to prove that $D(\chi)' \leq \ker(\chi)$ for all $\chi \in \mathrm{Irr}\,(G)$. Write $\chi = \lambda^G$ for a linear $\lambda \in \mathrm{Irr}\,(H)$ and $H \leq G$, then $D(\chi) \leq H$ by Proposition 16.1, and $D(\chi)' \leq \bigcap_{x \in G}(H')^x \leq \bigcap_{x \in G}(\ker(\lambda))^x = \ker(\chi)$.

16.2 Proposition. *Let G be solvable and let V be a completely reducible faithful G-module over possibly different finite fields. Then G has a faithful complex character ψ with $\psi(1) \leq \dim(V)$. (Here, $\dim(V)$ denotes the number of free generators of V.) Furthermore, it may be arranged that ψ is irreducible if and only if V is irreducible.*

Proof. We argue by induction on $\dim(V)$. If $V = V_1 \oplus V_2$ for proper G-submodules V_i, the inductive hypothesis implies the existence of $\psi_i \in \mathrm{Char}\,(G)$ with $\ker(\psi_i) = \mathbf{C}_G(V_i)$ and $\psi_i(1) \leq \dim(V_i)$. Let $\psi = \psi_1 + \psi_2$, so that $\ker(\psi) = \mathbf{C}_G(V_1) \cap \mathbf{C}_G(V_2) = 1$ and $\psi(1) \leq \dim(V)$. We may thus assume that V is irreducible over a field $\mathcal{F}$. Let $\mathcal{K}$ be an algebraically closed extension field of $\mathcal{F}$. Then $V \otimes_{\mathcal{F}} \mathcal{K} = V_1 \oplus \cdots \oplus V_t$ for distinct absolutely irreducible G-modules V_i that are Galois-conjugate by Proposition 0.4. Since V is faithful and the V_i are Galois-conjugate, V_1 is a faithful G-module. By the Fong–Swan Theorem (see Corollary 0.33), there exists $\psi \in \mathrm{Irr}\,(G)$ faithful with $\psi(1) = \dim_{\mathcal{K}}(V_1)$. Then $\psi(1) \leq \dim_{\mathcal{K}}(V \otimes_{\mathcal{F}} \mathcal{K}) = \dim_{\mathcal{F}}(V)$. $\qquad\square$

If V above has characteristic p, the proposition is still valid for p-solvable G. Note that if p is the smallest prime divisor of $|V|$, then $\psi(1) \leq \log_p(|V|)$.

16.3 Corollary. *Suppose* $\chi \in \mathrm{Irr}\,(G)$ *is faithful and primitive and* G *is solvable. Set* $F = \mathbf{F}(G)$ *and* $T = \mathbf{Z}(F)$. *Then*

(a) $T = \mathbf{Z}(G)$ *is cyclic;*

(b) $F/T = E_1/T \times \cdots \times E_m/T$ *where each* E_i/T *is an irreducible symplectic* G*-module;*

(c) G/F *acts faithfully on* F/T;

(d) $|F/T| \,\big|\, \chi(1)^2$; *and*

(e) F/T *has a complement in* G/T.

Proof. If $B \trianglelefteq G$ is abelian, then $\chi_B = e \cdot \beta$ for a faithful $\beta \in \mathrm{Irr}\,(B)$ and integer e, because χ is faithful and primitive. Consequently B is cyclic. Since β is G-invariant, linear and faithful, $B \leq \mathbf{Z}(G)$ (this uses that $\mathbb{C}$ is algebraically closed!). Thus every normal abelian subgroup of G is cyclic and central. The assertions now follow from Corollary 1.10, Corollary 2.6 and Lemma 1.11. ☐

16.4 Lemma. *Suppose that* $\chi \in \mathrm{Irr}\,(G)$ *is a faithful primitive character of a solvable group* G. *Set* $F = \mathbf{F}(G)$, $T = \mathbf{Z}(F) = \mathbf{Z}(G)$ *and* $K/F = \mathbf{O}_{2'}(G/F)$. *Assume that* $D(\chi) \not\leq F$. *Then*

(a) *there exists* $\rho \in \mathrm{Irr}\,(G/F)$ *faithful with* $\rho(1) = \chi(1)$;

(b) $K/F = \mathbf{F}(G/F)$ *is abelian and* $\chi(1) \,\big|\, |G/K|$;

(c) $D(\chi) \leq K$ *and* $D(\chi)''' = 1$;

(d) $\chi(1) = |F : T|^{1/2}$ *is* 2 *or* 2^2;

(e) F/T *is a faithful irreducible* G/F*-module; and*

(f) G/T *has a faithful irreducible character* ω *with* $\chi(1) < \omega(1) \leq (3/2) \cdot \chi(1)$.

Proof. Since χ is faithful and primitive, Corollary 16.3 applies. Since $D(\chi) \not\leq F$, we have that $F < G$ and hence $T < F$ (see Corollary 16.3 (c)). Proposition 16.2 implies that there exists a faithful $\rho \in \mathrm{Char}\,(G/F)$ satisfying $\rho(1) \leq \mathrm{rank}\,(F/T)$. Set $e = |F : T|^{1/2} \in \mathbb{Z}$ and let p be the smallest prime divisor of e. Then $\rho(1) \leq \text{rank}(F/T) \leq \log_p(e^2)$. Since $\ker(\rho) = F \not\leq D(\chi)$, some irreducible constituent ξ of ρ satisfies $\rho(1) \geq \xi(1) \geq \chi(1)$. Now $\chi(1) = et$ for an integer t (see Corollary 16.3 (d)). Thus

$$2 \cdot \log_p(e) \geq \rho(1) \geq \xi(1) \geq \chi(1) = et.$$

Then $e^2 \geq p^{et} \geq 2^e$. Since $p \mid e$, this can only occur when $p = 2$, $t = 1$, and e is 2 or 2^2. Hence, $\rho(1) = \xi(1) = \chi(1) = e$ and ρ is irreducible. It follows from Corollary 16.3 and the supplement of Proposition 16.2 that F/T is a faithful irreducible G/F-module. Since $T = \mathbf{Z}(G)$ and F/T is a 2-group, $\mathbf{O}_2(G/F) = 1$ and so $\mathbf{F}(G/F) \leq K/F$. Since $\rho \in \text{Irr}(G/F)$ is faithful of degree 2 or 2^2, $\rho_{K/F}$ is faithful and all irreducible constituents of $\rho_{K/F}$ are linear. Hence K/F is abelian and $K/F = \mathbf{F}(G/F)$. By Ito's Theorem [6.15 of Is], $\rho(1) \mid |G/K|$ and therefore $\chi(1) \mid |G/K|$. We have proven conclusions (a), (b), (d) and (e).

First suppose that $e = 2$ (recall $e = |F : T|^{1/2} = \chi(1)$). Since F/T is an irreducible and faithful G/F-module, $G/F \cong A_3$ or S_3. But $\rho \in \text{Irr}(G/F)$ has degree 2 and so $G/F \cong S_3$. Now $D(\chi) = G' \leq K$ and $K''' \leq F'' = 1$. If $1 \neq \lambda \in \text{Irr}(F/T)$, then each irreducible constituent of λ^G has degree 3 and kernel T. Parts (c) and (f) follow in this case.

Finally we may assume that $e = 2^2$. Now G/F acts faithfully, irreducibly and symplectically on F/T. A cyclic irreducible subgroup of G/F must have order dividing $2^2 + 1$ (see [Hu, II, 9.23]). Since $2^2 \mid |G/K|$, it follows from Corollary 2.15 applied to G/F acting on $V := F/T$ that

(i) $G/F \leq \Gamma(2^4)$, $|K/F| = 5$ and $|G/K| = 4$; or

(ii) $G/F \leq S_3 \,\text{wr}\, Z_2$, $K/F \cong Z_3 \times Z_3$ and G/K is abelian of order 4 or $G/K \cong D_8$.

Either $G' \leq K$ or G/K has a faithful irreducible character of degree 2. Thus $D(\chi) \leq K$ and $D(\chi)''' = 1$, proving (c).

In case (i), K/F induces three orbits of length 5 on $\text{Irr}(V)^{\#}$. We thus may find $\lambda \in \text{Irr}(V)$ such that $I_G(\lambda)/F$ is cyclic of order 4. Then λ extends to $\lambda^* \in \text{Irr}\,(I_G(\lambda)/T)$ by Proposition 0.11 and $(\lambda^*)^G \in \text{Irr}\,(G)$ of degree 5. Since F/T is the unique minimal normal subgroup of G/T, $T = \ker((\lambda^*)^G)$. Part (f) follows in this case, and we may assume that (ii) occurs.

Now $V = V_1 \oplus V_2$ for subspaces V_i permuted by G, and G/F has a subgroup M/F of index two that acts irreducibly on each V_i. If $C_i = \mathbf{C}_M(V_i)$, then $C_1 \cap C_2 = F$. Since $2 \mid |M/F|$, $2 \mid |M/C_i|$ for each i. Thus $M/C_1 \cong S_3$. Let $\alpha = \lambda \times 1 \in \text{Irr}\,(V)$ with $\lambda \neq 1$. Then $C_1 \leq I_M(\alpha) = I_G(\alpha)$ and $|I_G(\alpha)/C_1| = 2$. If $\alpha^* \in \text{Irr}\,(I_G(\alpha))$ were an extension of α, then $(\alpha^*)^G \in \text{Irr}\,(G)$ would have degree 6 and kernel T. To establish (f), it thus suffices to show that α extends to $I_G(\alpha)$. Since F/T has a complement H/T in G/T (by Corollary 16.3 (e)), and $T \leq \ker(\alpha) < F$, it follows that $F/\ker(\alpha)$ has a complement $J/\ker(\alpha)$ in $I_G(\alpha)/\ker(\alpha)$, namely $J = (H \cap I_G(\alpha)) \cdot \ker(\alpha) = I_H(\alpha) \cdot \ker(\alpha)$. Since J centralizes $F/\ker(\alpha)$, we have $I_G(\alpha)/\ker(\alpha) = F/\ker(\alpha) \times J/\ker(\alpha)$, and α trivially extends to $I_G(\alpha)$. $\qquad\square$

16.5 Theorem (Isaacs). *Let $\chi \in \text{Irr}\,(G)$ with G solvable. Then*

(a) $D(\chi)''' \leq \ker(\chi)$;

(b) *if $\chi(1)$ is odd, then $D(\chi)'' \leq \ker(\chi)$;*

(c) $\text{dl}\,(G) \leq 3 \cdot |\text{cd}\,(G)| - 2$; *and*

(d) *if $|G|$ is odd, then $\text{dl}\,(G) \leq 2 \cdot |\text{cd}\,(G)| - 1$.*

Proof. (a), (b) We argue by induction on $|G|$ and write $\chi = \theta^G$ for a primitive $\theta \in \text{Irr}\,(H)$, $H \leq G$. By Proposition 16.1, $D(\chi) \leq D(\theta)$. If $H < G$, the inductive hypothesis yields that $D(\chi)''' \leq D(\theta)''' \leq \ker(\theta)$. Since $D(\chi)''' \trianglelefteq G$, $D(\chi)''' \leq \bigcap_{g \in G}(\ker(\theta))^g = \ker(\theta^G) = \ker(\chi)$. Should $\chi(1)$ be odd, we also have $\theta(1)$ odd and argue inductively that $D(\chi)'' \leq \ker(\chi)$. So we may assume that χ is primitive. Let $F/\ker(\chi) = \mathbf{F}(G/\ker(\chi))$, so that $F'' \leq \ker(\chi)$, by Corollary 16.3. We can thus assume that $D(\chi) \not\leq F$. Then Lemma 16.4 implies that $\chi(1)$ is even and $D(\chi)''' \leq \ker(\chi)$. This proves (a) and (b).

(c), (d) Recall that $1 = f_1 < \cdots < f_l$ are the distinct irreducible character degrees of G and $D_i(G) = \bigcap\{\ker(\psi) \mid \psi \in \mathrm{Irr}(G),\ \psi(1) \leq f_i\}$. Parts (a) and (b) show $D_i(G)''' \leq D_{i+1}(G)$, and when G has odd order, $D_i(G)'' \leq D_{i+1}(G)$. Since $G/D_1(G) = G/G'$, we see that $\mathrm{dl}(G) \leq 3 \cdot l - 2$, and when $|G|$ is odd, $\mathrm{dl}(G) \leq 2 \cdot l - 1$. $\square$

Suppose that M is an elementary abelian p-group, on which G acts. The action of G on $\mathrm{Irr}(M)$ is given by $\lambda^g(m^g) = \lambda(m)$. If U is the subgroup of pth roots of unity in $\mathbb{C}$, then $\mathrm{Irr}(M)$ is just $\mathrm{Hom}(M, U)$. Writing M additively, M is a vector space over $\mathcal{F} := GF(p)$, and G acts on the dual space $M^* = \mathrm{Hom}_{\mathcal{F}}(M, \mathcal{F})$ by $f^g(m^g) = f(m)$. Since $\mathcal{F}$ is just the prime field, $M^* = \mathrm{Hom}(M, \mathbb{Z}_p)$. As $U \cong \mathbb{Z}_p$, it follows that M^* and $\mathrm{Irr}(M)$ are isomorphic G-modules. Indeed, an isomorphism is given by

$$\varphi \mapsto \exp((2\pi i/p)\varphi), \quad \varphi \in M^*.$$

16.6 Theorem (Berger). *Suppose that $\chi \in \mathrm{Irr}(G)$ is a faithful and primitive character for a group G of odd order. Then $D(\chi)' = 1$.*

Proof. We apply Corollary 16.3, and let $F = \mathbf{F}(G)$ and $T = \mathbf{Z}(F) = \mathbf{Z}(G)$. If $F = T$, then G is cyclic and we may assume that $F > T$.

By Corollary 16.3, $|F : T|^{1/2} \mid \chi(1)$, and $F/T = E_1/T \times \cdots \times E_m/T$ where each E_i/T is an irreducible symplectic G-module. Now $\mathrm{Irr}(F/T) = V_1 \oplus \cdots \oplus V_m$ where $V_i := \mathrm{Irr}(E_i/T)$ is an irreducible G-module as well (cf. Proposition 12.1). By the comments preceding the theorem, V_i is G-isomorphic to the dual space $(E_i/T)^*$ of E_i/T. Since E_i/T has a non-singular G-invariant form, $E_i/T \cong (E_i/T)^*$ (see [HB, VII, 8.10 (b)]). Thus each V_i is an irreducible symplectic G-module. By Theorem 8.4, there exists $1 \neq \lambda_i \in V_i$ such that

$$|G : I_G(\lambda_i)| \leq (|V_i|^{1/2} + 1)/2 < |V_i|^{1/2}.$$

Set $\alpha = \lambda_1 \cdots \lambda_m \in \mathrm{Irr}(F/T)$. Then

$$|G : I_G(\alpha)| \leq \prod_i |V_i|^{1/2} = |F/T|^{1/2},$$

and $|G : I_G(\alpha)| < \chi(1)$. Again by Corollary 16.3, F/T has a complement in G/T. Since $T \leq \ker(\alpha) \leq F$, it follows that $F/\ker(\alpha)$ has a complement $J/\ker(\alpha)$ in $I_G(\alpha)/\ker(\alpha)$. Since α is linear, J centralizes $F/\ker(\alpha)$ and so $I_G(\alpha)/\ker(\alpha) = F/\ker(\alpha) \times J/\ker(\alpha)$. Then α extends to $\alpha^* \in \mathrm{Irr}\,(I_G(\alpha)/\ker(\alpha))$, $(\alpha^*)^G \in \mathrm{Irr}\,(G)$ and $((\alpha^*)^G)(1) = |G : I_G(\alpha)| < \chi(1)$. Now $T \leq \ker((\alpha^*)^G)$, but $\ker((\alpha^*)^G) \cap F = T$, because $\alpha^*_{F/T} = \lambda_1 \cdots \lambda_m$ with $\lambda_i \neq 1$. By Corollary 16.3 (c), $F/T = \mathrm{socle}\,(G/T)$ and so $\ker((\alpha^*)^G) = T$. As $((\alpha^*)^G)(1) < \chi(1)$, we obtain $D(\chi) \leq T$ and $D(\chi)' = 1$, as desired. $\square$

16.7 Corollary. *Suppose that G has odd order. Then* $\mathrm{dl}\,(G) \leq |\mathrm{cd}\,(G)|$.

Proof. Arguing as in 16.5 (c), (d), it suffices to show that $D(\chi)' \leq \ker(\chi)$ for all $\chi \in \mathrm{Irr}\,(G)$. This is done by induction on $|G|$. If χ is primitive, our claim is an immediate consequence of Theorem 16.6. Otherwise, write $\chi = \theta^G$ for some $\theta \in \mathrm{Irr}\,(H)$ and $H < G$. By Proposition 16.1 and induction, $D(\chi)' \leq D(\theta)' \leq \ker(\theta)$, and $D(\chi)' \leq \bigcap_{g \in G}(\ker(\theta))^g = \ker(\chi)$. $\square$

Finally, we exploit the information in Lemma 16.4 to give Gluck's improvement that $\mathrm{dl}\,(G) \leq 2 \cdot |\mathrm{cd}\,(G)|$ for all solvable groups G.

16.8 Theorem (Gluck). *Suppose that G is solvable and that $D_{r-1}(G)'' \not\leq D_r(G)$ for some integer $r \geq 2$. Then $D_{r-1}(G)'''' \leq D_{r+1}(G)$.*

Proof. Recall that $1 = f_1 < \cdots < f_l$ are the distinct character degrees of G and $D_i(G) = \bigcap\{\ker(\psi) \mid \psi \in \mathrm{Irr}\,(G), \psi(1) \leq f_i\}$. Let $D = D_{r-1}(G)$. We may then assume that there exist χ, $\theta \in \mathrm{Irr}\,(G)$ such that $\chi(1) \leq f_r$, $\theta(1) \leq f_{r+1}$, $D'' \not\leq \ker(\chi)$ and $D'''' \not\leq \ker(\theta)$. By the definition of D, in fact $\chi(1) = f_r$, and so $D = D(\chi)$. By Theorem 16.5 (a), we may also assume that $\theta(1) = f_{r+1}$.

Step 1. We have $f_{r+1} \leq (3/2) \cdot f_r$.

Proof. Choose $H \leq G$ and a primitive character $\chi_0 \in \mathrm{Irr}\,(H)$ such that $\chi_0^G = \chi$. By Proposition 16.1, $D = D(\chi) \leq D(\chi_0) \leq H$. Let $K_0 = \ker(\chi_0)$, $T_0/K_0 = \mathbf{Z}(H/K_0)$ and $F_0/K_0 = \mathbf{F}(H/K_0)$. By Corollary 16.3, F_0/K_0 is metabelian. Now $D \not\leq F_0$, since otherwise $D'' \leq K_0 = \ker(\chi_0)$ and $D'' \leq \bigcap_{g \in G}(\ker(\chi_0))^g = \ker(\chi)$, a contradiction. By Lemma 16.4 (f), there exists a faithful $\omega \in \mathrm{Irr}\,(H/T_0)$ with $\chi_0(1) < \omega(1) \leq (3/2) \cdot \chi_0(1)$. Consequently, $\chi(1) < \omega^G(1) \leq (3/2) \cdot \chi(1)$. If ω^G is not irreducible, then there exists a constituent $\gamma \in \mathrm{Irr}\,(G)$ of ω^G such that $\gamma(1) < \chi(1)$. Thus $D \leq \ker(\gamma) \cap H = \ker(\gamma_H) \leq \ker(\omega) = T_0$, because ω is an irreducible constituent of γ_H. This contradicts $D \not\leq F_0$. Hence ω^G is irreducible, and $\chi(1) < \omega^G(1) \leq (3/2) \cdot \chi(1)$ implies that $f_{r+1} \leq (3/2) \cdot f_r$.

Step 2. Write $\theta = \mu^G$ for a primitive character $\mu \in \mathrm{Irr}\,(J)$ and $J \leq G$. Let $L = \ker(\mu)$. Then

(a) $\mu(1) > 1$;

(b) if $\alpha \in \mathrm{Irr}\,(J)$ and $\alpha(1) < (2/3) \cdot \mu(1)$, then $D \leq \ker(\alpha)$;

(c) $D \leq J'$; and

(d) $\mathrm{dl}\,(J/L) > \mathrm{dl}\,(DL/L) \geq 5$.

Proof. (a) Recall that $D = D(\chi)$, $\chi(1) = f_r$ and $\theta(1) = f_{r+1}$. By Theorem 16.5 (a), $D''' \leq D_r$. If μ is linear, then Lemma 16.1 (with μ in the role of θ) implies that $D_r' = D(\theta)' \leq J' \leq \ker(\mu)$, and therefore $D'''' \leq \bigcap_{g \in G}(\ker(\mu))^g = \ker(\theta)$, a contradiction. Hence $\mu(1) > 1$.

(b) We now have $\alpha^G(1) < (2/3) \cdot \mu^G(1) = (2/3) \cdot \theta(1) = (2/3) \cdot f_{r+1} \leq f_r$, by Step 1. Hence each irreducible constituent τ of α^G has degree less than f_r and so $D \leq \ker(\tau)$. Thus $D \leq \ker(\alpha^G) \leq \ker(\alpha)$.

(c) If $\lambda \in \mathrm{Irr}(J)$ is linear, then (a) implies $\lambda(1) < (2/3) \cdot \mu(1)$. By (b), $D \leq \ker(\lambda)$ and so $D \leq J'$.

(d) If $\mathrm{dl}\,(DL/L) \leq 4$, then $D'''' \leq \bigcap_{g \in G} L^g = \bigcap_{g \in G}(\ker(\mu))^g = \ker(\theta)$, a contradiction. By (c), it also follows that $\mathrm{dl}\,(J/L) > \mathrm{dl}\,(DL/L)$.

Step 3. Let $F/L = \mathbf{F}(J/L)$ and $T/L = \mathbf{Z}(J/L)$. Then F/T is a faithful irreducible symplectic J/F-module and $|F/T|^{1/2} = 8 = \mu(1)$. Also, J/F

has a faithful irreducible character ρ satisfying $\rho(1) = 6$.

Proof. By Corollary 16.3, $F/T = W_1 \oplus \cdots \oplus W_k$ for irreducible symplectic J/F-modules W_i. Also J/F acts faithfully on F/T. Write $|W_i| = e_i^2$ for integers e_i and set $e = e_1 \ldots e_k$. Then $e \mid \mu(1)$ and $e > 1$, by Corollary 16.3 and Step 2 (d). By Proposition 16.2, J/F has a faithful character σ such that $\sigma(1) \leq \operatorname{rank}(F/T)$. Since $F'' \leq L$, Step 2 (b, d) implies that there exists an irreducible constituent ρ of σ with $\rho(1) \geq (2/3) \cdot \mu(1) \geq (2/3) \cdot e$. Thus

$$\operatorname{rank}(F/T) \geq \sigma(1) \geq \rho(1) \geq (2/3) \cdot \mu(1) \geq (2/3) \cdot e. \qquad (16.1)$$

If p is the smallest prime divisor of e, then $(2/3) \cdot e \leq \operatorname{rank}(F/T) \leq \log_p(e^2)$. In particular, $e^3 \geq p^e \geq 2^e$. The only possibilities are $p = 3 = e$ or $p = 2$ and $e < 10$.

We claim that $e = e_1 = 8$. If not, then each e_i equals 2, 3 or 4. Now each W_i is an irreducible faithful symplectic $J/\mathbf{C}_J(W_i)$-module of order 2^2, 3^2 or 2^4. Observe that $GL(2,2) \cong S_3$ has derived length 2, $Sp(2,3) \cong SL(2,3)$ has derived length 3, and every solvable irreducible subgroup of $GL(4,2)$ has derived length at most three (cf. Corollary 2.15). Since $\bigcap_i \mathbf{C}_J(W_i) = F$, we have that $\operatorname{dl}(J/F) \leq 3$ and $\operatorname{dl}(J/L) \leq 5$, contradicting Step 2 (d). Thus $e = e_1 = 8$ and F/T is a faithful irreducible symplectic G/F-module of order $8^2 = 2^6$.

Since $\dim(F/T) = 6$, it follows from (16.1) that

$$6 \geq \sigma(1) \geq \rho(1) \geq (2/3) \cdot \mu(1) \geq (2/3) \cdot e = 16/3 > 5.$$

Since $8 \mid \mu(1)$, we have that $\mu(1) = 8$ and $\sigma = \rho$ is a faithful irreducible character of J/F of degree 6.

Step 4. We have $\rho^G \in \operatorname{Irr}(G)$ and $\rho^G(1) = \chi(1)$. In particular, $D = D(\rho^G)$.

Proof. If some irreducible constituent η of ρ^G satisfies $\eta(1) < \chi(1)$, then $D \leq \ker(\eta)$ and by Step 2 (c), $D \leq \ker(\eta) \cap J = \ker(\eta_J) \leq \ker(\rho) = F$, contradicting Step 2 (d). So every irreducible constituent of ρ^G has degree at least $\chi(1) = f_r$. But by Steps 1 and 3,

$$\rho^G(1) = (3/4) \cdot \mu^G(1) = (3/4) \cdot \theta(1) \leq (9/8) \cdot \chi(1) < 2 \cdot f_r.$$

Therefore, ρ^G is irreducible and $\rho^G(1) \geq \chi(1)$. Also $\rho^G(1) < \theta(1) = f_{r+1}$, whence $\rho^G(1) = \chi(1)$.

Step 5. Conclusion.

Proof. By Step 3, $\mathbf{O}_2(J/F) = 1$. Now $D \leq J$, and we let $E/F = \mathbf{F}(DF/F)$ so that $|E/F|$ is odd. Choose $F \leq Y \leq J$ and a primitive $\beta \in \mathrm{Irr}\,(Y)$ such that $\rho = \beta^J$. Since by Step 4, $\beta^G = \rho^G \in \mathrm{Irr}\,(G)$ and $D = D(\rho^G)$, Proposition 16.1 implies that $D \leq Y$. If $D'' \leq \ker(\beta)$, then $D'' \leq \ker(\beta^J) = F$ and $D'''' \leq L$, contradicting Step 2 (d). By Proposition 16.1, $D = D(\beta^G) \leq D(\beta)$, and therefore $D(\beta)'' \not\leq \ker(\beta)$. Applying Lemma 16.4 to the faithful primitive character β of $Y/\ker(\beta)$, we see that the Hall $2'$-subgroup of $\mathbf{F}(Y/\ker(\beta))$ is central in $Y/\ker(\beta)$. Recall that $E \leq DF \leq Y$ and $F \leq \ker(\beta)$. Thus $E \cdot \ker(\beta)/\ker(\beta)$ is nilpotent of odd order and central in $Y/\ker(\beta)$. Therefore, $[D, E] \leq \ker(\beta)$. Since $D, E \trianglelefteq J$, we also have that $[D, E] \leq \bigcap_{j \in J}(\ker(\beta))^j = \ker(\beta^J) = \ker(\rho) = F$. Since $E/F = \mathbf{F}(DF/F)$, in fact $E/F = DF/F$ is abelian. Then $D' \leq F$, $D''' \leq \ker(\mu)$ and $D''' \leq \ker(\mu^G) = \ker(\theta)$. The proof is complete. $\qquad\square$

16.9 Corollary. *If G is solvable, then $\mathrm{dl}\,(G) \leq 2 \cdot |\mathrm{cd}\,(G)|$.*

Proof. If, for some r, $\mathrm{dl}\,(D_r(G)/D_{r+1}(G)) > 2$, then Theorems 16.8 and 16.5 imply that $\mathrm{dl}\,(D_r(G)/D_{r+1}(G)) = 3$ and $\mathrm{dl}\,(D_{r+1}(G)/D_{r+2}(G)) \leq 1$ (where possibly $r + 1 = |\mathrm{cd}\,(G)|$). Since $G/D_1(G) = G/G'$ has derived length one, it follows that $\mathrm{dl}\,(G) \leq 2 \cdot |\mathrm{cd}\,(G)|$. $\qquad\square$

16.10 Examples. (a) We note that the assertion of Theorem 16.6 definitely does not hold for arbitrary solvable groups. Namely $G = SL(2,3)$ has a faithful primitive character $\chi \in \mathrm{Irr}\,(G)$ of degree 2. Thus

$$D(\chi) = \bigcap\{\ker(\lambda) \mid \lambda \in \mathrm{Irr}\,(G) \text{ and } \lambda(1) = 1\} = G' \cong Q_8$$

and $Q_8' \cong Z_2 \neq 1$.

(b) It is not known whether Seitz's conjecture $\mathrm{dl}\,(G) \leq |\mathrm{cd}\,(G)|$ would be best possible. Encouraged by several examples, one might rather believe in a logarithmic bound for $\mathrm{dl}\,(G)$ in terms of $|\mathrm{cd}\,(G)|$.

Let $P_m = \underset{m}{Z_p \,\mathrm{wr}\, \cdots \,\mathrm{wr}\, Z_p}$ be the m-fold iterated wreath-product of Z_p, with $P_1 = Z_p$. If p is odd, an easy induction argument yields that

$$\mathrm{cd}\,(P_m) = \{p^j \mid 0 \leq j \leq 1 + p + \cdots + p^{m-2}\}, \quad (m \geq 2).$$

On the other hand, $\mathrm{dl}\,(P_m) = m$ (see Proposition 3.10).

This section is based upon [Be 1], [Gl 2], and [Is 3].

§17   Huppert's ρ-σ-Conjecture

In this section, we are concerned with the "arithmetic complexity" of character degrees. To be more precise, we introduce some notation.

17.1 Definition. For a natural number n, we (as usual) let $\pi(n)$ be the set of distinct prime divisors of n and $\pi_0(n) = \pi(n) \setminus \{2, 3\}$. For a group G, we define

$$\rho(G) = \{p \text{ prime} \mid p \,|\, \chi(1) \text{ for some } \chi \in \mathrm{Irr}\,(G)\} \text{ and}$$

$$\sigma(G) = \max\{|\pi(\chi(1))| \mid \chi \in \mathrm{Irr}\,(G)\}.$$

Note that $\rho(G)$ is a set, whereas $\sigma(G)$ is an integer. Also observe that by the Ito–Michler Theorem (13.1, 13.13), $p \notin \rho(G)$ if and only if G has a normal abelian Sylow p-subgroup.

For solvable groups G, Huppert has asked the following questions:

(1) Is there a function f (independent of G) such that $|\rho(G)| \leq f(\sigma(G))$?

(2) Does even $|\rho(G)| \leq 2 \cdot \sigma(G)$ hold?

Before we proceed we show that (2) would be best possible.

17.2 Example. Let n be an integer and $p_1, \ldots, p_n$, $q_1, \ldots, q_n$ mutually distinct primes such that $p_i \mid q_i \pm 1$ $(i = 1, \ldots, n)$. Let E_i be extra-special of order q_i^3 and exponent q_i, Z_i cyclic of order p_i and $G_i = E_i \cdot Z_i$, where Z_i acts fixed-point-freely on $E_i / \mathbf{Z}(E_i)$ but trivially on $\mathbf{Z}(E_i)$. Since G_i has character degrees $\{1, p_i, q_i\}$ (see [Hu, V, 17.13]), it follows that $G := G_1 \times \cdots \times G_n$ satisfies $\rho(G) = \{p_1, \ldots, p_n, q_1, \ldots, q_n\}$ and $\sigma(G) = n$.

Whereas question (2) is still open, there are several results answering (1) in the affirmative (cf. [Is 7], [Gl 3], [GM 1]). Following [MW 3], we present a proof of the best function f known so far. We shall take advantage of the results of Section 11.

17.3 Proposition. Suppose that $K / \mathbf{F}(K)$ is nilpotent and $C \trianglelefteq K$. Then there exists $\mu \in \mathrm{Irr}(C)$ such that $\mu(1)$ is divisible by every prime divisor of $|C / (\mathbf{F}(K) \cap C)|$.

Proof. Since $C / (\mathbf{F}(K) \cap C)$ is nilpotent, there is no loss to assume that $C / (\mathbf{F}(K) \cap C)$ is abelian. Also $\mathbf{F}(C) = \mathbf{F}(K) \cap C$, because $\mathbf{F}(C) \trianglelefteq K$. Now the abelian group $C / \mathbf{F}(C)$ acts faithfully and completely reducibly on both $\mathbf{F}(C) / \Phi(C)$ and $V := \mathrm{Irr}(\mathbf{F}(C) / \Phi(C))$, by Theorem 1.12 and Proposition 12.1. Write $V = V_1 \oplus \cdots \oplus V_m$ for irreducible C-modules V_i. Since $C / \mathbf{F}(C)$ is abelian, $I_C(\lambda_i) = \mathbf{C}_C(V_i)$ for $1 \neq \lambda_i \in V_i$ $(i = 1, \ldots, m)$. Set $\lambda = \lambda_1 \cdots \lambda_m$ and $\mu = \lambda^C \in \mathrm{Irr}(C)$. Then $\mu(1) = |C / \mathbf{F}(C)| = |C / (\mathbf{F}(K) \cap C)|$. $\qquad\square$

17.4 Lemma. *Suppose that M is a normal elementary abelian subgroup of the solvable group G. Assume that $M = \mathbf{C}_G(M)$ is a completely reducible G-module (possibly of mixed characteristic). Set $V = \mathrm{Irr}(M)$ and write $V = V_1 \oplus \cdots \oplus V_m$ for irreducible G-modules V_i. For each i, write $V_i = Y_i^G$ for primitive modules Y_i. Assume that $\mathbf{N}_G(Y_i) / \mathbf{C}_G(Y_i)$ is nilpotent-by-nilpotent for each i. If $M \leq N \trianglelefteq G$, there exists $\theta \in \mathrm{Irr}(N)$ whose degree is divisible by at least half the primes of $\pi_0(N/M)$.*

Proof. We may write each V_i as a direct sum of the G-conjugates of Y_i, $i = 1, \ldots, m$. Consequently, $V = X_1 \oplus \cdots \oplus X_n$ for subspaces X_i of V permuted by G (not necessarily transitively) with $\{Y_1, \ldots, Y_m\} \subseteq \{X_1, \ldots, X_n\}$. Furthermore, if $N_i = \mathbf{N}_G(X_i)$, $C_i = \mathbf{C}_G(X_i)$ and $F_i/C_i = \mathbf{F}(N_i/C_i)$, then X_i is a primitive, faithful N_i/C_i-module and N_i/F_i is nilpotent.

Let $K = \bigcap_i N_i \trianglelefteq G$ be the kernel of the permutation representation of G on $\{X_1, \ldots, X_n\}$. Since $\bigcap_i C_i = M$, we have $\bigcap_i F_i/M = \mathbf{F}(K/M) \trianglelefteq G/M$. Let $H = \bigcap_i F_i$, so that $H/M = \mathbf{F}(K/M)$. Observe that K/H is nilpotent. Set $C = K \cap N$ and $F = H \cap N = C \cap H$. By Proposition 17.3, there exists $\theta \in \mathrm{Irr}\,(C/M)$ such that $\theta(1)$ is divisible by every prime divisor of $|C/F|$. Since $C \trianglelefteq N$, there exists $\tau \in \mathrm{Irr}\,(N)$ such that $\tau(1)$ is divisible by every prime divisor of $|C/F|$. Consequently it suffices to show there exists $\beta \in \mathrm{Irr}\,(N)$ with $\beta(1)$ divisible by each prime in $\pi_0(N/C) \cup \pi_0(F/M)$. To do this, we need just find some $\lambda \in V$ such that $\pi_0(N : \mathbf{C}_N(\lambda)) \supseteq \pi_0(N/C) \cup \pi_0(F/M)$.

By Corollary 5.7, we proceed to choose $\Delta \subseteq \{X_1, \ldots, X_n\}$ such that $\mathrm{stab}_N(\Delta)/(N \cap K) = \mathrm{stab}_N(\Delta)/C$ is a $\{2,3\}$-group. Furthermore, we can assume that Δ intersects each N-orbit non-trivially. Without loss of generality, $\Delta = \{X_1, \ldots, X_l\}$ for some $l \in \{1, \ldots, n\}$. Set $\lambda = \lambda_1 \cdots \lambda_l \in V$ for non-principal $\lambda_i \in X_i$. Finally suppose that $Q \in \mathrm{Syl}_q(N)$ for a prime $q \geq 5$, and Q centralizes λ. Thus $Q \leq \mathrm{stab}_N(\Delta)$. But $\mathrm{stab}_N(\Delta)/C$ is a $\{2,3\}$-group. Thus $Q \leq C$. For each i, $F_i \cap C/C_i \cap C$ is isomorphic to a normal nilpotent subgroup of N_i/C_i, and N_i/C_i acts irreducibly on X_i. Thus, for $i = 1, \ldots, l$, λ_i is not centralized by a non-trivial Sylow subgroup of $F_i \cap C/C_i \cap C$. Since $Q \cap F_i \in \mathrm{Syl}_q(F_i \cap C)$, we have that $q \nmid |F_i \cap C/C_i \cap C|$ for $i = 1, \ldots, l$. By our choice of Δ, each F_j/C_j $(j = 1, \ldots, n)$ is conjugate to some F_i/C_i with $i \in \{1, \ldots, l\}$. Hence $q \nmid |F_j \cap C/C_j \cap C|$ for all $j = 1, \ldots, n$. Since $\bigcap_i C_i = M$ and $\bigcap_i (F_i \cap C) = F$, we have that $q \nmid |F/M|$. We have already seen above that $Q \leq C$ and so $q \nmid |N/C|$. Thus $|N : \mathbf{C}_N(\lambda)|$ is divisible by every prime in $\pi_0(N/C) \cup \pi_0(F/M)$, as desired. $\quad\square$

17.5 Lemma. *Suppose that $M = \mathbf{C}_G(M)$ is a normal elementary abelian* subgroup of a solvable group G and a completely reducible G-module (possibly of mixed characteristic). Assume that G splits over M. Then there exists $\chi \in \mathrm{Irr}\,(G)$ such that $\chi(1)$ is divisible by at least half the primes in $\pi_0(G/M)$.

Proof. We proceed by induction on $|M|$. Write $M = M_1 \oplus \cdots \oplus M_n$ for $n \geq 1$ irreducible G-modules M_i. Set $V_i = \mathrm{Irr}\,(M_i)$ so that each V_i is an irreducible G-module and $V = V_1 \oplus \cdots \oplus V_n$ is a faithful G/M-module by Proposition 12.1. For each i, choose $H_i \leq G$ and X_i an irreducible primitive H_i-module with $X_i^G = V_i$. If $H_i/\mathbf{C}_{H_i}(X_i) \leq \Gamma(X_i)$ for each i, this lemma follows from Lemma 17.4. We assume without loss of generality that $H_1/\mathbf{C}_{H_1}(X_1) \not\leq \Gamma(X_1)$.

Let $K = \mathbf{C}_G(M_1) \trianglelefteq G$. Let H be a complement for M in G and let $J = NH$ where $N = M_2 \oplus \cdots \oplus M_n$. Then $J \cap M = N$. Now $J \cap K = N(H \cap K)$ acts on N and $\mathbf{C}_{J \cap K}(N) = N$. By induction, there exists $\tau \in \mathrm{Irr}\,(J \cap K)$ such that $\tau(1)$ is divisible by at least half the primes in $\pi_0((J \cap K)/N) = \pi_0(K/M)$, as $(J \cap K)/N \cong K/M$. Now $J \cap K \trianglelefteq J$ and centralizes $M/N \cong M_1$. Thus $J \cap K \trianglelefteq KJ = G$ and $K/N = M/N \times (J \cap K)/N$.

By the choice of M_1, Theorem 11.4 implies that there exists $\lambda \in V_1$ such that $\pi_0(G/K) = \pi_0(G : I_G(\lambda))$. Set $\beta = \lambda \cdot \tau \in \mathrm{Irr}\,(K)$. Now $I_G(\beta) \subseteq I_G(\lambda)$. Thus $\pi_0(G : I_G(\beta)) \supseteq \pi_0(G/K)$. If $\chi \in \mathrm{Irr}\,(G|\beta)$, then as $K \trianglelefteq G$, $\pi_0(\chi(1)) \supseteq \pi_0(G/K) \cup \pi_0(\tau(1))$. Since $\tau(1)$ is divisible by at least half the primes in $\pi_0(K/M)$, certainly $\chi(1)$ is divisible by at least half the primes in $\pi_0(G/M)$. $\qquad\square$

17.6 Theorem. If G is solvable, then there exists $\beta \in \mathrm{Irr}\,(G/\Phi(G))$ such that $\beta(1)$ is divisible by at least half the primes in $\pi_0(G/\mathbf{F}(G))$.

Proof. Apply Lemma 17.5 with $G/\Phi(G)$ and $\mathbf{F}(G)/\Phi(G)$ in the role of G and M, respectively. Note that Gaschütz's Theorem 1.12 guarantees the hypotheses of Lemma 17.5 are satisfied. $\qquad\square$

As already formulated above, Ito's Theorem 13.1 for solvable groups G states $p \notin \rho(G)$ if and only if G has a normal abelian Sylow p-subgroup. Thus $p \in \rho(G)$ if and only if $p \mid |G/\mathbf{F}(G)|$ or $\mathbf{F}(G)$ has a non-abelian Sylow p-subgroup, i.e. $p \in \rho(G)$ if and only if $p \mid |G/\mathbf{Z}(\mathbf{F}(G))|$. We let $\rho_0(G) = \rho(G) \setminus \{2, 3\}$.

17.7 Theorem. *Let G be solvable.*

(a) *There exists $\chi \in \mathrm{Irr}\,(G)$ such that $\chi(1)$ is divisible by at least one third of the primes in $\rho_0(G)$.*

(b) *Assume whenever r is a prime and $\mathbf{O}_r(G)$ is non-abelian, then also $r \mid |G/\mathbf{F}(G)|$. Then there exists $\chi \in \mathrm{Irr}\,(G)$ with $\chi(1)$ divisible by at least one half of the primes in $\rho_0(G)$.*

Proof. Let $\mathfrak{S}$ be the set of those primes s for which $\mathbf{O}_s(G)$ is non-abelian and $s \nmid |G/\mathbf{O}_s(G)|$. Now $\mathbf{F}(G)$ certainly has an irreducible character φ whose degree is divisible by all $s \in \mathfrak{S}$. Hence $\tau \in \mathrm{Irr}\,(G|\varphi)$ also satisfies $s \mid \chi(1)$ for all $s \in \mathfrak{S}$. By Theorem 17.6, there exists $\beta \in \mathrm{Irr}\,(G)$ with $\beta(1)$ divisible by at least half the primes in $\pi_0(G/\mathbf{F}(G))$. By the comments preceding this theorem, $\rho(G) = \pi(G/\mathbf{F}(G)) \dot\cup \mathfrak{S}$ holds.

Under the hypothesis of (b), $\mathfrak{S} = \varnothing$ and we just let $\chi = \beta$. To prove (a), we let $\chi = \beta$ if $|\rho_0(G)|/3 \geq |\mathfrak{S} \setminus \{2, 3\}|$, and let $\chi = \tau$ otherwise. $\quad\square$

We reformulate Theorem 17.7 in terms of Huppert's ρ-σ-conjecture. The summand "2" refers to the role of $\{2, 3\}$ above.

17.8 Corollary. *Let G be solvable.*

(a) $|\rho(G)| \leq 3 \cdot \sigma(G) + 2$; *and*

(b) $|\rho(G)| \leq 2 \cdot \sigma(G) + 2$ *if $r \mid |G/\mathbf{F}(G)|$ whenever $\mathbf{O}_r(G)$ is non-abelian.*

17.9 Remarks. (a) For non-solvable groups G, it is still unknown whether there exists a function f according to Huppert's question (1). Question (2) however, in general has a negative answer, e.g. $|\rho(A_5)| = 3$ and $\sigma(A_5) = 1$. Using the classification of finite simple groups, the following bound has been established by Alvis and Barry [AB 1] and Manz, Staszewski and Willems [MSW]:

$$\text{Let } G \text{ be simple. Then } |\rho(G)| \leq 3 \cdot \sigma(G).$$

It seems reasonable to ask whether this estimation holds in general, or whether even a factor 2 and an additive constant is the "right" answer.

(b) Returning to solvable groups G, the natural question arises whether also $\mathrm{dl}\,(G)$ can be bounded in terms of $\sigma(G)$. To see why this is not the case, we let p, q be distinct primes and consider the class $\mathfrak{C}$ of $\{p, q\}$-groups. If $G \in \mathfrak{C}$, then clearly $\sigma(G) \leq 2$ holds. On the other hand, it is well-known that within $\mathfrak{C}$ there is no universal bound for the derived length (nor even for the nilpotency length).

The following result (cf. [MS 1]) may serve as a substitute. It again relies on Theorem 12.9.

17.10 Theorem. *Let G be solvable. Then G has a characteristic series*

$$1 \leq N_0 \leq N_1 \leq N_2 \leq \cdots \leq N_{2k+1} \leq N_{2k} = G$$

with the following properties:

(i) $N_0' = 1$ *and* $A \leq N_0$ *for the normal abelian Hall $\rho(G)'$-subgroup A of G (A exists by Theorem 13.1);*

(ii) $(N_{2i+1}/N_{2i})' = 1$ *for $i = 0, \ldots, k - 1$;*

(iii) $|\pi(N_{2i}/N_{2i-1})| \leq \sigma(G)$ *for $i = 1, \ldots, k$; and*

(iv) $k \leq 2 \cdot \sigma(G)$.

Proof. We argue by induction on $s := \sigma(G)$. If $s = 0$, then $N_0 = G$ is abelian and $k = 0$. We may therefore assume that $s > 0$ and choose iterated commutator subgroups $N/A := (G/A)^{(j)}$ and $M/A := (G/A)^{(j+1)}$ such that

$\sigma(N) = s$ but $\sigma(M) < s$. By induction, there exists a characteristic series $1 \leq N_0 \leq N_1 \leq N_2 \leq \cdots \leq N_{2l-1} \leq N_{2l} = M$ of M with

(i)' $N_0' = 1$ and $A \leq B \leq N_0$ for the normal abelian Hall $\rho(M)'$-subgroup B of M;

(ii)' $(N_{2i+1}/N_{2i})' = 1$ for $i = 0, \ldots, l-1$;

(iii)' $|\pi(N_{2i}/N_{2i-1})| \leq \sigma(M) < s$ for $i = 1, \ldots, l$; and

(iv)' $l \leq 2 \cdot \sigma(M) \leq 2(s-1)$.

By their definition, N and M are characteristic in G and N/M is abelian. We may choose $\psi \in \mathrm{Irr}(N)$ with $|\pi(\psi(1))| = s$ and we denote the set of prime divisors of $\psi(1)$ by π. Let $\chi \in \mathrm{Irr}(G|\psi)$. Since $\psi(1) \mid \chi(1)$, each prime divisor of $\chi(1)$ belongs to π and $\chi(1)/\psi(1)$ is a π-number. By Theorem 12.9, G/N has an abelian Hall π'-subgroup. Consequently the π'-length of G/N is at most 1 by Lemma 0.19. We extend the above characteristic series of M to a characteristic series of G by setting $N_{2l+1} = N$, $N_{2l+2}/N_{2l+1} = \mathbf{O}_\pi(G/N_{2l+1})$, $N_{2l+3}/N_{2l+2} = \mathbf{O}_{\pi'}(G/N_{2l+2})$ and $N_{2l+4} = G$. Then G/N_{2l+3} is a π-group. With $k := l + 2$, properties (i)–(iv) are satisfied.   □

While $\sigma(G)$ is a measure for the number of different primes in the character degrees of G, $\tau(G)$ will measure the maximum multiplicity of the primes in the character degrees of G.

17.11 Definition. For a group G and a prime p, recall the definition of $e_p(G)$ as the smallest non-negative integer e such that $p^{e+1} \nmid \chi(1)$ for all $\chi \in \mathrm{Irr}(G)$ (see 14.1). We set

$$\tau(G) = \max\{e_p(G) \mid p \mid |G|\}.$$

Observe that the group G of Example 17.2 satisfies $\tau(G) = 1$, but $|\rho(G)| = n$. Therefore we cannot expect to estimate $|\rho(G)|$ in terms of $\tau(G)$. We finish this section with a result of Leisering and Manz [LM 1].

17.12 Theorem. *Let G be solvable. Then*

$$\mathrm{dl}(G) \leq 2 \cdot (\tau(G) + \log_2 \tau(G) + 3).$$

Proof. For a prime divisor p of $|G|$, we denote by r_p the p-rank of $G/\mathbf{O}_p(G)$, and by r the rank of $G/\mathbf{F}(G)$. It follows from 14.12 (c) that $r_p \leq 2 \cdot e_p(G)$ and therefore $r = \max\{r_p\} \leq \max\{2 \cdot e_p(G)\} = 2 \cdot \tau(G)$. We set $\bar{G} = G/\mathbf{F}(G)$. By Gaschütz's Theorem 1.12, $\mathbf{F}(\bar{G})/\Phi(\bar{G})$ is a faithful completely reducible $\bar{G}/\mathbf{F}(\bar{G})$-module (possibly of mixed characteristic). Write $\mathbf{F}(\bar{G})/\Phi(\bar{G}) = V_1 \oplus \cdots \oplus V_n$ with irreducible $\bar{G}$-modules V_i and $\bar{C}_i = \mathbf{C}_{\bar{G}}(V_i)$. Since $r \leq 2\tau(G)$, also $\dim(V_i) \leq 2 \cdot \tau(G)$ for $i = 1, \ldots, n$. It thus follows from Corollary 3.12 that

$$\mathrm{dl}\,(\bar{G}/\bar{C}_i) \leq 2 \cdot \log_2(2 \cdot \dim(V_i)) \leq 2 \cdot \log_2(4 \cdot \tau(G)).$$

Since $\bar{G}/\mathbf{F}(\bar{G}) \lesssim \prod_i \bar{G}/\bar{C}_i$, we also have $\mathrm{dl}\,(\bar{G}/\mathbf{F}(\bar{G})) \leq 2 \cdot \log_2(4 \cdot \tau(G))$.

Let P equal $\mathbf{O}_p(G)$ or $\mathbf{O}_p(\bar{G})$, respectively. Then P is normal in G or $\bar{G}$, respectively, and Clifford's Theorem implies that $\tau(P) \leq \tau(G)$. Since P is a p-group, it follows that $|\mathrm{cd}\,(P)| \leq \tau(P) + 1$, and Taketa's Theorem (cf. [Is, 5.12]) yields $\mathrm{dl}\,(P) \leq |\mathrm{cd}\,(P)| \leq \tau(P) + 1 \leq \tau(G) + 1$.

Altogether, we obtain

$$\mathrm{dl}\,(G) \leq \mathrm{dl}\,(\mathbf{F}(G)) + \mathrm{dl}\,(\mathbf{F}(\bar{G})) + \mathrm{dl}\,(\bar{G}/\mathbf{F}(\bar{G}))$$

$$\leq 2 \cdot (\tau(G) + 1) + 2 \cdot \log_2(4 \cdot \tau(G)) = 2 \cdot (\tau(G) + \log_2 \tau(G) + 3),$$

as required.   □

We note that Theorem 17.12 is far away from being best possible. If e.g. $\tau(G) = 1$ (i.e. all character degrees are squarefree), then the above estimate yields $\mathrm{dl}\,(G) \leq 8$. Best possible however in this case is $\mathrm{dl}\,(G) \leq 4$ (see [HM 1]).

§18   The Character Degree Graph

We construct a graph $\Gamma(G)$, whose vertices are the elements of $\rho(G)$, i.e. those primes q that divide the degree $\chi(1)$ of some irreducible character

$\chi \in \mathrm{Irr}\,(G)$. We draw an edge between distinct $q,\ r \in \Gamma(G)$ if and only if $qr \mid \chi(1)$ for some $\chi \in \mathrm{Irr}\,(G)$. A distance function $d(q,s) = d_G(q,s)$ is defined in the usual way:

$d(q,s)$ is the length of the shortest path between q and s.

In particular, $d(q,q) = 0$, $d(q,s) = \infty$ if q and s lie in different connected components and $d(q,s) = 1$ if and only if $q \neq s$ and $qs \mid \tau(1)$ for some $\tau \in \mathrm{Irr}\,(G)$. If Λ is a connected component of $\Gamma(G)$, then the diameter of Λ is defined by

$$\mathrm{diam}\,(\Lambda) = \max\{d(q,s) \mid q,s \in \Lambda\}.$$

Finally

$$\mathrm{diam}\,(\Gamma(G)) = \max\{\mathrm{diam}\,(\Lambda) \mid \Lambda \text{ a component of } \Gamma(G)\}.$$

The number of connected components of $\Gamma(G)$ will be denoted by $n(\Gamma(G))$. As usual, $\pi(G)$ is the set of prime divisors of $|G|$.

We will show that there are very limited configurations for the graph $\Gamma(G)$. If G is solvable, then the number of components of $\Gamma(G)$ is at most 2 (Theorem 18.4). If $\Gamma(G)$ has two components, then both components are regular graphs. More will be said about such groups in the next section. With further work, we show (Theorem 18.7) that whenever $\Delta \subseteq \Gamma(G)$ and $|\Delta| \geq 3$, there exists $\chi \in \mathrm{Irr}\,(G)$ with $\chi(1)$ divisible by at least two distinct primes in Δ. For $|\Delta| \geq 4$, this was proven in [MWW]. Modifications by Palfy [Pl 2] improved this to $|\Delta| \geq 3$. We will use Theorem 18.7 to prove for solvable G that $\mathrm{diam}\,(\Gamma(G)) \leq 3$. We will discuss graphs of non-solvable groups at the end of Sections 18 and 19.

18.1 Lemma. *Suppose that* $G/\mathbf{F}(G)$ *is abelian. Then there is* $\chi \in \mathrm{Irr}\,(G)$ *such that* $\chi(1) = |G : \mathbf{F}(G)|$.

Proof. As a consequence of Gaschütz's Theorem 1.12 and Proposition 12.1, $\mathrm{Irr}\,(\mathbf{F}(G)/\Phi(G))$ is a faithful and completely reducible $G/\mathbf{F}(G)$-module (possibly in different characteristics). Write

$$\mathrm{Irr}\,(\mathbf{F}(G)/\Phi(G)) = V_1 \oplus \cdots \oplus V_n$$

with irreducible G-modules V_i and set $C_i = \mathbf{C}_G(V_i)$. Since G/C_i is abelian, G/C_i acts fixed-point-freely on V_i. Take $1 \neq \lambda_i \in V_i$ $(i = 1, \ldots, n)$. Then

$$I(\lambda_1 \cdots \lambda_n) = \bigcap_i I(\lambda_i) = \bigcap C_i = \mathbf{F}(G)$$

and $\chi := (\lambda_1 \ldots \lambda_n)^G \in \mathrm{Irr}\,(G)$ has the desired property. $\square$

18.2 Lemma. Let $N \trianglelefteq G$ and $M/N = \mathbf{F}(G/N)$. Suppose that G/M is nilpotent. Then we have

(a) $q \in \Gamma(G)$ for all $q \in \pi(G/M)$ and

$d(q, q') = 1$ for different $q, q' \in \pi(G/M)$.

(b) If $v \in \Gamma(G)$ with $v \nmid |G/N|$, then either

$d(v, p) = 1$ for some $p \in \pi(M/N) \cap \Gamma(G)$, or

$d(v, q) = 1$ for all $q \in \pi(G/M)$.

Proof. (a) Replacing G by the complete preimage of $\mathbf{Z}(G/M)$ in G, we may assume that G/M is abelian. The assertion now follows from Lemma 18.1.

(b) Let $\chi \in \mathrm{Irr}\,(G)$ with $v \mid \chi(1)$. Suppose $p \nmid \chi(1)$ for all $p \in \pi(M/N)$. Let $\mathcal{Q} = \{q \mid q \text{ prime}, q \nmid \chi(1)\}$ and $H/M \in \mathrm{Hall}_\mathcal{Q}(G/M)$. Note that $H \trianglelefteq G$. Now choose $\psi \in \mathrm{Irr}\,(H)$ with $[\chi_H, \psi] \neq 0$. Then $\psi_N \in \mathrm{Irr}\,(N)$ and $v \mid \psi(1)$. By Lemma 18.1, there is $\tau \in \mathrm{Irr}\,(H/N)$ whose degree is divisible by all $q \in \pi(H/M)$. Then Lemma 0.10 implies $\tau\psi \in \mathrm{Irr}\,(H)$ and $vq \mid \tau\psi(1)$ for all $q \in \pi(H/M)$. Now the degree of $\rho \in \mathrm{Irr}\,(G|\tau\psi)$ yields $d(v, q) = 1$ for all $q \in \pi(H/M)$ and the degree of χ leads to $d(v, q) = 1$ for all $q \in \pi(G/H)$. $\square$

18.3 Lemma. *Suppose that $N \trianglelefteq G$ is maximal such that G/N is solvable but non-abelian. Then one of the following two cases occurs.*

  (i) G/N *is a non-abelian p-group and $d(v, p) \leq 1$ for all $v \in \Gamma(G)$; or*

  (ii) G/N *is a Frobenius group. The Frobenius kernel M/N is an elementary abelian r-group and G/M is abelian. Furthermore if $v \in \Gamma(G)$, then $d(v, r) \leq 1$ or $d(v, q) \leq 1$ for all $q \in \pi(G/M)$.*

Proof. If G/N is a p-group and $\chi \in \mathrm{Irr}\,(G)$ satisfying $p \nmid \chi(1)$, then $\chi_N \in \mathrm{Irr}\,(N)$ and $\sigma\chi \in \mathrm{Irr}\,(G)$ for all $\sigma \in \mathrm{Irr}\,(G/N)$ by Lemma 0.10. Conclusion (i) then holds. We may assume that G/N is not a p-group. It follows from the hypotheses and Lemma 12.3 of [Is] that G/N is a Frobenius group whose Frobenius kernel M/N is an elementary abelian r-group with $M/N = (G/N)' = \mathbf{F}(G/N)$. Conclusion (ii) follows with help of Lemma 18.2 (b). $\qquad\square$

Our first result requires solvability of only a small factor group of G.

18.4 Theorem. *Assume that G has a non-abelian solvable factor group. Then one of the following occurs.*

  (i) $n(\Gamma(G)) = 1$ *and* $\mathrm{diam}\,(\Gamma(G)) \leq 4$; *or*

  (ii) $n(\Gamma(G)) = 2$ *and* $\mathrm{diam}\,(\Gamma(G)) \leq 2$.

Proof. Choose $N \trianglelefteq G$ maximal such that G/N is solvable but non-abelian. If G/N is a p-group, Lemma 18.3 gives that $d(v, p) \leq 1$ for all $v \in \Gamma(G)$ and thus $n(\Gamma(G)) = 1$ and $\mathrm{diam}\,(\Gamma(G)) \leq 2$. We may therefore assume that G/N is a Frobenius group, with kernel $M/N \trianglelefteq G/N$ an elementary abelian r-group, and G/M is a $\mathcal{Q}$-group for a set of primes $\mathcal{Q}$. Furthermore, if $v \in \Gamma(G)$, we have $d(v, r) \leq 1$ or $d(v, q) \leq 1$ for all $v \in \mathcal{Q}$. Thus $n(\Gamma(G)) \leq 2$ and if $\Gamma(G)$ has 2 components, its diameter is at most 2. Assume $\Gamma(G)$ has 1 component. We may assume that $r \in \Gamma(G)$, since otherwise $\Gamma(G)$ has diameter at most 2. Let $q_0 \in \mathcal{Q}$ and let

$$\bullet\!\!-\!\!\!-\!\!\bullet\!\!-\!\!\!-\!\!\bullet \cdots \bullet\!\!-\!\!\!-\!\!\bullet$$
$$q_0 \quad t_1 \quad t_2 \qquad t_{d-1}r$$

be a shortest path of length d between q_0 and r. Assume that $d \geq 3$ and choose $\psi \in \mathrm{Irr}\,(G)$ such that $t_1 t_2 \mid \psi(1)$, but q_0 and r do not divide $\psi(1)$. Since G/N has an irreducible character divisible by every $q \in \mathcal{Q}$ and $d(t_2, q_0) = 2$, $t_2 \notin \mathcal{Q}$. Let Q/M be a Sylow q_0-subgroup of G/M and let τ be an irreducible constituent of ψ_Q. Then $t_2 \mid \tau(1)$ and $\tau_N \in \mathrm{Irr}\,(N)$. Now Q/N has an irreducible character β with $q_0 \mid \beta(1)$. Therefore $\beta\tau \in \mathrm{Irr}\,(Q)$, $q_0 t_2 \mid \beta\tau(1)$, and $Q \trianglelefteq G$. Thus $d(q_0, t_2) = 1$, a contradiction. Hence $d(q, r) \leq 2$ for all $q \in \mathcal{Q}$. If $v \in \Gamma(G)$, we have either $d(v, q) \leq 1$ for all $q \in \mathcal{Q}$ or $d(v, r) \leq 1$. Hence $\mathrm{diam}\,(\Gamma(G)) \leq 4$. $\square$

If G is actually solvable, we may strengthen both (i) and (ii) (see Corollary 18.8 below). This however requires some very technical preparations. The next result we are aiming at is that if $\pi \subseteq \Gamma(G)$ with $|\pi| \geq 3$, then $d(p, q) = 1$ for some $p, q \in \pi$ (Theorem 18.7).

18.5 Lemma. *Let G be solvable, and π be a non-empty set of prime divisors of G. Suppose that V is a faithful G-module and $|V| = p^m$ for some prime p. Assume that $\mathbf{C}_G(v)$ contains a Hall π-subgroup of G for each $v \in V$. Furthermore suppose that V_N is homogeneous for all N char G. Then*

(a) *There exists $\chi \in \mathrm{Irr}\,(G)$ such that $q \mid \chi(1)$ for all $q \in \pi$.*

(b) *If G is an $\{s\} \cup \pi$-group for some $s \notin \pi$, then $|\pi| = 1$ and a Hall π-subgroup of G has prime order.*

(c) *If $|V| \neq 3^2$, then each prime in π divides m.*

Proof. The hypothesis on centralizers implies that $\mathbf{O}_\pi(G)$ centralizes V and so $\mathbf{O}_\pi(G) = 1$. In particular, $\mathbf{F}(G) \neq 1$ is a π'-group. Note, without loss of generality, that we may assume $G = \mathbf{O}^{\pi'}(G)$. Corollary 10.6 applies here and we may conclude that V is an irreducible G-module and one of the following occurs:

(i) $G \leq \Gamma(V)$ and $G/\mathbf{F}(G)$ is cyclic;

(ii) $|V| = 3^2$, $G \cong SL(2, 3)$, and $\pi = \{3\}$; or

(iii) $|V| = 2^6$, $\pi = \{2\}$, $|G/\mathbf{F}(G)| = 2$, and $|\mathbf{F}(G)| = 3^3$.

In cases (ii) and (iii), all three conclusions (a), (b), and (c) hold. We thus assume that $G \leq \Gamma$ where $\Gamma = \Gamma(V)$.

Since Γ' is cyclic, so is G'. Since $\mathbf{O}^{\pi'}(G) = G$, we have that $\mathbf{F}(G) = G'$ is a cyclic π'-group and $G/\mathbf{F}(G)$ is a cyclic π-group. Because $\Gamma_0(V) \cap G \leq \mathbf{F}(G)$, $|G/\mathbf{F}(G)| \mid |\Gamma(V)/\Gamma_0(V)|$. Hence $|G/\mathbf{F}(G)| \mid m$, proving (c). By Lemma 18.1, there exists $\chi \in \mathrm{Irr}\,(G)$ with $\chi(1) = |G/\mathbf{F}(G)|$, proving (a). To prove (b), we may assume that $\mathbf{F}(G)$ is a cyclic s-group for a prime s.

Let $S \leq \mathbf{F}(G)$ with $|S| = s$. Because $\mathbf{F}(G) = \mathbf{C}_G(\mathbf{F}(G))$ is a cyclic s-group and $G/\mathbf{F}(G)$ is an s'-group, it follows that $\mathbf{C}_G(S) = \mathbf{F}(G)$. Consequently, whenever $\mathbf{F}(G) < H \leq G$, H is a Frobenius group. Because $H \trianglelefteq G$, each $0 \neq v \in V$ is centralized by a Hall π-subgroup of H. But $\mathbf{C}_{\mathbf{F}(G)}(v) = 1$ and $H/\mathbf{F}(G)$ is a π-group. Hence $\mathbf{C}_H(v) \in \mathrm{Hall}_\pi(H)$. If $R \in \mathrm{Hall}_\pi(H)$, then $|V^\#| = |\mathbf{C}_V(R)^\#||\mathrm{Hall}_\pi(H)| = |\mathbf{C}_V(R)^\#||\mathbf{F}(G)|$.

In particular $|\mathbf{C}_V(R)|$ is independent of the choice of H for $\mathbf{F}(G) < H \leq G$, and $R \in \mathrm{Hall}_\pi(H)$. But $\dim(\mathbf{C}_V(R)) = \dim(V)/|H : \mathbf{F}(G)|$ by Lemma 0.34. Thus it follows that $G/\mathbf{F}(G)$ has prime order and $|\pi| = 1$. $\square$

18.6 Lemma. *Suppose that V is finite faithful irreducible G-module for a solvable group G. Assume that π is a non-empty set of prime divisors of $|G|$ and $\mathbf{C}_G(v)$ contains a Hall π-subgroup of G for each $v \in V$. Then*

(a) *There exists $\chi \in \mathrm{Irr}\,(G)$ such that*

$$q \mid \chi(1) \quad \text{for all} \quad q \in \pi.$$

(b) *If G is an $\{s\} \cup \pi$-group for some $s \notin \pi$, then $|\pi| = 1$.*

Proof. Observe that if $K \trianglelefteq G$, then $\mathbf{C}_K(v)$ contains a Hall π-subgroup of K. In particular, $\mathbf{O}_\pi(G)$ centralizes V, whence $\mathbf{O}_\pi(G) = 1$ and $\mathbf{F}(G)$ is a π'-group.

We may assume that $|\pi| \geq 2$, since otherwise (b) is trivial and part (a) follows from Ito's Theorem (see Theorem 13.1). Let $H = \mathbf{O}^{\pi'}(G)$ and write

$V_H = W_1 \oplus \cdots \oplus W_t$ for (possibly isomorphic) irreducible H-modules W_i. Let $D_i = \mathbf{C}_H(W_i)$, so that $\bigcap_i D_i = 1$ and the D_i are G-conjugate. Now $|H/D_i|$ is divisible by every prime in π. Suppose that $H < G$. By the inductive hypothesis applied to the action of H/D_1 on W_1, we have that in (b), $|\pi| = 1$ and in (a) there exists $\tau \in \mathrm{Irr}\,(H)$ with $q \mid \tau(1)$ for all $q \in \pi$. Part (a) is then completed by choosing $\chi \in \mathrm{Irr}\,(G|\tau)$. We thus assume that $G = \mathbf{O}^{\pi'}(G)$.

By Lemma 18.5, the result is valid if V_N is quasi-primitive. Choose $C \trianglelefteq G$ maximal with respect to V_C not homogeneous. Since $\mathbf{O}^{\pi'}(G) = G$, we will now fix a prime $q \in \pi$ with $q \mid |G/C|$. Next apply Lemma 9.2 and Theorem 9.3 to conclude that
$$V_C = V_1 \oplus \cdots \oplus V_n$$
for homogeneous components V_i of V_C that are faithfully and primitively permuted by G/C. Also

(1) $n = 3$, 5 or 8;

(2) $q = 2$, 2 or 3 (respectively);

(3) G/C is isomorphic to D_6, D_{10} or $A\Gamma(2^3)$ (respectively);

(4) $C/\mathbf{C}_C(V_i)$ acts transitively on $V_i \setminus \{0\}$; and

(5) if $q = 2$, then $\mathrm{char}\,V = 2$.

(As usual D_m denotes the dihedral group of order m. Recall that $A\Gamma(2^3)$ has a unique minimal normal subgroup B. Also $A\Gamma(2^3)/B$ is non-abelian of order 21 and $|B| = 8$.)

To prove (b), assume that G is a $\pi \cup \{s\}$-group. The last paragraph shows that $|G/C|$ is divisible by exactly one prime in π. Since $A\Gamma(2^3)$ has order $8 \cdot 3 \cdot 7$, we thus assume that $q = 2 = \mathrm{char}\,V$, $G/C \cong D_6$ or D_{10} and $s = 3$ or 5. Now (4) above and the hypothesis that each $v \in V$ is centralized by a Hall π-subgroup imply that $|V_i| - 1 = s^j$ for some j. Since $s = 3$ or 5 and $\mathrm{char}\,(V_i) = 2$, we must have $|V_i| = 4$ and $s = 3$ by Proposition 3.1. Then $C/\mathbf{C}_C(V_i)$ is a $\{2,3\}$-group and G is a $\{2,3\}$-group. This proves (b).

We set $C_i = \mathbf{C}_C(V_i)$ and note $\bigcap_i C_i = 1$. Since the C_i are G-conjugate, every prime in π, except possibly q, divides $|C/C_1|$. Observe that the hypotheses imply that $\mathbf{C}_C(w)/C_1$ contains a Hall π-subgroup ($\neq 1$) of C/C_1 for each $w \in V_1$. Since C/C_1 acts transitively on $V_1 \setminus \{0\}$, C/C_1 acts quasi-primitively on V_1. Now Lemma 18.5 implies that there exists $\tau \in \mathrm{Irr}\,(C)$ with $C_1 \leq \ker(\tau)$ and $\tau(1)$ divisible by every prime of π dividing $|C|$. Since $C \trianglelefteq G$, we may assume that $q \nmid |C|$. Let $\pi_0 = \pi \setminus \{q\}$ and recall that $q = 2$ or 3. We have that $\pi_0 \neq \varnothing$, no prime in π_0 divides $|G/C|$ and $2 \notin \pi_0$. If $|V_i| = 3^2$, then C/C_i would be a $\{2, 3\}$-group, which is not possible because $q \leq 3$, $q \nmid |C|$, and $\mathbf{O}_\pi(G) = 1$. By Lemma 18.5 (c), every prime in π_0 divides m, where $|V_1| = p^m$, $p = \mathrm{char}\,V$. Also, no divisor of $p^m - 1$ is in π_0, by transitivity. As $2 \notin \pi_0$, it thus follows that $m \geq 3$, $m \neq 4$ and $p^m \neq 2^6$. By Corollary 6.6, we have a series $C_i < R_i \leq F_i \leq C$ of normal subgroups of C such that $F_i/C_i = \mathbf{F}(C/C_i)$ is a cyclic π'-group, C/F_i is cyclic, R_i/C_i is a Sylow r-subgroup of C/C_i for a Zsigmondy prime divisor r of $p^m - 1$, and $F_i/C_i = \mathbf{C}_{C/C_i}(R_i/C_i)$. We let $F = \bigcap_i F_i$ and $R = \bigcap_i R_i$. Then $F = \mathbf{F}(C) = \mathbf{C}_C(R)$, $R \trianglelefteq G$ is a Sylow r-subgroup of C and both R and F are abelian (see Proposition 9.5). Also F is a π'-group, as $\mathbf{F}(G)$ is.

Let $C_i \leq M_i < R_i$ with $|R_i/M_i| = r$ and let $M = \bigcap_i M_i$, so that R/M is elementary abelian. Since R_i/C_i is cyclic, the M_i are all the G-conjugates of M_1 and hence $M \trianglelefteq G$. Since $r \nmid |C/F_i|$, then R_i/M_i is a faithful C/F_i-module. Furthermore, $(M_i \cap R)/M$ is a C-submodule of R/M of codimension 1, and $R/(M_i \cap R)$ is C-isomorphic to R_i/M_i. It follows that R/M is a faithful C/F-module and, as $r \nmid |C/F|$, $R/M = A_1 \oplus \cdots \oplus A_l$, where each A_j is C-isomorphic to R_i/M_i for some i.

We next show that R/M is in fact a faithful G/F-module. Since C/F acts faithfully on R/M, we have that $C \cap \mathbf{C}_G(R/M) = F$ and thus $\mathbf{C}_G(R/M)$ centralizes C/F. Since $m \neq 2$ or 4 and since $C > F$ (as $\pi_0 \neq \varnothing$), Lemma 9.10 yields that $C/F = \mathbf{C}_{G/F}(C/F)$. Thus $\mathbf{C}_G(R/M) \leq C \cap \mathbf{C}_G(R/M) = F$. So R/M is a faithful G/F-module.

We claim that for each $1 \neq \tau \in \mathrm{Irr}\,(R/M)$, $|C : I_C(\tau)|$ is divisible by every prime in π_0. In particular, we can then assume that $q \nmid |G : I_G(\tau)|$ for all

$\tau \in \mathrm{Irr}\,(R/M)$, since otherwise the conclusion (a) of the lemma is satisfied with $\chi \in \mathrm{Irr}\,(G|\tau)$. For this claim, we write $\tau = \tau_1 \times \cdots \times \tau_l$ and observe that $I_C(\tau) = \bigcap_i I_C(\tau_i)$. Without loss of generality, $\tau = \tau_1 \times 1 \times \cdots \times 1$ and $1 \neq \tau_1 \in \mathrm{Irr}\,(A_1)$. Since A_1 is C-isomorphic to, say, R_1/M_1, $I_C(\tau) = F_1$. Since C/F_1 is divisible by all primes in π_0, the claim holds.

By the last paragraph, $q \nmid |G : I_G(\tau)|$ for all $\tau \in \mathrm{Irr}\,(R/M)$. Now $\mathrm{Irr}\,(R/M)$ is a faithful G/F-module. If $q = 2$, then $r = 2$ by Lemma 9.2. This is a contradiction, because $q \nmid |C|$. Hence $q = 3$. Since each prime in $\pi_0 \neq \varnothing$ divides m and $2 \notin \pi_0$, either $m \geq 7$ or $m = 5$. In particular, the Zsigmondy prime divisor r of $p^m - 1$ is not 2, 3, or 7. Thus $r \nmid |G/C|$ and $r \nmid |G/F|$.

We now have that $q = 3$ and $G/C \cong A\Gamma(2^3)$. Let $S = \mathbf{N}_G(C_1) \geq \mathbf{N}_G(V_1) > C$. Then $S/C \cong A\Gamma(2^3)$ or $\Gamma(2^3)$. Let $D = \mathbf{C}_S(V_1)$ so that $D \cap C = C_1$ and S/D acts transitively on $V_1 \setminus \{0\}$, because $C \leq S$. Recall that $|C/C_1|$ is divisible by all primes in π_0. By Lemma 6.5, we may choose a faithful linear $\lambda \in \mathrm{Irr}\,(F_1/C_1)$ such that $\lambda^C \in \mathrm{Irr}\,(C)$ and hence $\lambda^C(1)$ is divisible by every prime in π_0. Since λ^C has kernel C_1, $I_G(\lambda^C) \leq S$. Furthermore, $\{\lambda^C\} = \mathrm{Irr}\,(C|\lambda)$. Thus it suffices to show that there exists $\delta \in \mathrm{Irr}\,(S|\lambda) = \mathrm{Irr}\,(S|\lambda^C)$ with $3 \mid \delta(1)$, because then $\delta(1)$ is divisible by all primes in π and δ^G is irreducible.

Since $DF_1/D \cong F_1/C_1$, λ may be considered as a linear character of $DF_1/D \leq \mathbf{F}(S/D)$. But S/D acts transitively on $V_1 \setminus \{0\}$, and thus applying Lemma 6.5, λ extends to $\lambda^* \in \mathrm{Irr}\,(\mathbf{F}(S/D))$ with $(\lambda^*)^S$ irreducible. Note that for $v = (x, 0, \ldots, 0) \in V \setminus \{0\}$, we have $\mathbf{C}_G(v) \leq \mathbf{N}_G(V_1) \leq S$, hence each $x \in V_1$ is centralized by a Sylow 3-subgroup of S/D. Thus $3 \nmid |\mathbf{F}(S/D)|$ and we obtain either $3 \mid (\lambda^*)^S(1)$ or $3 \mid |D/C_1|$. By the last paragraph, we are done in the first case and we thus assume that $3 \mid |D/C_1|$. Now $DC/C \trianglelefteq S/C$ and $S/C \cong A\Gamma(2^3)$ or $\Gamma(2^3)$. Since $3 \mid |DC/C|$, $DC = S$ and $S/C_1 = C/C_1 \times D/C_1$. Since $D/C_1 \cong S/C$, there exists $\mu \in \mathrm{Irr}\,(D/C_1)$ with $3 \mid \mu(1)$. Now $\delta := \lambda^C \times \mu \in \mathrm{Irr}\,(S|\lambda)$ and $3 \mid \delta(1)$. $\qquad\square$

18.7 Theorem. *Let G be solvable and let π be a set of primes contained in $\Gamma(G)$. Assume that $|\pi| \geq 3$. Then there exist distinct $u, v \in \pi$ such that $uv \mid \chi(1)$ for some $\chi \in \mathrm{Irr}(G)$.*

Proof. We proceed by induction on $|G|$. Let $F = \mathbf{F}(G)$. Now

$$\Gamma(G) = \{p \mid G \text{ does not have a normal abelian Sylow } p\text{-subgroup}\}$$

$$= \{p \mid p \text{ divides } |G/F| \text{ or } \mathbf{O}_p(G) \text{ is non-abelian}\}.$$

Note that $F' \leq \Phi(G) \leq F$ and $\mathbf{F}(G/\Phi(G)) = F/\Phi(G)$ (see Theorem 1.12). Arguing by induction on $|G|$, we may assume that

Step 1. If $P \in \mathrm{Syl}_p(F)$ for a prime p, then either

   (i) P is elementary abelian, or

   (ii) $p \in \pi$, $p \nmid |G/F|$, and P' is a minimal normal subgroup of G.

Step 2. If M is a minimal normal subgroup of G, then G/M has a normal abelian Sylow s-subgroup S/M for some $s \in \pi$. Furthermore, S is either a non-abelian s-group or M is a non-trivial S/M-module.

Proof. Arguing by induction, we may assume that there is some $s \in \pi$ such that $s \notin \Gamma(G/M)$. Thus G/M has a normal abelian Sylow s-subgroup S/M. Since G does not have a normal abelian Sylow s-subgroup, S is non-abelian and $S' = M$. This step follows.

Step 3. If F is abelian, then F is the unique minimal normal subgroup of G.

Proof. If F is abelian, then $\Phi(G) = 1$, by Step 1. By Theorem 1.12, F is a completely reducible G-module. For this step, we may assume there exist distinct minimal normal subgroups M, N of G. By Step 2, G/M and G/N have a normal abelian Sylow s-subgroup S/M and a normal abelian Sylow t-subgroup T/N, respectively, for primes $s, t \in \pi$. (We do not assume that s and t are distinct!) Furthermore, S and T are non-abelian. Since F is abelian, $s \nmid |M|$ and $t \nmid |N|$. Now $MN = M \times N$. Let $1 \neq \mu \in \mathrm{Irr}\,(M)$ and $1 \neq \nu \in \mathrm{Irr}\,(N)$. Choose $\theta \in \mathrm{Irr}\,(ST|\mu\nu)$. Now μ is a constituent of θ_M and thus there exists $\alpha \in \mathrm{Irr}\,(S|\mu)$ with $[\theta_S, \alpha] \neq 0$. Since M is a non-trivial module for the s-group $S/M \trianglelefteq G/M$, and M is an irreducible G/M-module, $\mathbf{C}_M(S/M) = 1$. Since $\mu \neq 1$, $s \mid \alpha(1)$ and $s \mid \theta(1)$. Likewise, $t \mid \theta(1)$. If $\chi \in \mathrm{Irr}\,(G|\theta)$, then $st \mid \chi(1)$, since $ST \trianglelefteq G$. The conclusion of the theorem is satisfied unless $s = t$. Let $S_0 \in \mathrm{Syl}_s(G)$ and observe $S = S_0 M$, $T = S_0 N$, and S_0 is abelian. Now MS_0/M and MN/M are normal in G/M and so $[S_0, MN] \leq M$. Likewise $[S_0, MN] \leq M \cap N = 1$. Thus $S = S_0 \times M$ and G have a normal abelian Sylow s-subgroup, a contradiction to $s \in \Gamma(G)$.

Step 4. F is non-abelian.

Proof. By Steps 2 and 3, we can assume that F is the unique minimal normal subgroup of G and G/F has a normal abelian Sylow s-subgroup $S/F \neq 1$ for a prime $s \in \pi$. If $1 \neq \lambda \in \mathrm{Irr}\,(F)$ and $\tau \in \mathrm{Irr}\,(S|\lambda)$, then $s \mid \tau(1)$. Hence $s \mid \beta(1)$ for all $\beta \in \mathrm{Irr}\,(G|\lambda)$. Thus we may assume that $|G : I_G(\lambda)|$ is not divisible by each prime in $\pi \setminus \{s\}$. Applying Lemma 18.6 (a) to the action of G/F on F, there exists $\gamma \in \mathrm{Irr}\,(G/F)$ such that $\gamma(1)$ is divisible by all primes in $\pi \setminus \{s\}$. Since $|\pi| \geq 3$, the conclusion of the theorem is satisfied.

Step 5. F is a non-abelian Sylow r-subgroup of G for some $r \in \pi$.

Proof. Now F has an irreducible character whose degree is divisible by every prime p for which the Sylow p-subgroup of F is non-abelian. Thus by Steps 1 and 4, there is a unique prime r for which F has a non-abelian Sylow r-subgroup R. Furthermore, $r \in \pi$ and $R \in \mathrm{Syl}_r(G)$. For this step, we may assume that G has a minimal normal subgroup M with $M \not\leq R$. By Step 2, G/M has a normal abelian Sylow s-subgroup S/M for some $s \in \pi$ and S is non-abelian. Since $M \not\leq R$, the uniqueness of r implies that $s \nmid |M|$. Hence a Sylow s-subgroup of G is abelian, whence $s \neq r$. Now $RS = R \times S \trianglelefteq G$ and RS has a character of degree divisible by rs. In this case, the conclusion of the theorem is satisfied. This step follows.

For convenience, say $\pi = \{r, s, t\}$. Also let $S \in \mathrm{Syl}_s(G)$ and $T \in \mathrm{Syl}_t(G)$ such that $ST = TS$. By Steps 1 and 5, F' is the unique minimal normal subgroup of G. Let $C = \mathbf{C}_G(F') \geq F$.

Step 6.

(i) Without loss of generality, $S \leq C$ and $t \nmid |C|$.

(ii) SF/F is a normal abelian Sylow s-subgroup of G/F.

(iii) For all $1 \neq \alpha \in \mathrm{Irr}(F')$, $|G : I_G(\alpha)|$ is coprime to st and $I_G(\alpha)/F$ contains normal abelian Sylow s- and Sylow t-subgroups. In particular, ST is an abelian Hall $\{s, t\}$ subgroup of G.

Proof. Let $1 \neq \alpha \in \mathrm{Irr}(F')$ and observe that $r \mid \delta(1)$ for all $\delta \in \mathrm{Irr}(F|\alpha)$. Thus $r \mid \sigma(1)$ for all $\sigma \in \mathrm{Irr}(G|\alpha)$. Consequently, $I_G(\alpha)$ contains a Hall $\{s, t\}$-subgroup of G for all $\alpha \in \mathrm{Irr}(F')$. Applying Lemma 18.6 (a) we may assume that G/C is divisible by at most one prime in $\pi \setminus \{r\} = \{s, t\}$. Without loss of generality, S is contained in C.

Since C/F is an r'-group, there exists $\beta \in \mathrm{Irr}(F|\alpha)$ that is invariant in C by Lemma 0.17 (d). Again by coprimeness, β extends to $\beta^* \in \mathrm{Irr}(C)$. As $\xi\beta^* \in \mathrm{Irr}(C|\beta)$ for all $\xi \in \mathrm{Irr}(C/F)$ and $r \mid \beta(1)$, we have $(st, \xi(1)) = 1$ for all $\xi \in \mathrm{Irr}(C/F)$. Thus C/F has normal abelian Sylow subgroups SF/F and $(T \cap C)F/F$. Clearly, SF/F then is a normal abelian Sylow subgroup of G/F.

By Lemma 18.1, there exists $\eta \in \mathrm{Irr}(C)$ such that $\eta(1)$ is divisible by all prime divisors of $|\mathbf{F}(C/F)|$. In particular, $s \mid \eta(1)$ and we may assume then that $t \nmid |\mathbf{F}(C/F)|$. But C/F has a normal Sylow t-subgroup. Thus $t \nmid |C|$. Since $(|I_G(\alpha)/F|, |F|) = 1$, the same argument as in the previous paragraph shows that $I_G(\alpha)/F$ has normal abelian Sylow s- and Sylow t-subgroups and $(st, |G : I_G(\alpha)|) = 1$.

Step 7. If $F \leq N \trianglelefteq G$ with $st \mid |N|$, then $N = G$.

Proof. Since $F = \mathbf{F}(N)$, we have $\pi \subseteq \Gamma(N)$. If $N < G$, the theorem follows by induction. Thus $N = G$.

Step 8. $G/F = S_0/F \times H/F$ for $S_0 := SF$ and some $H \trianglelefteq G$. Also $|S_0/F| = |S| = s$.

Proof. Now $S_0/F \lhd G$ and $ST \leq \mathbf{C}_G(S)$ by Step 6. Thus $\mathbf{O}^{\{s,t\}}(G/F) \leq \mathbf{C}_G(S_0/F)$. By Step 7, $S_0/F \leq \mathbf{Z}(G/F)$. Thus $G/F = SF/F \times H/F$ for some $H \leq G$. Now $S_0/F \cong S$ since $s \nmid |F|$. Finally Step 7 implies that $|S_0/F| = s$.

Step 9. H/F $(\cong G/S_0)$ has a unique maximal normal subgroup L/F. Also $|H/L| = t$ and $t \nmid |L/F|$.

Proof. If $S_0 \leq J \lhd G$, then Step 7 implies that $t \nmid |J/S_0|$, whence $t \mid |G/J|$. By the solvability of G/S_0, it follows that G/S_0 has a unique maximal normal subgroup K/S_0, $t \nmid |K/S_0|$, and $t = |G : K|$. Since $H/F \cong G/S_0$, this step follows with $L = H \cap K$.

Step 10. (a) If $\lambda \in \mathrm{Irr}\,(F/F')$ is not S-invariant, then $I_G(\lambda)/F$ contains exactly one Sylow t-subgroup of G/F.
(b) If $1 \neq \varphi \in \mathrm{Irr}\,(F')$, then $I_G(\varphi)/F$ contains exactly one Sylow t-subgroup of G/F.

Proof. (a) We note that λ is S-invariant if and only if λ is S_0-invariant because $FS = S_0$. Assuming that λ is not S-invariant, we have that every $\tau \in \mathrm{Irr}\,(S_0|\lambda)$ satisfies $s \mid \tau(1)$. We may thus assume that $t \nmid \chi(1)$ for all $\chi \in \mathrm{Irr}\,(G|\lambda)$, as otherwise the conclusion of the theorem is satisfied. Since $r \nmid |G/F|$, we have that λ extends to $\lambda^* \in \mathrm{Irr}\,(I_G(\lambda)|\lambda)$ (see Theorem 0.13). Consequently Gallagher's and Clifford's Theorems (0.8 and 0.9) imply that $\beta \mapsto (\beta\lambda^*)^G$ is a bijection from $\mathrm{Irr}\,(I_G(\lambda)/F)$ onto $\mathrm{Irr}\,(G|\lambda)$. Thus $t \nmid |G : I_G(\lambda)|$ and $I_G(\lambda)$ contains a Sylow t-subgroup of G. Furthermore, $t \nmid \beta(1)$ for all $\beta \in \mathrm{Irr}\,(I_G(\lambda)/F)$ and Ito's Theorem 13.1 yields that $I_G(\lambda)/F$ has a normal Sylow t-subgroup. This proves (a).

(b) Let $1 \neq \varphi \in \mathrm{Irr}(F')$. Then $F \leq I_G(\varphi)$ because $F' \leq \mathbf{Z}(F)$. If $\theta \in \mathrm{Irr}(F|\varphi)$, then $r \mid \theta(1)$. Consequently $t \nmid \chi(1)$ for all $\chi \in \mathrm{Irr}(G|\theta)$. Repeat the argument of the last paragraph to conclude that $I_G(\theta)/F$ contains exactly one Sylow t-subgroup of G/F. Observe that $I_G(\theta) \leq I_G(\varphi)$ because $F' \leq \mathbf{Z}(F)$. Because $r \nmid |G/F|$, we may apply Lemma 0.17 to conclude there exists $\theta^* \in \mathrm{Irr}(F|\varphi)$ such that $I_G(\theta^*) = I_G(\varphi)$. Part (b) follows.

Step 11.

(a) Suppose that $F/F' = A/F' \times B/F'$ with $A, B \trianglelefteq G$. Then $A = F'$ or $B = F'$.

(b) $F = F'[F, S]$

(c) $F/\Phi(F)$ is a faithful irreducible G/F-module.

Proof. (a) Suppose not. By Gaschütz's Theorem 1.12, $S \cong S_0/F$ acts faithfully on $F/\Phi(F)$ and also on F/F'. Without loss of generality, S does not centralize B/F' and we may choose $\beta \in \mathrm{Irr}(B/F')$ that is not S-invariant. By Step 10, $I_G(1_A \times \beta)/F = I_G(\beta)/F$ contains a unique Sylow t-subgroup T_0/F of G/F. For $\alpha \in \mathrm{Irr}(A/F')$, $I_G(\alpha \times \beta) = I_G(\alpha) \cap I_G(\beta)$. Thus $\alpha \times \beta$ is not S-invariant, and by Step 10, we must have $T_0/F \leq I_G(\alpha)$. Thus T_0 fixes all $\alpha \in \mathrm{Irr}(A/F')$. Then T_0/F centralizes A/F by Proposition 12.1. Hence $H/F = \mathbf{O}^t(G/F) \leq \mathbf{C}_{G/F}(A/F')$.

If S does not centralize A/F', the same argument repeated with A and B interchanged implies that $H/F \leq \mathbf{C}_{G/F}(B/F')$. This then implies that $1 \neq H/F$ centralizes F/F', contradicting Gaschütz's Theorem 1.12. So S and H/F centralize $A/F' \cong F/B$. Consequently, $F/B \leq \mathbf{Z}(G/B)$ and $[F, S]F' \leq B$.

Since $F' < [F, S]F' \leq B$, we may apply Fitting's Lemma 0.6 to assume without loss of generality that $B = [F, S]F'$ and $A/F' = \mathbf{C}_{F/F'}(S) = \mathbf{C}_{F/F'}(S_0/F)$. Since S centralizes $F' \leq \mathbf{Z}(F)$, we see that $[A, F, S] \leq [F', S] = 1$ and $[S, A, F] \leq [F', F] = 1$. By the Three Subgroups Lemma, $[F, S, A] = 1$. Since $B = [F, S]F' \leq [F, S]\mathbf{Z}(F)$, we have that $[A, B] = 1$.

We next observe that $B \leq \mathbf{Z}(F)$. Since $F/B \leq \mathbf{Z}(G/B)$, it follows that $G/B = F/B \times J/B$ where $J/B \in \mathrm{Hall}_{r'}(G/B)$. Since $|G/J|$ is a power of r, certainly s, $t \in \Gamma(J)$. If B is non-abelian, then $r \in \Gamma(J)$ because $B \trianglelefteq J$. Then the inductive hypothesis yields $\rho \in \mathrm{Irr}\,(J)$ with $\rho(1)$ divisible by at least two primes in $\{r, s, t\}$. The desired conclusion would then follow because $J \trianglelefteq G$. Hence B is abelian. Because $[A, B] = 1$ and $F = AB$, it follows that $B \leq \mathbf{Z}(F)$.

Now $[F, H] \leq B \leq \mathbf{Z}(F)$ and consequently $[F, H, F] = 1 = [H, F, F]$. By the Three Subgroups Lemma, $[F, F, H] = 1$, i.e. H centralizes F', contradicting Step 6 (i). This contradiction yields part (a).

(b) By Fitting's Lemma 0.6, write $F/F' = \mathbf{C}_{F/F'}(S) \times [F/F', S]$. Since $[F/F', S] = F'[F, S]/F'$, part (b) follows from part (a) or S centralizes F/F'. But $S \cong S_0/F$ does not centralizes F/F' by Theorem 1.12 and so (b) follows.

(c) Now $F/\Phi(F)$ is a faithful completely reducible G/F-module by Theorem 1.12. To prove (c), we may assume that $F/\Phi(F) = A_0/\Phi(F) \times B_0/\Phi(F)$ for A_0, $B_0 \trianglelefteq G$ and A_0, $B_0 > \Phi(F)$. Repeating the arguments of the first two paragraphs of part (a), we may assume that S centralizes $A_0/\Phi(F)$. Then $F'[F, S] \leq B_0 < F$, contradicting (b) and completing this step.

Step 12. F/F' is an irreducible H/F-module. In particular, $\Phi(F) = F' = \mathbf{C}_F(S) = \mathbf{Z}(F)$.

Proof. Assume not and let $F' < E < F$ with $E \trianglelefteq H$. Choose $1 \neq \xi \in \mathrm{Irr}\,(E/F')$. Since $r \nmid |H/F|$, Lemma 0.17 implies there exists an extension $\xi^* \in \mathrm{Irr}\,(F/F')$ of ξ such that ξ^* is invariant in $I_H(\xi)$. Then $I_H(\xi) = I_H(\xi^*)$.

By Step 11 and Fitting's Lemma, the principal character is the only S-invariant irreducible character of F/F'. By Step 10 (a), $I_G(\xi^*)/F$ contains exactly one Sylow t-subgroup T_1/F of G/F. By Step 8, $T_1/F \leq \mathbf{O}^{t'}(G/F) = H/F$. If $\lambda \in \mathrm{Irr}\,(F/E)$, then $\lambda\xi^*$ extends ξ and $I_H(\lambda\xi^*) \leq I_H(\xi) = I_H(\xi^*)$.

By Step 10, we must have that T_1 stabilizes $\lambda\xi^*$ and also λ. Since this is true for all $\lambda \in \mathrm{Irr}\,(F/E)$, T_1/F centralizes F/E. Since $\mathbf{O}^{t'}(H/F) = H/F$, in fact H/F centralizes F/E. Then $\mathbf{C}_{F/F'}(H/F) \neq 1$. Now $H/F \unlhd G/F$, so that Fitting's Lemma 0.6 and Step 11 (a) imply that H/F centralizes F/F'. This contradicts Theorem 1.12, because $H/F \neq 1$. This step is complete.

Step 13. (a) F/F' is a faithful irreducible L/F-module.

(b) H/F is a Frobenius group with cyclic Frobenius kernel L/F.

Proof. Now $V := \mathrm{Irr}\,(F/F')$ is a faithful irreducible H/F-module by Step 12 and Proposition 12.1. Let $0 \neq W$ be an L-submodule of V and let $0 \neq w \in W$. By Step 12, w is not S-invariant and so, by Step 10 (a), w is centralized by a Sylow t-subgroup T_2/F of H/F. Then W is stabilized by $L\,T_2 = H$ and so $W = V$. Thus V is an irreducible L-module. Part (a) follows via Proposition 12.1.

Now each $\lambda \in V$ is centralized by a Sylow t-subgroup of H/F and $\mathbf{O}^{t'}(H/F) = H/F$. If V is a quasi-primitive module, Theorem 10.4 implies conclusion (b) or that $t = 3 = \mathrm{char}\,(V) = r$. Since t and r are distinct, we may assume that V is not quasi-primitive. By Theorem 9.3, there exists $D/F \unlhd H/F$ such that $V_D = V_1 \oplus \cdots \oplus V_n$ $(n > 1)$ for homogeneous components V_i of V_D that are transitively permuted by H/D. Furthermore, D/F transitively permutes the elements of $V_1 \setminus \{0\}$. So $|V_1 \setminus \{0\}| \mid |H/F|$. Since S centralizes H/F, S permutes the V_i. By Glauberman's Lemma 0.14, S fixes V_1. Since $\mathbf{C}_V(S) = \mathbf{C}_V(S_0/F) = 1$, $s \mid |V_1 \setminus \{0\}|$. Then $s \mid |H/F|$, a contradiction. Part (b) holds.

Step 14. $|F/F'| = |F'|$.

Proof. Since F' is a minimal normal subgroup of G and $S_0 \leq \mathbf{C}_G(F')$, in fact F' is an irreducible H/F-module, as is $\mathrm{Irr}\,(F')$. For $1 \neq \lambda \in \mathrm{Irr}\,(F')$, $I_H(\lambda)/F$ contains a unique Sylow t-subgroup of H/F by Step 10 (b). Since H/F is a Frobenius group, we must have that $I_H(\lambda)/F \in \mathrm{Syl}_t(H/F)$. Since $\mathbf{O}^{t'}(H/F) = H/F \neq 1$, H/F acts faithfully on F'. Since $I_H(\lambda)/F \in$

$\mathrm{Syl}_t(H/F)$ and $L/F = \mathbf{O}^t(H/F)$, in fact F' is a faithful irreducible L/F-module (just as in Step 13 (a)). Now F/F' and F' are faithful irreducible L/F-modules in characteristic r. Since L is cyclic, $|F/F'| = |F'|$ by Example 2.7.

Step 15. If $U < F'$, then $F'/U = \mathbf{Z}(F/U)$.

Proof. Set $Z/U = \mathbf{Z}(F/U) \geq F'/U$. Since U is S-invariant, so is Z. Let $1 \neq \varphi \in \mathrm{Irr}(F'/U)$. For $\theta \in \mathrm{Irr}(F|\varphi)$, we have that $r \mid \theta(1)$. Then θ is S-invariant, since otherwise the conclusion of the theorem is satisfied. Thus the unique irreducible constituent of θ_Z is S-invariant and extends φ. Applying Gallagher's Theorem, every $\alpha \in \mathrm{Irr}(Z/F')$ is S-invariant. Thus S and S_0/F centralize Z/F'. Step 12 implies that $Z = F'$, as desired.

Step 16. Conclusion.

Proof. Fix $x \in F \setminus F'$ and let $Y = [F, x]$. Now Yx is central in F/Y and $Y \leq F'$. By Step 15, $Y = F'$. Since $F' \leq \mathbf{Z}(F)$, the map $g \mapsto [g, x]$ is a homomorphism of F onto $F' = [F, x]$. Thus $|F/\mathbf{C}_F(x)| = |F'| = |F/F'|$ by Step 14. This is a contradiction because $F' = \mathbf{Z}(F) < \mathbf{C}_F(x)$. This completes the proof of the Theorem.   □

The assertion of the theorem above is wrong if G is not solvable. To see this observe that $\Gamma(PSL(2, 2^f))$, $f \geq 2$, has three components. Namely the ordinary character degrees of $PSL(2, 2^f)$ are $1, 2^f - 1, 2^f$ and $2^f + 1$ (see [HB, Theorem XI, 5.5]). As another example, consider the group $PSL(2, 11)$, which has the following graph (see [HB, Theorem XI, 5.7]):

$$3 \text{—} 2 \text{—} 5 \qquad\qquad 11$$

18.8 Corollary. *Assume that G is solvable.*

(a) *Then* $\mathrm{diam}(\Gamma(G)) \leq 3$.

(b) *If* $\Gamma(G)$ *has two components* Γ_1 *and* Γ_2, *then both are regular graphs.*

Proof. (a) If not, $\Gamma(G)$ contains a shortest path of length 4, say

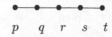

$$p \quad q \quad r \quad s \quad t$$

for (distinct) primes p, q, r, s, t. By Theorem 18.7, two of the vertices p, r and t must be connected, a contradiction.

(b) Let $p \in \Gamma_1$ and q, $r \in \Gamma_2$. By Theorem 18.7, $d(q,r) \leq 1$. Thus Γ_2 is regular. Likewise, Γ_1 is regular.   □

Solvable groups with two components cannot be too complicated. For such G, the nilpotence length must be between 2 and 4. We will prove this in Theorem 19.6, whose proof gives much more information about such G.

On the other extreme, we do not know of a solvable group whose graph has diameter 3. But the simple Janko group J_1 does have diameter 3. The graph for J_1 is:

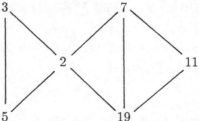

But even this seems rare. Consulting The Atlas of Finite Groups [CCNPW], many simple groups have regular graphs.

The graphs $\Gamma(A_5)$ and $\Gamma(PSL(2,8))$ each have 3 components. It is true that $n(\Gamma(G)) \leq 3$ for arbitrary G, a theorem due to Manz, Staszewski, and Willems [MSW]. In Corollary 19.8, we show that a minimal counterexample to this theorem is simple.

§19   Coprime Group Actions and Character Degrees

We begin this section by proving an interesting Theorem 19.3 of Isaacs [Is 9]. Suppose that H acts non-trivially on N, $(|H|, |N|) = 1$ and H fixes every non-linear character of N. Then N is solvable (without appeal to the classification of simple groups) and the nilpotence length of N is at most two. More can be said regarding the structure of N, see Theorem 19.3.

We apply Isaacs' Theorem 19.3 and Theorem 12.9 to study solvable groups G whose graph $\Gamma(G)$ has exactly 2 components, i.e. there exists a "non-trivial" set π of primes such that each $\chi(1)$, $\chi \in \mathrm{Irr}(G)$, is a π-number or π'-number. The structure of such groups G is very limited (e.g. $\mathrm{dl}(G/\mathbf{F}(G)) \le 4$ and the nilpotence length $n(G)$ is 2, 3, or 4).

We close this section with another application of Theorem 19.3. If G is non-solvable, then $n(\Gamma(G)) \le \max_E n(\Gamma(E))$ as E ranges over non-solvable composition factors of G. This result employs the Ito–Michler Theorem 13.13 for non-solvable groups and thus relies upon the classification of simple groups.

We begin with two lemmas. In Lemma 19.1, we do not assume that P acts faithfully on Q.

19.1 Lemma. Let P be a p-group of class $\mathrm{cl}(P) \le 2$ and suppose that P acts on some non-trivial p'-group Q such that $\mathbf{C}_P(x) \le P'$ for all $1 \ne x \in Q$. Then P acts fixed-point-freely and is either cyclic or isomorphic to Q_8.

Proof. We argue by induction on $|P|$. First assume that P does not act fixed-point-freely on Q. Then there exists $1 \ne x \in Q$ such that $Z := \mathbf{C}_P(x) \ne 1$. Since $Z \le P' \le \mathbf{Z}(P)$, Z is normal in P. Let $C = \mathbf{C}_Q(Z)$. Then $x \in C$ and P/Z acts on C. Now if $y \in C \setminus \{1\}$, then

$$\mathbf{C}_{P/Z}(y) = \mathbf{C}_P(y)/Z \le P'/Z = (P/Z)',$$

and the action of P/Z on C satisfies the hypotheses of the Lemma. By the inductive hypothesis, P/Z acts fixed-point-freely on C. Also P/Z is cyclic or $P/Z \cong Q_8$. In the first case, clearly P has to be abelian, contradicting $1 \neq Z \leq P'$. In the second case, we may take subgroups A and B of P, such that $P = AB$, $Z \leq A \cap B$ and such that both A/Z and B/Z are cyclic of order 4. Consequently, A and B are abelian, $A \cap B \leq \mathbf{Z}(P)$ and $|P : \mathbf{Z}(P)| \leq 4$. This forces $|P'| \leq 2$ and since $1 \neq Z \leq P'$, we conclude $Z = P'$ and $Q_8 \cong P/Z$ is abelian, a contradiction. Hence P acts fixed-point-freely on Q. Because every abelian normal subgroup of P is cyclic and $\mathrm{cl}(P) \leq 2$, either P is cyclic or $P \cong Q_8$ (see Corollary 1.3). $\qquad \square$

The next result generalizes earlier results of A. Camina [Cm 1].

19.2 Lemma (Isaacs [Is 9]). *Let K be a proper normal subgroup of G and assume that G/K is nilpotent. Suppose that each conjugacy class of G outside of K is a union of cosets of K. Then*

(i) *G is a Frobenius group with kernel K; or*

(ii) *G/K is a p-group for some prime p; also G has a normal p-complement M and $\mathbf{C}_G(m) \leq K$ for all $1 \neq m \in M$.*

Proof. (1) For $g \in G \setminus K$, we claim that $|\mathbf{C}_G(g)| = |\mathbf{C}_{G/K}(Kg)|$. To see this, note that by our hypothesis, the conjugacy class $\mathrm{cl}_G(g)$ is the union of exactly those cosets of K which constitute the class $\mathrm{cl}_{G/K}(Kg)$. It follows that

$$|G : \mathbf{C}_G(g)| = |\mathrm{cl}_G(g)| = |K||\mathrm{cl}_{G/K}(Kg)| = |K||(G/K) : \mathbf{C}_{G/K}(Kg)|,$$

which at once yields the claim.

(2) Let $g \in G \setminus K$ and let $k \in \mathbf{C}_K(g)$. Then g is conjugate to gk and so $o(kg) = o(g)$. Since $kg = gk$, $o(k) \mid o(g)$.

(3) Suppose first that G splits over K, i.e. there is some subgroup U of G such that $G = KU$ and $K \cap U = 1$. If $u \in U \setminus \{1\}$ then (1) yields

$$|\mathbf{C}_U(u)| = |\mathbf{C}_{G/K}(Ku)| = |\mathbf{C}_G(u)|,$$

and so $\mathbf{C}_G(u) \leq U$. It follows that G is a Frobenius group with kernel K and assertion (i) holds.

(4) Suppose now that G does not split over K. We wish to establish assertion (ii) and we assume that G/K is not a p-group for any prime p. We can choose, therefore, $z \in G \setminus K$ such that $Kz \in \mathbf{Z}(G/K)$ and $o(Kz) = pq$ for primes $p \neq q$. We may assume that $z = xy = yx$ with x a p-element, y a q-element and $x, y \in \langle z \rangle \cap [G \setminus K]$. If $k \in \mathbf{C}_K(z)$, then (2) implies that $o(k) \mid (o(x), o(y))$. Thus $\mathbf{C}_K(z) = 1$. Let $C = \mathbf{C}_G(z)$. Then $K \cap C = 1$ and $|C| = |\mathbf{C}_{G/K}(Kz)| = |G/K|$, where (1) is used again and the fact that $Kz \in \mathbf{Z}(G/K)$. Consequently $G = KC$ and G splits over K, a contradiction. This shows that G/K is a p-group.

To prove the second assertion of (ii), we fix $P \in \mathrm{Syl}_p(G)$ and claim that $P \cap K \leq P'$. Let $z \in P \setminus K$ with $Kz \in \mathbf{Z}(G/K)$ and let $Q = [P, z] \lhd P$. Since $[G, z] \leq K$, we have $Q \leq P \cap K$. As $|P/Q| = |\mathbf{C}_{P/Q}(Qz)| \leq |\mathbf{C}_P(z)|$ (see [Is, Corollary 2.24]), we obtain by (1) that

$$|P/Q| \leq |\mathbf{C}_P(z)| \leq |\mathbf{C}_G(z)| = |\mathbf{C}_{G/Z}(Kz)| = |G/K| = |P/(P \cap K)| \leq |P/Q|,$$

and hence $Q = P \cap K$. We have shown that

$$P \cap K = Q = [P, z] \leq P',$$

and the claim holds. We have thus established the hypothesis of Tate's Theorem (see [Hu, IV, 4.7]) and G has a normal p-complement M.

Finally, let $1 \neq m \in M$ and suppose that $\mathbf{C}_G(m) \not\subseteq K$. Choose $g \in G \setminus K$ centralizing m and note that $|\mathbf{C}_G(g)| = |\mathbf{C}_{G/K}(Kg)|$ is a p-power. Thus $\mathbf{C}_G(g)$ is a p-group and cannot contain m. $\qquad\square$

Suppose that H acts on N and $(|H|, |N|) = 1$. It is a consequence of the Glauberman–Isaacs correspondence that the number of H-invariant conjugacy classes of N equals the number of H-invariant irreducible characters of N. When H is solvable (i.e. when the Glauberman correspondence applies), this is Theorem 13.24 of [Is]. The more general case appears as Lemma 5.5 of [Wo 2]. The next theorem of Isaacs does not really require that H be solvable (and Isaacs did not assume that). We only use the solvability of H to quote [Is] for the aforementioned consequence of character correspondence.

19.3 Theorem (Isaacs [Is 9]). *Suppose H acts non-trivially on N and fixes every non-linear irreducible character of N. Assume $(|N|, |H|) = 1$. Set $M = [N, H]$. Assume H is solvable. Then*

(a) $N' = M'$.

(b) *One of the following occurs:*

   (i) N *is abelian;*

   (ii) M *is a p-group of class 2 and $N' \leq \mathbf{Z}(NH)$; or*

   (iii) M *is a Frobenius group with Frobenius kernel M'.*

In all cases, N' is nilpotent by Thompson's Theorem [Hu, V, 8.7].

Proof. Since $N'M$ is H-invariant, H permutes $\mathrm{Irr}\,(N'M)$.

(1) If $\alpha \in \mathrm{Irr}\,(N'M)$ and $N' \not\leq \ker(\alpha)$, we claim that α is H-invariant. To see this, let $\chi \in \mathrm{Irr}\,(N|\alpha)$ and note that $\chi(1) > 1$. By our hypothesis, χ is H-invariant. Since H centralizes $N/N'M$, Lemma 0.17 (c) shows that α is H-invariant.

(2) Let $1 \neq \nu \in \mathrm{Irr}\,(N')$. We show that ν cannot extend to $\nu^* \in \mathrm{Irr}\,(N'M)$. Otherwise Gallagher's Theorem 0.9 and (1) imply that ν^* and $\lambda\nu^*$ are H-invariant for all $\lambda \in \mathrm{Irr}\,(N'M/N')$. Consequently, H fixes all $\lambda \in \mathrm{Irr}\,(N'M/N')$ and H centralizes $N'M/N'$ (see Proposition 12.1). By Fitting's Lemma 0.6, $N/N' = N'M/N' \times \mathbf{C}_{N/N'}(H)$. Hence $N'M/N' = 1$ and $M \leq N'$. Then H fixes every linear character of N and thus all characters of N. This implies that H centralizes N (see Lemma 12.2), a contradiction. The claim holds.

(3) We next prove assertion (a). Observe that $N'/(N' \cap M)$ is a direct factor of $N'M/(N' \cap M)$ and therefore every $\nu \in \mathrm{Irr}\,(N')$ with $N' \cap M \leq \ker(\nu)$

extends to $N'M$. By (2), this can only happen for the trivial character of N' and thus $N' \cap M = N'$, i.e. $N' \leq M$. Finally, if $\nu \in \mathrm{Irr}\,(N')$ with $M' \leq \ker(\nu)$, then ν extends to $M = N'M$, since M/M' is abelian. Again by (2), the only possibility is $\nu = 1$, and this shows $M' = N'$, proving (a).

(4) We next prove (b). Because $N' = M'$ and $N'M = M$, step (1) yields that every non-linear character of M is H-invariant (i.e. the action of H on M satisfies the hypotheses). Because $(|N|,|H|) = 1$, we have that $[M,H] = [N,H,H] = [N,H] = M$. Hence $\mathbf{C}_{M/M'}(H) = 1$ and 1_M is the only H-invariant linear character of M. Thus the number of irreducible characters of M that are not H-invariant is $|M : M'| - 1$. By a consequence of the Glauberman correspondence (see Theorem 13.24 of [Is]), $|M : M'| - 1$ is also the number of conjugacy classes of M that are not H-invariant. If $x \notin M'$, then $\mathrm{cl}_M(x) \subseteq M'x$ because M/M' is abelian. Since $\mathbf{C}_{M/M'}(H) = 1$, we see that neither $M'x$ nor $\mathrm{cl}_M(x)$ is H-invariant. Since M has only $|M : M'| - 1$ non H-invariant conjugacy classes, it follows that $M'x$ is a single conjugacy class of M whenever $x \notin M'$.

If $M' = 1$ then M is abelian and assertion (b) holds. If, on the other hand, $M = M'$, then (1) yields that H fixes all irreducible characters of M. By Lemma 12.2, H centralizes M, contradicting the non-triviality of the action of H on N. We may therefore assume that $1 < M' < M$.

We have now established the hypotheses of Lemma 19.2 (with M and M' in place of G and K). If M is a Frobenius group with kernel M', we are done. We may therefore assume that M/M' is a p-group for some prime p, that M has a normal p-complement $Q \leq M'$ and that $\mathbf{C}_M(x) \subseteq M'$ for $1 \neq x \in Q$.

We claim that $[M',H] \leq Q$. To see this, work in the semi-direct product $G = MH$ and consider a chief factor U/V of G with $Q \leq V < U \leq M'$. Since M/Q is a p-group, $U/V \leq \mathbf{Z}(M/V)$. Let $1 \neq \lambda \in \mathrm{Irr}\,(U/V)$, hence $\lambda(1) = 1$. Let further $\chi \in \mathrm{Irr}\,(M|\lambda)$. Then χ_U is a multiple of λ and we have $M' \nsubseteq \ker \chi$. Thus χ is non-linear and hence H-invariant. It follows that λ is

H-invariant and therefore H fixes all elements of Irr(U/V). Consequently, $[U, H] \leq V$. Since H thus centralizes all chief factors of G between Q and M', the coprime action of H on M implies that $[M', H] \leq Q$, as claimed.

Since we now have $[M', H, M] \leq [Q, M] \leq Q$ and $[M, M', H] \leq [M', H] \leq Q$, it follows by the Three Subgroups Lemma [Hu, III, 1.10] that $[M', M] = [H, M, M'] \leq Q$. Let $P \in \mathrm{Syl}_p(M)$, so that $P \cong M/Q$ has class at most 2.

We next prove the result when $Q = 1$. In this case, M is a class two p-group and so $N' = M' \leq \mathbf{Z}(M)$. By Fitting's Lemma 0.6, $N/M' = M/M' \times C/M'$, where $C/M' = \mathbf{C}_{M/M'}(H)$. Since $[M, C] \leq M' \leq \mathbf{Z}(M)$, we have that $[M, C, M] = 1 = [C, M, M]$. By the Three Subgroups Lemma [Hu, III, 1.10], $M' \leq \mathbf{Z}(C)$. Since $N = MC$, indeed $N' = M' \leq \mathbf{Z}(N)$. By the next to last paragraph, $[M', H] \leq Q = 1$. Thus $M' \leq \mathbf{Z}(NH)$. Conclusion (b) holds. We thus assume that $Q \neq 1$.

Now $Q \leq M' \leq M$, $QP = M$ and $P \cap Q = 1$. Hence $P' = M' \cap P$. For $1 \neq x \in Q$, we have that $\mathbf{C}_M(x) \leq M'$ and so $\mathbf{C}_P(x) \leq P'$. Now Lemma 19.1 applies to the action of P on Q. Thus P acts fixed-point-freely on Q and either P is cyclic or $P \cong Q_8$.

First assume that P is cyclic. Thus $M' = Q \geq \mathbf{C}_M(x)$ for all $1 \neq x \in Q$. Since $Q \neq 1$, then M is a Frobenius group with kernel M'. Conclusion (b) is then satisfied. So we now assume that $P \cong Q_8$.

Now M'/Q has order two. Clearly we may assume that H acts faithfully on N and also on M. If $1 \neq H_0 < H$, the action of H_0 on N satisfies the hypotheses of the theorem. The proof of part (a) shows that $N' = M' = [N, H_0]'$. But $|M'/Q| = 2$ and $M/Q \cong Q_8$. It thus follows that $[N, H_0] = M$ and H_0 acts non-trivially on M/M'. Hence H acts faithfully on M/M'. Since $M/M' \cong Z_2 \times Z_2$ and $(|H|, |M|) = 1$, we have that $|H| = 3$ and $G/Q = MH/Q \cong SL(2, 3)$.

Since P acts fixed-point-freely on Q, Thompson's Theorem [Hu, V, 8.7], shows that Q is nilpotent. Let Q/L be a chief factor of G. Since $P \cong M/Q$ acts fixed-point-freely on Q, Q/L is a faithful irreducible module of G/Q. Let $1 \neq \lambda \in \mathrm{Irr}\,(Q/L)$ and $\theta \in \mathrm{Irr}\,(M|\lambda)$. Since $M' \nleq \ker(\theta)$, θ is H-invariant. By Lemma 0.17, some M-conjugate of λ is H-invariant. Thus λ is fixed by one of the four Sylow 3-subgroups of G/Q. Indeed, $I_G(\lambda)/Q \in \mathrm{Syl}_3(G/Q)$ for each $1 \neq \lambda \in \mathrm{Irr}\,(Q/L)$. Now $|Q:L| = q^m$ and $|\mathbf{C}_{Q/L}(H)| = q^l$ for a prime q and integers m, l. Since each $1 \neq \lambda \in \mathrm{Irr}\,(Q/L)$ is fixed by exactly one Sylow 3-subgroup of G/Q, we have that $q^m - 1 = 4(q^l - 1)$. Now $l \mid m$ and it easily follows that $q^m = 9$. This is a contradiction, because $(|H|, |M|) = 1$. The proof is complete. $\qquad\square$

As mentioned before Theorem 19.3, the solvability of H is not really necessary in that theorem. In our first application, Theorem 19.6, H will be abelian. Our second application uses the next corollary.

19.4 Corollary. *Suppose that H acts non-trivially on N and fixes every non-linear irreducible character of N. If $(|H|, |N|) = 1$, then N is solvable. In fact, N' is nilpotent.*

Proof. Without loss of generality, $H \neq 1$ acts faithfully on N. The hypotheses are met by every non-trivial subgroup of H and so we may assume H to be cyclic. Now Theorem 19.3 shows that N' is nilpotent. $\qquad\square$

We next apply Isaacs' Theorem 19.3 along with Theorem 12.9 to investigate solvable groups whose graphs have two components. Initial study of groups whose graphs have more than one component was initiated by Manz [Mz 1, 2]. Manz and Staszewski [MS 1] showed solvable groups G with two components must have nilpotence length $n(G) \leq 5$. Palfy, in correspondence, has announced $n(G) \leq 4$ and when $n(G) = 4$, $G/\mathbf{Z}(G)$ is a $\{2,3\}$-group. We have yet to see Palfy's proof, but we prove below (Theorem 19.6) that $n(G) \leq 4$ and $\mathrm{dl}(G/\mathbf{F}(G)) \leq 4$. More information about G is evident from the proof of Theorem 19.6. We have seen above (Corollary

18.8) that the two components of $\Gamma(G)$ are regular.

We also mention one example before proceeding with the proof. Let H be the semi-direct product of $Z_3 \times Z_3$ and $GL(2,3)$. If A is abelian and $G = H \times A$, then $\Gamma(G)$ has two components, namely $\{2\}$ and $\{3\}$. Also $n(G) = 4 = \mathrm{dl}(G/\mathbf{F}(G))$. It is convenient to first isolate, in a lemma, one of the arguments needed in the proof of Theorem 19.6.

19.5 Lemma. *Suppose that M is a normal Hall π-subgroup of $G = \mathbf{O}^\pi(G)$ and G/M is abelian. Assume that M is a Frobenius group with Frobenius kernel M'. If V is a finite faithful irreducible G-module such that $\mathbf{C}_G(v)$ contains a Hall π'-subgroup of G for each $v \in V$, then there exists $0 \neq w \in V$ such that $\mathbf{C}_G(w)$ does not have a normal Hall π'-subgroup.*

Proof. Since M' is the Frobenius kernel of M, then $M' = \mathbf{F}(M)$ and also $(|M : M'|, |M'|) = 1$. Let H/M' be a Hall π'-subgroup of G/M', so that $G = MH$ and $M' = M \cap H$. Since $\mathbf{F}(G) \cap M = M'$, it follows $\mathbf{F}(G)/M'$ is a π'-group. The hypothesis on centralizers implies that $\mathbf{O}_{\pi'}(G)$ centralizes V, whence $\mathbf{F}(G)$ is a π-group. Thus $\mathbf{F}(G) = M'$. Now M/M' and G/M are non-trivial abelian groups. Because $\mathbf{O}^\pi(G) = G$, $H \ntrianglelefteq G$ and $n(G) = 3$.

Now V is a faithful irreducible G-module and $\mathbf{C}_G(v)$ contains a Hall π'-subgroup of G for each $v \in V$. Since $n(G) = 3$ and $\mathbf{O}^\pi(G) = G$, Corollary 10.6 implies that V is not a quasi-primitive G-module. Choose $C \triangleleft G$ maximal such that V_C is not homogeneous and write $V_C = V_1 \oplus \cdots \oplus V_n$ for homogeneous components V_i of V_C with $n > 1$. By Proposition 0.2, G/C primitively and faithfully permutes $\{V_1, \ldots, V_n\}$. Since $\mathbf{O}^\pi(G) = G$, we apply Theorem 9.3 to conclude that:

(i) $G/C \cong D_6$, D_{10} or $A\Gamma(2^3)$;

(ii) $n = 3, 5$, or 8 (respectively); and

(iii) $p = 2, 2$, or 3 (respectively) is the unique π'-prime divisor of $|G/C|$.

Furthermore $\mathrm{char}(V) = 2$ if $p = 2$, by Lemma 9.2.

We claim that C is a π-group. For $0 \neq v_1 \in V_1$, we may assume that $\mathbf{C}_G(v_1)$ contains a Hall π'-subgroup S of G and that $S \trianglelefteq \mathbf{C}_G(v_1)$. Now $S \cap C$ is the unique Hall π'-subgroup of $\mathbf{C}_C(v_1)$ and is a Hall π'-subgroup of C. If $v_i \in V_i^{\#}$ for $i \geq 2$, then $\mathbf{C}_C(v_1 + \cdots + v_n) \leq \mathbf{C}_C(v_1)$ and $\mathbf{C}_C(v_1 + \cdots + v_n)$ contains a Hall π'-subgroup of C. Hence $S \cap C$ must centralize $v_2 + \cdots + v_n$. Since these were arbitrarily chosen, $S \cap C$ centralizes $V_2 + \cdots + V_n$. Now fix $0 \neq v_2 \in V_2$. Then, repeating the above arguments, $S \cap C$ is the unique Hall π'-subgroup of C centralizing v_2 and $S \cap C$ centralizes V_1. Thus $S \cap C$ centralizes V, whence $S = 1$ and C is a π-group.

Now $C \leq M$ and it follows from (i), (ii), (iii) above that M/C is the unique maximal normal subgroup of G/C and $|G : M| = p$. Also G/C has a unique minimal normal subgroup L/C with $|L/C| = n$. Also $M = L$ when $p = 2$, and $|M/L| = 7$ when $p = 3$.

We designate a prime t by letting $t = 7$ when $p = 3$ and letting $t = n$ when $p = 2$. We now prove the lemma in the case $t^2 \nmid |G|$. Then a Sylow t-subgroup T of G permutes $\{V_1, \ldots, V_n\}$ non-trivially and we may find a non-zero vector $w = (w_1, \ldots, w_t, 0, \ldots, 0)$ that is centralized by T. But w is also centralized by a Hall π'-subgroup of G. When $p = 3$, $L\mathbf{C}_G(w) = G$ and when $p = 2$, $C\mathbf{C}_G(w) = G$. Thus $\mathbf{C}_G(w)$ has a non-abelian factor group of order 21 when $p = 3$, or $\mathbf{C}_G(w)$ has a factor group isomorphic to D_{2n} when $p = 2$. Since p is the only π'-divisor of $|G|$, it follows that $\mathbf{C}_G(w)$ does not have a normal Hall π'-subgroup, as desired. The lemma holds when $t^2 \nmid |G|$.

Set $|V_1| = q^m$ for a prime q and integer m. We next prove the lemma when $q^m = 3^2$ or 3^4. Here $7 \nmid |GL(m, q)|$, whence $7 \nmid |C/\mathbf{C}_C(V_i)|$ for each i, and $7 \nmid |C|$. Since $\mathrm{char}(V) \neq 2$, $p = 3$ and thus $t = 7$ is an exact divisor of $|G|$. In this case, the lemma follows from the last paragraph. So we assume that $q^m \neq 3^2$ or 3^4.

First suppose that $p = 2$. Then $M' \leq C$. Since $M' = \mathbf{F}(G)$, indeed $M' = \mathbf{F}(C)$. Now $C < M \leq \mathbf{C}_G(M/\mathbf{F}(C)) \leq \mathbf{C}_G(C/\mathbf{F}(C))$, and Lemma 9.10 (b) yields that $C = \mathbf{F}(C) = M'$. Since $(|M : M'|, |M'|) = 1$, then

$t = n = |M : M'|$ is an exact divisor of $|G|$ and the lemma follows when $p = 2$.

Finally, we assume that $p = 3$. Here $(M/C)' = L/C$ and $|M : L| = 7 = t$. If $M' = L$, we argue as in the last paragraph that t is an exact divisor of $|G|$, as desired. So we thus assume that $M' < L \leq M$. Since M/C is non-abelian, $M' \not\leq C$. Then $L = M'C$ because L/C is a chief factor of G, and hence $M' \cap C < M'$. Because $M' \cap C < M' < L \leq M$ and M' is the Frobenius kernel of M, the group $L/M' \cap C$ must be a Frobenius group with kernel $M'/M' \cap C$. This is a contradiction, because $L/(M' \cap C) = C/(M' \cap C) \times M'/(M' \cap C)$. $\qquad\square$

We define characteristic subgroups $\mathbf{F}_i(G)$ iteratively by letting $\mathbf{F}_0(G) = 1$ and $\mathbf{F}_{i+1}(G)/\mathbf{F}_i(G) = \mathbf{F}(G/\mathbf{F}_i(G))$. So $\mathbf{F}_1(G) = \mathbf{F}(G)$ and $n(G)$ is the smallest n for which $\mathbf{F}_n(G) = G$.

Let $Y/\mathbf{F}(G) = \mathbf{Z}(\mathbf{F}_2(G)/\mathbf{F}(G))$. By Lemma 18.1, there exists $\eta \in \mathrm{Irr}(Y)$ with $\eta(1) = |Y : \mathbf{F}(G)|$. If $\tau \in \mathrm{Irr}(\mathbf{F}_2(G) \mid \eta)$, then τ is divisible by every prime divisor of $|\mathbf{F}_2(G)/\mathbf{F}(G)|$. Hence, for each $i \geq 2$, there exists $\tau_i \in \mathrm{Irr}(\mathbf{F}_i(G))$ such that $\tau_i(1)$ is divisible by each prime divisor of $|\mathbf{F}_i(G)/\mathbf{F}_{i-1}(G)|$ and $\mathbf{F}_{i-2}(G) \leq \ker(\tau_i)$.

19.6 Theorem. *Suppose G is a solvable group, whose graph has (exactly) two components. Then*

　(i) $2 \leq n(G) \leq 4$, *and*

　(ii) $\mathrm{dl}(G/\mathbf{F}(G)) \leq 4$.

Proof. We let $F_i = \mathbf{F}_i(G)$ and set $F = F_1$.

Step 1. (i) G is not nilpotent.

(ii) Choose a prime $p \mid |F_2/F|$. Then $p \in \pi$ for a component $\pi \subseteq \Gamma(G)$.

(iii) For each $i \geq 2$, F_i/F_{i-1} is a π-group or a π'-group. In particular F_2/F is a π-group.

Proof. Whenever J is a nilpotent group, $\Gamma(J)$ is a regular graph. Since $n(\Gamma(G)) = 2$, G is not nilpotent. Thus $F_2 > F$ and we choose a prime p dividing $|F_2/F|$. Since $p \mid |G/F|$, indeed $p \in \Gamma(G)$. Parts (i) and (ii) follow.

By definition of π, the degree of an irreducible character of G is a π-number or π'-number. Also, if $N \trianglelefteq G$ and $\theta \in \mathrm{Irr}(N)$, then $\theta(1)$ is a π-number or π'-number. This fact is used repeatedly in the proof.

For $i \geq 2$, it follows from Lemma 18.1 (see comments preceding the theorem) that there exists $\beta_i \in \mathrm{Irr}(F_i)$ such that $\beta_i(1)$ is divisible by every prime divisor of $|F_i/F_{i-1}|$. Thus F_i/F_{i-1} is a π-group or π'-group.

Step 2. Choose m maximal such that F_m/F is a π-group. Then

   (i) F_{m+1}/F_m is an abelian π'-group; and

   (ii) G/F_{m+1} is an abelian π-group.

Proof. Since $m \geq 2$ (see Step 1), there exists a non-linear $\tau \in \mathrm{Irr}(F_m)$ with $\tau(1)$ a π-number (by Lemma 18.1). If $\xi \in \mathrm{Irr}(G|\tau)$, then $\xi(1)$ must be a π-number. By Theorem 12.9, G/F_m has an abelian Hall π'-subgroup. Since F_{m+1}/F_m is a π'-group by Step 1 and since $\mathbf{C}_G(F_{m+1}/F_m) \leq F_{m+1}$ by Lemma 0.19, F_{m+1}/F_m must be an abelian Hall π'-subgroup of G. This proves (i) and that G/F_{m+1} is a π-group.

By Lemma 18.1, there exists a non-linear $\eta \in \mathrm{Irr}(F_{m+1})$ with $\eta(1)$ a π'-number. Since G/F_{m+1} is a π-group, each $\mu \in \mathrm{Irr}(G|\eta)$ must extend η. It follows from Gallagher's Theorem 0.9 that G/F_{m+1} is abelian.

Step 3. Let $P \in \mathrm{Hall}_\pi(F)$ and $Q \in \mathrm{Hall}_{\pi'}(F)$.

   (i) Then P or Q is abelian.

   (ii) If Q is non-abelian, then F_2/Q is an abelian π-group, and G/F_2 is an abelian π'-group. Also Q is a class two group.

Proof. Part (i) is immediate, because F is nilpotent and the degree of every irreducible character of F must be a π-number or π'-number.

We assume that Q is non-abelian and choose $\theta \in \mathrm{Irr}(Q)$ with $\theta(1) \neq 1$. Every $\chi \in \mathrm{Irr}(G|\theta)$ must be of π'-degree and so Theorem 12.9 implies that G/Q has an abelian Hall π-subgroup. Since $Q \leq F \leq F_2$, since F_2/F is a π-group and $\mathbf{C}_G(F_2/F) \leq F_2/F$ (Lemma 0.19), it follows that F_2/F contains a Hall π-subgroup of G/F. So F_2/Q is an abelian Hall π-subgroup of G/Q. By Step 2, G/F_2 is an abelian π'-group.

We must prove that $\mathrm{cl}(Q) = 2$. Let $D \in \mathrm{Hall}_\pi(F_2)$ so that D is abelian. Then $D \cong F_2/Q$ acts on Q and fixes every non-linear character of Q. Now Q is nilpotent and thus contains no Frobenius group. Since Q is non-abelian, Theorem 19.3 implies Q is a class two group, as desired, or that F_2/Q acts trivially on Q. In the latter case, $F_2 = D \times Q$. Since D is abelian, F_2 is nilpotent, a contradiction as $F_2 > F$. This step is complete.

Step 4. If G/F is a π-group, then G/F is abelian. In this case, $\mathrm{dl}(G) \leq 3$.

Proof. Since $n(\Gamma(G)) = 2$ and G/F is a π-group, Q must be non-abelian. Now Step 3 (ii) shows that G/Q is an abelian π-group and $\mathrm{dl}(Q) \leq 2$.

Step 5. (i) $m = 2$.
(ii) F_2/F is an abelian group or is a class 2 nilpotent group.

Proof. By Steps 4 and 2, we may assume that $m \geq 3$ and F_{m+1}/F_m is a non-trivial abelian π'-group. Since $\tau(1)$ is π-number or π'-number for $\tau \in \mathrm{Irr}(F_{m+1})$ and since F_m/Q is a π-group, every non-linear irreducible character of F_m/Q is invariant in F_{m+1}. Applying Theorem 19.3 to the action of a Hall π'-subgroup H/Q of F_{m+1}/Q on F_m/Q, we conclude that one of the following occurs:

(a) F_m/Q is abelian.

(b) F_m/Q is nilpotent of class two, or

(c) There exist normal subgroups $Q \leq K \leq M \leq F_m$ of G such that M/Q is a Frobenius group with kernel $K/Q = (M/Q)' = (F_m/Q)'$.

Also $M/Q = [F_m/Q, H/Q]$.

In cases (a) and (b), F_m/Q is nilpotent. But $Q \leq F \leq F_2 \leq F_m$. Thus $m = 2$ and this step follows in these cases. We assume (c) holds and adopt that notation.

We let $J = MH$. Now $F_{m+1}/M = F_m/M \times J/M$ and $J = \mathbf{O}^\pi(F_{m+1})$. Also $K < M$ and $[M/K, HK/K] = M/K$ because the order of $HK/K \cong H/Q$ is coprime to that of M/K. Now K/Q is the Frobenius kernel of M/Q and thus $K/Q = \mathbf{F}(M/Q)$. Also $K/Q = \mathbf{F}(J/Q)$ because F_2/Q is a π-group.

By Step 3 (ii), we can assume that Q is abelian. If $Q \leq \mathbf{Z}(K)$, then K is indeed nilpotent, whence $K \leq F$. Since F_m/K is abelian, both conclusions hold in this case. We may thus assume that $K \not\leq \mathbf{C}_M(Q)$. Since $(|Q|, |K/Q|) = 1$, we may choose a chief factor Q/Q_1 of G that is not central in K. Let $V = \mathrm{Irr}(Q/Q_1)$. Since $K \not\leq \mathbf{C}_M(V)$ and K/Q is the Frobenius kernel of M/Q, we also have that $M/\mathbf{C}_M(V)$ is a Frobenius group with $K\mathbf{C}_M(V)/\mathbf{C}_M(V) = (M/\mathbf{C}_M(V))'$ and that $\mathbf{O}^\pi(J/\mathbf{C}_J(V)) = J/\mathbf{C}_J(V)$ because $\mathbf{O}^\pi(J/Q) = J/Q$. If $1 \neq \lambda \in V$, then λ extends to $I_M(\lambda)$, because $(|M : Q|, |Q|) = 1$. But $I_M(\lambda) < M$ and so each $\eta \in \mathrm{Irr}(J|\lambda)$ has degree divisible by a prime in π. So $\eta(1)$ is a π-number for all $\eta \in \mathrm{Irr}(J|\lambda)$. Hence $I_J(\lambda)$ contains a Hall π-subgroup of J and λ extends to $\alpha \in \mathrm{Irr}(I_J(\lambda))$ (see Proposition 0.12). Then $\beta\alpha \in \mathrm{Irr}(I_J(\lambda))$ for all $\beta \in \mathrm{Irr}(I_J(\lambda)/Q)$. Thus $\beta(1)$ is a π-number for all $\beta \in \mathrm{Irr}(I_J(\lambda)/Q)$, whence $I_J(\lambda)/Q$ has a normal Hall π'-subgroup. So $I_J(\lambda)/\mathbf{C}_J(V)$ has a normal Hall π'-subgroup $U/\mathbf{C}_J(V) \in \mathrm{Hall}_{\pi'}(J/\mathbf{C}_J(V))$ for all non-zero $\lambda \in V$. Applying Lemma 19.5, we get a contradiction, completing this step.

Step 6. Conclusion.

Proof. By Step 1, $n(G) \geq 2$. By Step 5, $m = 2$ and $\mathrm{dl}(F_2/F) \leq 2$. By Step 2, $n(G/F_2) \leq \mathrm{dl}(G/F_2) \leq 2$. Hence $n(G) \leq 4$ and $\mathrm{dl}(G/F) \leq 4$. $\qquad\square$

We give another application of Theorem 19.3 in the following reduction theorem for graphs of non-solvable groups. The proof is dependent upon the classification of simple groups, because the Ito–Michler Theorem 13.13 is applied to a non-solvable group.

19.7 Theorem. *Suppose that* $K, L \trianglelefteq G$ *and* $1 \neq K/L$ *is a direct product of simple non-abelian groups. If* G/K *is solvable, then* $n(\Gamma(G)) \leq n(\Gamma(K/L))$.

Proof. We argue by induction on $|G|$. Since $\mathbf{Z}(K/L) = 1$, we have that $K\mathbf{C}_G(K/L)/\mathbf{C}_G(K/L)$ is G-isomorphic to K/L. Without loss of generality, we may assume that $\mathbf{C}_G(K/L) = L$.

First assume that $K = G$. It suffices to show that each $q \in \Gamma(G) \backslash \Gamma(G/L)$ is connected in $\Gamma(G)$ to some prime in $\Gamma(G/L)$. By Remark 13.13, $\Gamma(G/L) = \pi(G/L)$. So, if $\theta \in \mathrm{Irr}(G)$ with $q \mid \theta(1)$, we can assume that θ_L is irreducible. Choose $\beta \in \mathrm{Irr}(G/L)$ non-linear. Then $\beta\theta \in \mathrm{Irr}(G)$ and q is connected in $\Gamma(G)$ to some prime in $\Gamma(G/L)$. Hence we may assume that $K < G$.

We may choose $K \leq N \trianglelefteq G$ with $|G/N| = p$, a prime. If $\Gamma(G) = \Gamma(N)$, then $n(\Gamma(G)) \leq n(\Gamma(N))$ and we apply the inductive hypothesis. So we can assume that $\Gamma(G) = \Gamma(N) \cup \{p\}$, $p \notin \Gamma(N)$, and that $\chi(1) = p$ whenever $\chi \in \mathrm{Irr}(G)$ and $p \mid \chi(1)$. In particular, G/N fixes every non-linear character of N. Since $p \notin \Gamma(N)$, the Ito–Michler Theorem 13.13 implies that $\mathbf{O}_p(N) \in \mathrm{Syl}_p(N)$. A Sylow p-subgroup P of G acts non-trivially on $N/\mathbf{O}_p(N)$, because $\mathbf{C}_G(K/L) = L$. Applying Theorem 19.3 to the action of P on $N/\mathbf{O}_p(N)$, we get that $N/\mathbf{O}_p(N)$ is solvable, a contradiction.   □

19.8 Corollary [MSW]. *If* G *is non-solvable, then*

$$n(\Gamma(G)) \leq \max_E n(\Gamma(E)),$$

as E *ranges over the non-solvable composition factors of* G.

In addition to the above reduction, Manz, Staszewski and Willems used the classification of simple groups to show that $n(\Gamma(S)) \leq 3$ for every simple group S. We do not prove that here, but just state the consequence of that result and Corollary 19.8.

19.9 Theorem. *For every group G, $n(\Gamma(G)) \leq 3$.*

§20   Brauer Characters — the Modular Degree Graph

In this section, we again investigate Brauer characters with respect to a prime p. We construct a graph $\Gamma_p(G)$ with vertex set

$$\{q \mid q \text{ prime}, q \mid \varphi(1) \text{ for some } \varphi \in \mathrm{IBr}_p(G)\}$$

and make a graph by connecting distinct q_1, $q_2 \in \Gamma_p(G)$ if $q_1 q_2 \mid \eta(1)$ for some $\eta \in \mathrm{IBr}_p(G)$. As in the previous sections, we denote by $d(\cdot,\cdot)$ the natural distance function on $\Gamma_p(G)$. Also $\mathrm{diam}(\Gamma_p(G))$ denotes the diameter and $n(\Gamma_p(G))$ the number of connected components of $\Gamma_p(G)$.

We start with an easy result, which is similar to Theorem 18.4. We will implicitly use Proposition 14.4 throughout this section (including the next Proposition).

20.1 Proposition. *Suppose that G has a non-abelian solvable p'-factor group G/K. Then one of the following occurs.*

(i) *$n(\Gamma_p(G)) = 1$ and $\mathrm{diam}(\Gamma_p(G)) \leq 4$; or*

(ii) *$n(\Gamma_p(G)) = 2$ and $\mathrm{diam}(\Gamma_p(G)) \leq 2$.*

Proof. Choose $K \leq N \trianglelefteq G$ maximal such that G/N is non-abelian. Then mimic the proof of Theorem 18.4, using that $\mathrm{IBr}_p(G/N) = \mathrm{Irr}(G/N)$. $\square$

To obtain bounds for $n(\Gamma_p(G))$ and $\mathrm{diam}(\Gamma_p(G))$, where G is an arbitrary solvable group, we need the following modular analogue of [Is, 12.3] (see also Lemma 18.3).

20.2 Proposition. *Assume that $N \trianglelefteq G$ is maximal with respect to G/N having a non-linear irreducible p-Brauer character. Then the following hold:*

(i) *G/N has a unique minimal normal subgroup M/N and M/N is not a p-group.*

(ii) *G/N has a normal subgroup $K/N \geq M/N$ such that K/M is a p-group and G/K is an abelian p'-group.*

(iii) *If M/N is non-solvable, then $\mathbf{C}_{G/N}(M/N) = 1$.*

(iv) *If M/N is solvable, then M/N is an elementary abelian r-group for a prime $r \neq p$. Furthermore, G/M acts faithfully on M/N or G/N is a non-abelian r-group.*

Proof. First observe that every irreducible Brauer character of a group H is linear if and only if H' is a p-group. If $N < L \trianglelefteq G$, then each $\varphi \in \mathrm{IBr}_p(G/L)$ is linear and so $\mathbf{O}^p(G') \leq L$. But G/N itself has a non-linear Brauer character, so that $\mathbf{O}^p(G') \not\leq N$. Hence G/N has a unique minimal normal subgroup M/N (namely $\mathbf{O}^p((G/N)')$) and M/N is not a p-group. Set $K/M \in \mathrm{Syl}_p(G/M)$. Then $K/N \trianglelefteq G/N$, and G/K is an abelian p'-group. This establishes assertions (i) and (ii).

Let $C/N = \mathbf{C}_{G/N}(M/N) \trianglelefteq G/N$. By the first paragraph, either $C = N$ or $M \leq C$. Part (iii) thus follows and we may assume that M/N is an elementary abelian r-group for a prime $r \neq p$. If J/M is a characteristic subgroup of C/M and $r \nmid |J/M|$, then $J/N = M/N \times D/N$ for some $D \trianglelefteq G$. By the uniqueness of M/N, $D = N$ and $J = M$. Hence r divides the order of every non-trivial characteristic subgroup of C/M. Since G/M has a normal p-subgroup K/M with abelian factor group G/K, it follows that C/M is an r-group. To prove (iv), we may assume that G/C is not an r-group. Consequently, the normal subgroup $H/M \in \mathrm{Hall}_{r'}(G/M)$ is non-trivial. Observe that H/M acts coprimely on C/N and faithfully on M/N and $[M/N, H/M] = M/N$. On the other hand, H/M centralizes C/M. Therefore $M/N \not\leq \Phi(C/N)$ and C/N is abelian. By Lemma 0.6, $C/N = M/N \times \mathbf{C}_{C/N}(H/M)$. By the uniqueness of M, we conclude $\mathbf{C}_{C/N}(H/M) =$

1, whence $C = M$.   □

Our next result shows that for solvable G, $n(\Gamma_p(G)) \le 2$ and $\mathrm{diam}(\Gamma_p(G))$ ≤ 5. In fact, the sum of the diameters of components is at most 5.

20.3 Theorem. *Let $N \trianglelefteq G$ be maximal such that G/N has a non-linear irreducible p-Brauer character. If G/N is solvable, then*

(a) $n(\Gamma_p(G)) \le 2$;

(b) $\mathrm{diam}(\Gamma_p(G)) \le 5$; and

(c) *if $n(\Gamma_p(G)) = 2$, then $\mathrm{diam}(\Gamma_p(G)) \le 3$ and at most one component has diameter 3.*

Proof. We may assume that every p'-factor group of G is abelian, since otherwise the result follows from Proposition 20.1. By Proposition 20.2, G has normal subgroups M and K with $N \le M \le K$ such that the conclusions of Step 1 are satisfied.

Step 1. (a) M/N is the unique minimal normal subgroup of G/N and is an elementary abelian r-group for $r \ne p$.

(b) K/M is a non-trivial p-group.

(c) G/K is an abelian p'-group, and

(d) G/M acts faithfully on M/N.

Step 2. $p \in \Gamma_p(K/N)$.

Proof. The assertion is an immediate consequence of Step 1 (a, b, d).

Step 3. Let $\theta \in \mathrm{IBr}_p(G)$ with $(\theta(1), pr) = 1$, and let y be a prime such that $y \mid \theta(1)$. Then $d(y, p) \le 2$, and if $y \nmid |G/K|$, then even $d(y, p) = 1$.

Proof. Let η be an irreducible constituent of θ_K and observe that $\eta_N \in \mathrm{IBr}_p(N)$. By Step 2, there exists $\tau \in \mathrm{IBr}_p(K/N)$ with $p \mid \tau(1)$. It follows from Lemma 0.9 that $\tau\eta \in \mathrm{IBr}_p(K)$. Thus $d(p,z) = 1$ for all prime divisors z of $\eta(1)$ (if any). In particular, $d(p,y) = 1$ if $y \nmid |G/K|$. If $\eta(1) \neq 1$, there exists a prime divisor z_0 of $\eta(1)$ such that $d(y,z_0) \leq 1$ and $d(z_0,p) = 1$. To prove that $d(y,p) \leq 2$, we need just show that η is non-linear.

Assume that $\eta(1) = 1$, so that $K/\ker(\eta)$ is an abelian p'-group. Since the irreducible constituents of θ_K are G-conjugate, $K/\ker(\theta_K)$ is an abelian p'-group. Since $K/M \neq 1$ is a p-group and M/N is the unique minimal normal subgroup of G/N, this forces $\ker(\theta_K) = K$, i.e. $K \leq \ker(\theta)$. Since $G' \leq K$, θ is linear. This contradicts the assumption that $y \mid \theta(1)$. Step 3 now follows.

Step 4. There is at most one prime q_0 such that $q_0 \mid |G/K|$ and $d(q_0,p) > 1$.

Proof. Let $\pi = \{q \mid q \, \vert|G/K|$ and $d(q,p) > 1\}$. If $1 \neq \lambda \in \mathrm{IBr}_p(M/N) = \mathrm{Irr}(M/N)$, then $p \mid \xi(1)$ for every $\xi \in \mathrm{IBr}_p(G|\lambda)$, because G/M has a normal p-subgroup $K/M \neq 1$ that acts faithfully on the irreducible G/M-module M/N. In particular, λ is centralized by a Hall π-subgroup of G/M.

Let $H/K = \mathbf{O}^{\pi'}(G/K) = \mathbf{O}_\pi(G/K)$. Then $\mathrm{Irr}(M/N) = M_1 \oplus \cdots \oplus M_l$ for irreducible H/M-modules M_i. Let $C_i/M = \mathbf{C}_{H/M}(M_i)$. Then $\cap_{i=1}^l C_i = M$ by Step 1 (d). Since the C_i are G-conjugate, $|H/C_i|$ is divisible by every prime in π. By the last paragraph, each $\lambda \in M_i$ is fixed by a Hall π-subgroup of H/C_i. Since H/C_i is a $\pi \cup \{p\}$-group, Lemma 18.6 (b) implies $|\pi| \leq 1$.

Step 5. We may assume that $\Gamma_p(G)$ is connected, that $r \in \Gamma_p(G)$ and $\mathrm{diam}(\Gamma_p(G)) \leq 3 + d(p,r)$.

Proof. Let $y \in \Gamma_p(G)$. By Step 3, $d(y,r) \leq 1$ or $d(y,p) \leq 2$. In particular, $\Gamma_p(G)$ has at most two components, and assertion (a) of the theorem holds.

If $y_0 \in \Gamma_p(G)$ is chosen such that $d(y_0,r) > 1$ and $d(y_0,p) = 2$, then Step 3 yields that $y_0 \mid |G/K|$. Hence Step 4 implies that there exists at most one such prime y_0.

If $\Gamma_p(G)$ has two components Γ_1 and Γ_2, we may assume without loss of generality that $r \in \Gamma_1$ and $p \in \Gamma_2$. Consequently $\mathrm{diam}(\Gamma_1) \leq 2$ and $\mathrm{diam}(\Gamma_2) \leq 3$, by the last paragraph. This proves assertion (c) of the theorem.

To complete the proof of the theorem (i.e. part (b)), we may assume that $n(\Gamma_p(G)) = 1$. If $r \notin \Gamma_p(G)$, then Steps 3 and 4 yield that $\mathrm{diam}(\Gamma_p(G)) \leq 3$. We may finally assume that $r \in \Gamma_p(G)$. By Step 3, $\mathrm{diam}(\Gamma_p(G)) \leq 2 + d(p,r) + 1$. This step is complete.

Step 6. Conclusion.

Proof. It remains to prove that $\mathrm{diam}(\Gamma_p(G)) \leq 5$. We may assume by Step 5 that $n(\Gamma_p(G)) = 1$, but $d(p,r) > 2$. Consider a shortest path between p and r, say

$$p \qquad u_1 \qquad u_2 \qquad u_{l-1}\, r$$

where $d(p,r) = l$. By Step 5, we assume that $l \geq 3$. Suppose at first that $l \geq 4$ and choose $\theta \in \mathrm{IBr}_p(G)$ such that $u_2 u_3 \mid \theta(1)$. Then $(\theta(1), pr) = 1$ and Step 3 yields $d(p, u_3) \leq 2$, a contradiction. Hence $l = 3$, and it remains to show that each $t \in \Gamma_p(G) \setminus \{p, r, u_1, u_2\}$ (if any) has distance 1 either to p or to r. If not, then Step 3 yields $d(t,p) = 2$ and $t \mid |G/K|$. Choose $\eta \in \mathrm{IBr}_p(G)$ such that $u_1 u_2 \mid \eta(1)$. Therefore $(\eta(1), pr) = 1$ and Step 3 implies $u_2 \mid |G/K|$, a contradiction to Step 4. This completes the proof of the theorem. $\qquad\square$

We next bound $n(\Gamma_p(G))$ for p-solvable groups G. Our proof however relies on Remark 13.13 and Theorem 19.9, which have not been proven here.

20.4 Corollary. *Let G be p-solvable. Then $n(\Gamma_p(G)) \leq 3$.*

Proof. We take a maximal normal subgroup N of G and assume by induction that $\Gamma_p(N)$ has at most three connected components. Suppose that

$n(\Gamma_p(G)) > 3$.

We first consider the case $|G/N| = r$ for some prime r. If $r \notin \Gamma_p(G)$ or $r \in \Gamma_p(N)$, we are clearly done. We may also assume that r is an isolated point in $\Gamma_p(G)$. Hence there exists $\beta \in \mathrm{IBr}_p(G)$ with $\beta(1) = r$ and so $\beta_N = \lambda_1 + \cdots + \lambda_r$, where $\lambda_i(1) = 1$. Set $L/N' = \mathbf{O}_p(N/N')$. Then $L \leq \ker(\beta)$ and G/L is non-abelian. Since N/L is abelian and G/L is not, we may choose $L \leq M \trianglelefteq G$ with N/M a q-group for a prime q with G/M non-abelian. By definition of L, $q \neq p$. Possibly $q = r$. Thus there exists $\tau \in \mathrm{IBr}_p(G/M)$ of degree r. By assumption, there is a component Δ of $\Gamma_p(G)$ such that $r, q \notin \Delta$ and we consider $\psi \in \mathrm{IBr}_p(G)$ such that the prime divisors of $\psi(1)$ belong to Δ. As $(\psi(1), |G/M|) = 1$, we conclude that $\psi_M \in \mathrm{IBr}_p(M)$ and $\psi\tau \in \mathrm{IBr}_p(G)$ (see Lemma 0.9). This shows that $r \in \Delta$, a contradiction.

We may thus assume that $S := G/N$ is a simple non-abelian p'-group. As $\mathrm{IBr}_p(S) = \mathrm{Irr}(S)$, Theorem 19.9 yields that $n(\Gamma_p(S)) \leq 3$ and Remark 13.13 shows that $\Gamma_p(S)$ consists of the set of prime divisors of $|S|$. Should $n(\Gamma_p(G)) > 3$, there exists $\tau \in \mathrm{IBr}_p(G)$ such that $\tau(1) > 1$ and $(\tau(1), |S|) = 1$. Then τ_N is irreducible and we again use Lemma 0.9 to derive a contradiction and complete the proof. $\square$

If G is not p-solvable, there is no universal bound for $n(\Gamma_p(G))$ independent of G and p.

20.5 Example. For an odd prime p, the p-Brauer characters of $SL(2,p)$ have degrees $\{1, 2, 3, \ldots, p-1, p\}$ (see [HB, VII, 3.10]). It follows from the Prime Number Theorem that the number of primes between $p/2$ and p tends to infinity. Hence

$$\limsup_{p \to \infty} n(\Gamma_p(SL(2,p))) = \infty.$$

This section is based on [MWW].

Chapter VI

π-SPECIAL CHARACTERS

§21 Factorization and Restriction of π-Special Characters

For p-solvable groups G, the Fong–Swan Theorem (Cor. 0.33) states that each $\varphi \in \mathrm{IBr}_p(G)$ can be lifted to some $\chi \in \mathrm{Irr}\,(G)$. Isaacs [Is 4, 5] showed the existence of a "canonical" lift. An important role in this lift is played by "p'-special" characters. For a set of primes π, we may likewise define "π-special" characters, which were developed by Gajendragadkar [Ga 1] and Isaacs [Is 6]. Let G be π-separable. Then any primitive $\psi \in \mathrm{Irr}\,(G)$ necessarily factors $\psi = \psi_1 \psi_2$ as a product of a π-special character ψ_1 and π'-special character ψ_2. Furthermore for $H \in \mathrm{Hall}_\pi(G)$, restriction defines an injection from the set of π-special characters of G into $\mathrm{Irr}\,(H)$. This concept yields a powerful tool for studying problems in the character theory of solvable groups. Then namely any primitive character χ factors $\chi = \prod_p \chi_p$ as a product of p-special characters χ_p, each of which very much "behaves" like an irreducible character of a p-group. We use this approach in Section 22, where we give a proof of a conjecture of W. Feit.

Recall that when $\eta \in \mathrm{Char}\,(G)$, then the order of the linear character $\det(\eta)$ is denoted by $o(\eta)$. For convenience, we restate Theorem 0.13, which is of central importance here.

21.1 Lemma. *Suppose that* $N \trianglelefteq G$, $\theta \in \mathrm{Irr}\,(N)$ *is invariant in* G *and* $(o(\theta) \cdot \theta(1), |G/N|) = 1$. *Then there is a unique extension* χ *of* θ *to* G *satisfying* $(o(\chi), |G/N|) = 1$. *In fact,* $o(\chi) = o(\theta)$.

Recall that χ is called the *canonical extension* of θ to G. Whereas uniqueness of χ is easy to see, it is more difficult to prove existence.

We now state the definition of these "magical" π-special characters.

21.2 Definition. We say that $\chi \in \text{Irr}\,(G)$ is π-*special* if

(i) $\chi(1)$ is a π-number, and if

(ii) $o(\theta)$ is a π-number for all subnormal $S \trianglelefteq\trianglelefteq G$ and all irreducible constituents θ of χ_S.

We write $X_\pi(G)$ to denote the set of π-special characters of G.

21.3 Remarks. For arbitrary G and $\chi \in X_\pi(G)$, the following facts are immediate:

(i) If $N \trianglelefteq\trianglelefteq G$ and $\varphi \in \text{Irr}\,(N)$ is a constituent of χ_N, then $\varphi \in X_\pi(N)$.

(ii) $X_\pi(G) \cap X_{\pi'}(G) = \{1_G\}$.

(iii) If $N \trianglelefteq G$ and $\theta \in X_\pi(N)$, then $\theta^g \in X_\pi(N)$ for all $g \in G$.

(iv) If $E \supseteq \mathbb{Q}(\chi) \supseteq \mathbb{Q}$ is a Galois extension of $\mathbb{Q}$ and $\sigma \in \text{Gal}\,(E/\mathbb{Q})$, then $\chi^\sigma \in X_\pi(G)$.

(v) If G is a π-group, then $X_\pi(G) = \text{Irr}\,(G)$.

(vi) If G is a π'-group, then $X_\pi(G) = \{1_G\}$.

The following lemma is a generalization of (v) and (vi) above.

21.4 Lemma. *Let* $M \trianglelefteq G$, *let* $\theta \in X_\pi(M)$ *and* $\chi \in \text{Irr}\,(G|\theta)$. *Assume that*

(1) G/M *is a* π-*group, or*

(2) G/M *is a* π'-*group and* $o(\chi) \cdot \chi(1)$ *is a* π-*number.*

Then $\chi \in X_\pi(G)$.

Proof. We first show that also under hypothesis (1), $o(\chi) \cdot \chi(1)$ is a π-number. Write $\chi_M = e(\theta_1 + \cdots + \theta_t)$ for an integer e and characters $\theta_i \in \text{Irr}\,(M)$ that are G-conjugate to θ. By 21.3 (iii), $\theta_i \in X_\pi(M)$. Now $\chi(1) = et\theta(1) \mid |G/M|\theta(1)$, and $\chi(1)$ is a π-number. Also the order of $\det(\chi)_M = \det(\chi_M) = \prod_{i=1}^{t}(\det(\theta_i))^e$ is a π-number. Thus both $M/\ker(\det(\chi)_M)$ and G/M are π-groups, and therefore $o(\chi) = |G/\ker(\det \chi)|$ is a π-number as well. This establishes the claim.

Observe that by the last paragraph it suffices to show that whenever N is a maximal normal subgroup of G and $\varphi \in \text{Irr}\,(N)$ is a constituent of χ_N, then in fact φ is π-special. Since N is a maximal normal subgroup of G, either $M \leq N$ or $MN = G$. We argue by induction on $|G|$ and assume first that $M \leq N$. If hypothesis (1) holds, then induction immediately implies that $\varphi \in X_\pi(N)$. Suppose next that hypothesis (2) is valid. Since $\chi(1)$ is a π-number, χ extends θ and φ. Consequently, $o(\varphi)\varphi(1)$ is a π-number, and induction again yields that $\varphi \in X_\pi(N)$. We may assume that $MN = G$.

Without loss of generality, $\theta_{M \cap N}$ and $\varphi_{M \cap N}$ have a common irreducible constituent μ; otherwise namely replace φ by a suitable G-conjugate and apply 21.3 (iii). Since $\theta \in X_\pi(M)$, also $\mu \in X_\pi(M \cap N)$. Under hypothesis (1), $N/(M \cap N) \cong G/M$ is a π-group, and induction implies that $\varphi \in X_\pi(N)$, as desired. It thus remains to consider hypothesis (2), where $N/(N \cap M)$ is a π'-group.

Now $o(\mu) \cdot \mu(1)$ is a π-number, because μ is π-special. Also $\varphi(1) \mid \chi(1)$ is a π-number, and since $N/(N \cap M)$ is a π'-group, $\varphi_{M \cap N} = \mu$. By Lemma 21.1, we let $\hat{\mu}$ be the canonical extension of μ to N. By Lemma 0.9, $\varphi = \lambda \hat{\mu}$ for some linear character $\lambda \in \text{Irr}\,(N/(M \cap N))$. Note that M centralizes $N/(M \cap N)$, and so $\varphi^m = \lambda(\hat{\mu})^m$ for all $m \in M$. This implies that $\chi_N = f\lambda(\mu_1 + \cdots + \mu_l)$, where $f \cdot l$ is a π-number, and $\mu_i \in \text{Irr}\,(N)$ are M-conjugates of $\hat{\mu}$. In particular, $\mu_i(1) = \mu(1)$ and $o(\mu_i) = o(\hat{\mu}) = o(\mu)$. Now

$$\det(\chi_N) = \lambda^{l \cdot \mu(1) \cdot f} \cdot \prod_{i=1}^{l} \det(\mu_i)^f.$$

Since $o(\mu_i)$, $f \cdot l \cdot \mu(1)$ and $o(\chi_N)$ are π-numbers, so is $o(\lambda)$. But $\lambda \in \text{Irr}\,(N/(M \cap N))$ and $N/(M \cap N)$ is a π'-group. Consequently, $\lambda = 1$, $\varphi = \hat{\mu}$ and $o(\varphi)$ is a π-number. As $\mu \in X_\pi(M \cap N)$, the inductive hypothesis implies $\varphi \in X_\pi(N)$, as desired. $\qquad\square$

We next sum up what we know about restriction and induction of π-special characters. If $N \trianglelefteq G$ and $\theta \in \text{Irr}\,(N)$, we set $X_\pi(G|\theta) = \text{Irr}\,(G|\theta) \cap X_\pi(G)$.

21.5 Proposition. Let $N \trianglelefteq G$, $\chi \in X_\pi(G)$ and $\varphi \in X_\pi(N)$.

(a) Every irreducible constituent of χ_N is π-special.

(b) If G/N is a π'-group, then χ_N is irreducible and χ is the canonical extension of χ_N.

(c) If G/N is a π-group, then $X_\pi(G|\varphi) = \mathrm{Irr}\,(G|\varphi)$.

(d) If G/N is a π'-group, then $X_\pi(G|\varphi)$ is non-empty if and only if φ is G-invariant. In this case, $X_\pi(G|\varphi) = \{\hat{\varphi}\}$, where $\hat{\varphi}$ is the canonical extension of φ to G.

Proof. Part (a) was already mentioned as Remark 21.3 (i), (b) is an immediate consequence of the definition of π-special characters and (c) follows from Lemma 21.4.

For (d), first suppose that $\psi \in X_\pi(G|\varphi)$. Since G/N is a π'-group and $\psi(1)$ is a π-number, $\psi_N = \varphi$ and $I_G(\varphi) = G$. Conversely suppose that $\varphi \in X_\pi(N)$ is G-invariant. Since $o(\varphi) \cdot \varphi(1)$ is a π-number, φ has a canonical extension $\hat{\varphi}$ to G with $o(\hat{\varphi}) = o(\varphi)$, by Lemma 21.1. Since $o(\hat{\varphi}) \cdot \hat{\varphi}(1)$ is a π-number, $\hat{\varphi} \in X_\pi(G|\varphi)$, by Lemma 21.4. As we have seen above, every $\psi \in X_\pi(G|\varphi)$ must extend φ, and clearly satisfies $(o(\psi), |G/N|) = 1$. By the uniqueness statement of Lemma 21.1, $X_\pi(G|\varphi) = \{\hat{\varphi}\}$. $\qquad\square$

We continue with a fact which was first observed by Gajendragadkar, and which was the starting-point for many rather recent results about π-special characters.

21.6 Theorem (Gajendragadkar). Let G be a π-separable group with $\alpha, \alpha_1 \in X_\pi(G)$ and $\beta, \beta_1 \in X_{\pi'}(G)$. Then

(a) $\alpha\beta \in \mathrm{Irr}\,(G)$, and

(b) if $\alpha\beta = \alpha_1\beta_1$, then $\alpha = \alpha_1$ and $\beta = \beta_1$.

Proof. We argue by induction on $|G|$. Let M be a maximal normal subgroup of G. Without loss of generality, G/M is a π-group. In particular, β

and β_1 must restrict irreducibly to M and so β_M, $(\beta_1)_M \in X_{\pi'}(M)$.

Let φ be an irreducible constituent of α_M, so that $\varphi \in X_\pi(M)$. By the inductive hypothesis, $\varphi\beta_M \in \mathrm{Irr}\,(M)$. Clearly, $I_G(\varphi) \leq I_G(\varphi\beta_M)$. For $x \in I_G(\varphi\beta_M)$, we have that $\varphi^x \in X_\pi(M)$ by Remark 21.3 (iii), and also $\varphi^x\beta_M = (\varphi\beta_M)^x = \varphi\beta_M$. Applying the uniqueness part of the inductive hypothesis, $\varphi^x = \varphi$, and so $I_G(\varphi) = I_G(\varphi\beta_M)$. By Lemma 0.10, $\alpha\beta \in \mathrm{Irr}\,(G)$.

Say $\varphi = \varphi_1, \ldots, \varphi_t$ are the distinct irreducible constituents of α_M. Then, by induction, the $\varphi_i\beta_M$ are irreducible and distinct. Observe that the $\varphi_i\beta_M$ $(1 \leq i \leq t)$ are the distinct irreducible constituents of $(\alpha\beta)_M$. Likewise, if $\gamma_1, \ldots, \gamma_s$ are the distinct irreducible constituents of $(\alpha_1)_M$, then $\gamma_j(\beta_1)_M$ $(1 \leq j \leq s)$ are the distinct irreducible constituents of $(\alpha_1\beta_1)_M$. Since $\alpha\beta = \alpha_1\beta_1$, it follows that $\{\varphi_i\beta_M \mid 1 \leq i \leq t\} = \{\gamma_j(\beta_1)_M \mid 1 \leq j \leq s\}$. Of course φ_i, $\gamma_j \in X_\pi(M)$ and β_M, $(\beta_1)_M \in X_{\pi'}(M)$. It thus follows by induction that $\beta_M = (\beta_1)_M$. Since β, $\beta_1 \in X_{\pi'}(G|\beta_M)$ and G/M is a π-group, we obtain $\beta_1 = \beta$, by Proposition 21.5 (d). Now $\alpha\beta = \alpha_1\beta$ and α, $\alpha_1 \in \mathrm{Irr}\,(G|\varphi)$. Therefore Lemma 0.10 yields $\alpha = \alpha_1$, proving (b). $\square$

We next show that a primitive character χ of a π-separable group has a unique factorization $\chi = \alpha\beta$ with $\alpha \in X_\pi(G)$ and $\beta \in X_{\pi'}(G)$. In order to give a straightforward inductive argument, we use a weaker primitivity condition in the hypothesis and so get a stronger statement.

21.7 Theorem. Let $\chi \in \mathrm{Irr}\,(G)$, and suppose that there is a normal series $1 = M_0 < M_1 < M_2 \cdots < M_n = G$ such that M_i/M_{i-1} is a π- or π'-group and that χ_{M_i} is homogeneous for all i. Then χ factors uniquely as $\chi = \alpha\beta$ with $\alpha \in X_\pi(G)$ and $\beta \in X_{\pi'}(G)$.

Proof. We argue by induction on $|G|$ and set $M = M_{n-1}$. Now $\chi_M = e\mu$ for an integer e and $\mu \in \mathrm{Irr}\,(M)$. Because μ_{M_i} is homogeneous for all $i \leq n-1$, the inductive hypothesis implies that $\mu = \gamma\delta$ where $\gamma \in X_\pi(M)$ and $\delta \in X_{\pi'}(M)$. Furthermore, this factorization is unique. Because μ is

G-invariant, so are γ and δ (see 21.3 (iii)).

Without loss of generality, G/M is a π-group. Since $\delta \in X_{\pi'}(M)$ is G-invariant, there exists a unique $\beta \in X_{\pi'}(G|\delta)$ and $\beta_M = \delta$ (see Proposition 21.5 (d)). By Lemma 0.10, $\chi = \alpha\beta$ for some $\alpha \in \mathrm{Irr}\,(G|\gamma)$. Since G/M is a π-group, Proposition 21.5 (c) implies $X_\pi(G|\gamma) = \mathrm{Irr}\,(G|\gamma)$. This gives the existence of the factorization. Uniqueness follows from Theorem 21.6.   □

By a trivial induction argument, we can deduce the following result from Theorem 21.7. Note that by Berger's Theorem [Is, 11.33], the notions of primitivity and quasi-primitivity coincide for solvable groups.

21.8 Corollary. *Let G be solvable and let $\chi \in \mathrm{Irr}\,(G)$ be primitive. Then $\chi = \prod_p \chi_p$ with uniquely determined $\chi_p \in X_p(G)$.* ·

21.9 Proposition. *Suppose that $G = MH$ with $M \trianglelefteq G$. Assume that $\varphi \in \mathrm{Irr}\,(M)$ and $\varphi_{M \cap H} \in \mathrm{Irr}\,(M \cap H)$. If $I_G(\varphi) \cap H = I_H(\varphi_{M \cap H})$, then $\chi \mapsto \chi_H$ defines a bijection from $\mathrm{Irr}\,(G|\varphi)$ onto $\mathrm{Irr}\,(H|\varphi_{M \cap H})$.*

Proof. Let $I = I_G(\varphi)$, so that $I \cap H = I_H(\varphi)$. Note that $M \cap (I \cap H) = M \cap H$ and $M(I \cap H) = I \cap MH = I$. In particular, if $I < G$, induction yields that $\psi \mapsto \psi_{I \cap H}$ defines a bijection from $\mathrm{Irr}\,(I|\varphi)$ onto $\mathrm{Irr}\,(I \cap H|\varphi_{M \cap H})$. By our hypothesis, $I_H(\varphi_{M \cap H}) = I \cap H$, and Clifford correspondence implies that $\psi \mapsto (\psi_{I \cap H})^H$ is a bijection from $\mathrm{Irr}(I|\varphi)$ onto $\mathrm{Irr}(H|\varphi_{M \cap H})$. Because $IH = G$, $(\psi_{I \cap H})^H = (\psi^G)_H$ (cf. [Is, Ex. 5.2]). So $\psi \mapsto (\psi^G)_H$ is a bijection from $\mathrm{Irr}\,(I|\varphi)$ onto $\mathrm{Irr}\,(H|\varphi_{M \cap H})$. Again by Clifford correspondence, $\psi \mapsto \psi^G$ defines a bijection from $\mathrm{Irr}\,(I|\varphi)$ onto $\mathrm{Irr}\,(G|\varphi)$, and the result follows when $I < G$. Thus we assume that φ is G-invariant and $\varphi_{M \cap H}$ is H-invariant.

Because $\varphi \in \mathrm{Irr}\,(M)$ is G-invariant,

$$[\varphi^G, \varphi^G] = [\varphi_M^G, \varphi] = [|G : M| \cdot \varphi,\, \varphi] = |G/M|.$$

Likewise, we have $[(\varphi_{M \cap H})^H, (\varphi_{M \cap H})^H] = |H/(M \cap H)| = |G/M|$. Write $\varphi^G = \sum_i a_i \chi_i$ with $a_i > 0$ and distinct $\chi_i \in \text{Irr}(G|\varphi)$. Then

$$|G/M| = [\varphi^G, \varphi^G] = \sum_i a_i^2.$$

Since $G = MH$, $\varphi_H^G = (\varphi_{M \cap H})^H$ (cf. [Is, Ex. 5.2]). Thus

$$\sum_i a_i^2 = |G/M| = [(\varphi_{M \cap H})^H, (\varphi_{M \cap H})^H] = [\varphi_H^G, \varphi_H^G]$$

$$= [\sum_i a_i (\chi_i)_H, \sum_j a_j (\chi_j)_H] = \sum_i \sum_j a_i a_j [(\chi_i)_H, (\chi_j)_H].$$

Consequently, $[(\chi_i)_H, (\chi_j)_H] = \delta_{ij}$, i.e. $(\chi_i)_H \in \text{Irr}(H)$ and $(\chi_i)_H \neq (\chi_j)_H$ for $i \neq j$. Since $(\varphi_{M \cap H})^H = \sum_i a_i (\chi_i)_H$, the map $\chi \mapsto \chi_H$ is a bijection from $\text{Irr}(G|\varphi)$ onto $\text{Irr}(H|\varphi_{M \cap H})$. □

21.10 Theorem. *Suppose that G is π-separable and $H \leq G$ has π'-index. Then $\chi \mapsto \chi_H$ defines a 1–1 map from $X_\pi(G)$ into $X_\pi(H)$.*

Proof. We may assume that H is a maximal subgroup of G. Set $M = \mathbf{O}^{\pi'}(G)$ By Proposition 21.5 (d), $\chi \mapsto \chi_M$ is an injection from $X_\pi(G)$ into $X_\pi(M)$. Now $M \cap H$ has π'-index in M, because $M \trianglelefteq G$. If $M < G$, we apply the inductive hypothesis to $X_\pi(M)$ and see that $\chi \mapsto \chi_{M \cap H}$ is an injection from $X_\pi(G)$ into $X_\pi(M \cap H)$. In particular, $\chi \mapsto \chi_H$ is an injection from $X_\pi(G)$ into $\text{Irr}(H)$. For $\chi \in X_\pi(G)$, the character $\chi_H \in \text{Irr}(H)$ extends $\chi_{M \cap H} \in X_\pi(M \cap H)$. Since $H/(M \cap H)$ is a π'-group and $o(\chi_H) \mid o(\chi)$ is a π-number, in fact $\chi_H = \widehat{(\chi_{M \cap H})}$ is the canonical extension of $\chi_{M \cap H}$. By Proposition 21.5 (d), $\chi_H \in X_\pi(H)$, and the theorem holds should $M < G$.

We may now choose $N \trianglelefteq G$ with G/N a non-trivial π-group. Then $NH = G$ and $N \cap H$ has π'-index in N. By the inductive hypothesis, $\varphi \mapsto \varphi_{N \cap H}$ is an injection from $X_\pi(N)$ into $X_\pi(N \cap H)$. In particular, $I_G(\varphi) \cap H = I_H(\varphi) = I_H(\varphi_{N \cap H})$. By Proposition 21.9, $\chi \mapsto \chi_H$ defines a bijection from $\text{Irr}(G|\varphi)$ onto $\text{Irr}(H|\varphi_{N \cap H})$. Since $G/N \cong H/(N \cap H)$

is a π-group, Proposition 21.5 (c) yields that $\mathrm{Irr}\,(G|\varphi) = X_\pi(G|\varphi)$ and $\mathrm{Irr}\,(H|\varphi_{N\cap H}) = X_\pi(H|\varphi_{N\cap H})$. Consequently $\chi \mapsto \chi_H$ defines a 1–1 map from $X_\pi(G|\varphi)$ into $X_\pi(H|\varphi_{N\cap H})$, given $\varphi \in X_\pi(N)$. By Proposition 21.5 (a) and the injectivity of $\varphi \to \varphi_{N\cap H}$, restriction is a 1–1 map from $X_\pi(G)$ into $X_\pi(H)$. $\square$

We let $\mathbb{Q}_f$ be the field $\mathbb{Q}(\varepsilon)$ obtained by adjoining a primitive f^{th} root of unity ε to $\mathbb{Q}$. This is of course independent of the choice of ε. For $\psi \in \mathrm{Char}\,(G)$, we have $\mathbb{Q}(\psi) \subseteq \mathbb{Q}_g$ where $g = \exp(G)$.

21.11 Corollary. Let G be π-separable, $H \in \mathrm{Hall}_\pi(G)$ and $\chi \in X_\pi(G)$. Then $\mathbb{Q}(\chi) = \mathbb{Q}(\chi_H) \subseteq \mathbb{Q}_h$ where $h = \exp(H)$.

Proof. The Galois group $\mathrm{Gal}\,(\mathbb{Q}_{|G|}/\mathbb{Q})$ permutes both $\mathrm{Irr}\,(G)$ and $X_\pi(G)$ (see 21.3 (iv)). If $\sigma \in \mathrm{Gal}\,(\mathbb{Q}(\chi)/\mathbb{Q}(\chi_H))$, then $(\chi^\sigma)_H = (\chi_H)^\sigma = \chi_H$. Since by Theorem 21.10, $\chi \mapsto \chi_H$ is an injection from $X_\pi(G)$ into $X_\pi(H)$, it follows that $\chi^\sigma = \chi$. Thus $\sigma = 1$ and $\mathbb{Q}(\chi) = \mathbb{Q}(\chi_H) \subseteq \mathbb{Q}_h$. $\square$

§22   Some Applications — Character Values and
Feit's Conjecture

As in the previous section, we let $\mathbb{Q}_n$ denote the cyclotomic extension of $\mathbb{Q}$ generated by a primitive n^{th} root of unity. If $\chi \in \mathrm{Irr}\,(G)$, then $\mathbb{Q}(\chi) \subseteq \mathbb{Q}_g$ where g is the exponent of G. Set $f = f(\chi)$ to be the smallest integer with $\mathbb{Q}(\chi) \subseteq \mathbb{Q}_f$. Generalizing a question of R. Brauer, W. Feit conjectured that G necessarily has an element of order f. While Gow [Go 3] established the conjecture for groups of odd order, later Amit and Chillag [AC 1] extended this to solvable groups. Ferguson and Turull [FT 1] then gave another proof using factorization of primitive characters. With further use of the results of Section 21, Isaacs found an even slicker proof, which we present here with his permission. Due to further results of Ferguson and Turull, Feit's conjecture also holds provided that the non-solvable composition factors of G satisfy certain conditions. For arbitrary G however this conjecture is still open.

A related question is how large can $|\mathbb{Q}_f : \mathbb{Q}(\chi)|$ be, where again $f = f(\chi)$? One would expect this to be small. For linear characters χ, obviously this is 1. Cram [Cr 1] proved for solvable G and arbitrary $\chi \in \text{Irr}(G)$, that $|\mathbb{Q}_f : \mathbb{Q}(\chi)|$ divides $\chi(1)$. By using p-special characters, he gave a second and much shorter proof [Cr 2], which we present. This result certainly does not extend to arbitrary G, as is evidenced by A_5 (see discussion following Theorem 22.3).

We start by considering Feit's conjecture. In a minimal counterexample, $\chi \in \text{Irr}(G)$ is rather easily seen to be primitive and thus the factorization techniques of Section 21 apply.

22.1 Theorem. Let $\chi \in \text{Irr}(G)$ with G solvable. Let $f = f(\chi)$. Then G has an element of order f.

Proof. We argue by induction on $|G|$. If $\chi = \beta^G$ for some $\beta \in \text{Irr}(H)$ and $H < G$, then $\mathbb{Q}(\chi) \subseteq \mathbb{Q}(\beta) \subseteq \mathbb{Q}_{f(\beta)}$. In particular, $\mathbb{Q}_f \subseteq \mathbb{Q}_{f(\beta)}$ and thus $f \mid f(\beta)$. By the inductive hypothesis, there exists $y \in H$ with $o(y) = f(\beta)$. Since $f \mid f(\beta)$, we may choose $x \in \langle y \rangle$ with $o(x) = f$. We may thus assume that χ is primitive.

By Corollary 21.8, we may uniquely factor $\chi = \prod_{p \in \Delta} \alpha_p$ where the product is taken over the set Δ of all prime divisors of $|G|$ and where each α_p is p-special. Let g be the exponent of G. For $\sigma \in \text{Gal}(\mathbb{Q}_g/\mathbb{Q}_f)$,

$$\chi = \chi^\sigma = (\prod_{p \in \Delta} \alpha_p)^\sigma = \prod_{p \in \Delta} (\alpha_p)^\sigma.$$

Since $(\alpha_p)^\sigma \in X_p(G)$ (cf. 21.3 (iv)), it follows from the uniqueness of the factorization that $(\alpha_p)^\sigma = \alpha_p$ for all $p \in \Delta$. Hence $\mathbb{Q}(\alpha_p) \subseteq \mathbb{Q}_f$ and $f(\alpha_p) \mid f$. By Corollary 21.11, $\mathbb{Q}(\alpha_p) \subseteq \mathbb{Q}_{h_p}$ where h_p is the exponent of a

Sylow p-subgroup of G. Consequently $f(\alpha_p) \mid h_p$ is a p-power. If π is the set of prime divisors of f, then $\pi \subseteq \Delta$ because $f \mid |G|$. For $p \in \Delta \setminus \pi$, it follows that $f(\alpha_p) = 1$ and α_p is rational-valued. We may clearly assume that $\pi \neq \varnothing$.

Let $p \in \pi$. Observe that the group $L := \mathrm{Gal}(\mathbb{Q}_f/\mathbb{Q}_{f/p})$ is cyclic. In fact, $|L| = p$ when $p^2 \mid f$. When $p^2 \nmid f$, then $L \cong \mathrm{Gal}(\mathbb{Q}_p/\mathbb{Q})$ and is cyclic of order $p - 1$. Let $\langle \sigma_p \rangle = L \leq \mathrm{Gal}(\mathbb{Q}_f/\mathbb{Q})$. Because $\mathbb{Q}(\chi) \nsubseteq \mathbb{Q}_{f/p}$, we have $\chi^{\sigma_p} \neq \chi$ and so $(\alpha_q)^{\sigma_p} \neq \alpha_q$ for some $q \in \Delta$. By the last paragraph, $f(\alpha_q)$ is a q-power dividing f. If $q \neq p$, then $\mathbb{Q}(\alpha_q) \subseteq \mathbb{Q}_{f(\alpha_q)} \subseteq \mathbb{Q}_{f/p}$, a contradiction because $\alpha_q^{\sigma_p} \neq \alpha_q$ and $\langle \sigma_p \rangle = \mathrm{Gal}(\mathbb{Q}_f/\mathbb{Q}_{f/p})$. So $q = p$ and $(\alpha_p)^{\sigma_p} \neq \alpha_p$.

Next let $g \in G$. We may assume that $f \nmid o(g)$ and thus $(o(g), f) \mid f/p$ for some $p \in \pi$ dependent on g. Now $\mathbb{Q}(\alpha_p(g)) \subseteq \mathbb{Q}_{o(g)} \cap \mathbb{Q}_f \subseteq \mathbb{Q}_{f/p}$. Thus $(\alpha_p)^{\sigma_p}(g) = \alpha_p(g)$ for some $p \in \pi$ dependent on g.

Now $\Psi := \prod_{p \in \pi}((\alpha_p)^{\sigma_p} - \alpha_p)$ is identically zero on G. Recall that for each $p \in \pi$, α_p and $(\alpha_p)^{\sigma_p}$ are p-special characters. Expanding the generalized character Ψ, we get $\Psi = \pm\theta_1 \pm \theta_2 \cdots \pm \theta_l$ with $l = 2^{|\pi|}$ characters θ_i, each being a product $\theta_i = \prod_{p \in \pi} \beta_p$ of p-special characters β_p. By Theorem 21.6, $\theta_i \in \mathrm{Irr}\,(G)$ for all i. Furthermore, $\theta_i \neq \theta_j$ for $i \neq j$, because $(\alpha_p)^{\sigma_p} \neq \alpha_p$ for all $p \in \pi$ and because of the uniqueness of factorization. But Ψ is identically zero on G, violating the linear independence of the elements of $\mathrm{Irr}\,(G)$. This contradiction proves the result. $\qquad\qquad\square$

22.2 Proposition. *Let $N \trianglelefteq G$, $\theta \in \mathrm{Irr}\,(N)$, $I = I_G(\theta)$, $\psi \in \mathrm{Irr}(I|\theta)$ and $\chi = \psi^G$. Let $T = \{g \in G \mid \theta^g \text{ is Galois-conjugate to } \theta\}$. Then*

   (a) $\mathbb{Q}(\chi)\mathbb{Q}(\theta) = \mathbb{Q}(\psi)$;

   (b) $\mathbb{Q}(\chi) = \mathbb{Q}(\psi^T)$; *and*

   (c) $I \trianglelefteq T$ *and* $|\mathbb{Q}(\psi) : \mathbb{Q}(\chi)| \mid |T/I|$.

Proof. Clearly $\mathbb{Q}(\chi)\mathbb{Q}(\theta) \subseteq \mathbb{Q}(\psi)$. If $\sigma \in \mathrm{Gal}\,(\mathbb{Q}(\psi)/\mathbb{Q}(\chi)\mathbb{Q}(\theta))$, then $\psi^\sigma \in$

Irr$(I|\theta)$ and $(\psi^\sigma)^G = \chi$. Hence $\psi^\sigma = \psi$ and $\mathbb{Q}(\psi) = \mathbb{Q}(\chi)\mathbb{Q}(\theta)$. This proves part (a).

Now $\mathbb{Q}(\chi) \subseteq \mathbb{Q}(\psi^T)$ because $(\psi^T)^G = \chi$. If $\tau \in \mathrm{Gal}\,(\mathbb{Q}(\psi^T)/\mathbb{Q}(\chi))$, then $[\chi, \theta^\tau] \neq 0$ and hence $\theta^\tau = \theta^t$ for some $t \in T$. Now $(\psi^T)^\tau$ lies over θ^t and also over θ. Since $((\psi^T)^\tau)^G = \chi = (\psi^T)^G$, we have that $(\psi^T)^\tau = \psi^T$. Hence $\mathbb{Q}(\chi) = \mathbb{Q}(\psi^T)$, proving (b).

For $t \in T$, θ^t is Galois-conjugate to θ and hence $I = I_G(\theta^t)$ for all $t \in T$. Hence $I \trianglelefteq T$ and T/I faithfully and regularly permutes the set of Galois conjugates of θ, although the action is not necessarily transitive. For $\alpha \in \mathrm{Gal}\,(\mathbb{Q}(\psi)/\mathbb{Q}(\chi))$, θ^α is a constituent of χ_N and so $\theta^\alpha = \theta^{t(\alpha)}$ for some $t(\alpha) \in T$. Now $t(\alpha)$ is uniquely determined (mod I). Since $\mathbb{Q}(\psi) \subseteq \mathbb{Q}_{|G|}$, then $\mathrm{Gal}\,(\mathbb{Q}(\psi)/\mathbb{Q}(\chi))$ is abelian. It is then easy to see that the map $\alpha \mapsto I\,t(\alpha)$ is a homomorphism. If α is in the kernel of the homomorphism, then α centralizes $\mathbb{Q}(\theta)\mathbb{Q}(\chi) = \mathbb{Q}(\psi)$ by part (a) and $\alpha = 1$. So $\alpha \mapsto I\,t(\alpha)$ is a 1–1 homomorphism, whence $|\mathbb{Q}(\psi) : \mathbb{Q}(\chi)| = |\mathrm{Gal}\,(\mathbb{Q}(\psi)/\mathbb{Q}(\chi))|$ divides $|T : I|$. This proves (c). $\qquad\square$

22.3 Theorem. *Let $\chi \in \mathrm{Irr}\,(G)$ and $f = f(\chi)$. If G is solvable, then $|\mathbb{Q}_f : \mathbb{Q}(\chi)| \mid \chi(1)$.*

Proof. We argue by induction on $|G|$. First suppose that $N \trianglelefteq G$, $\theta \in \mathrm{Irr}\,(N)$ is an irreducible constituent of χ_N and that $I_G(\theta) < G$. Choose $\psi \in \mathrm{Irr}\,(I_G(\theta)|\theta)$ with $\psi^G = \chi$. By the inductive hypothesis, $|\mathbb{Q}_{f(\psi)} : \mathbb{Q}(\psi)|$ divides $\psi(1)$. By Proposition 22.2, $|\mathbb{Q}(\psi) : \mathbb{Q}(\chi)|$ divides $|G : I_G(\theta)| = \chi(1)/\psi(1)$. Thus $|\mathbb{Q}_{f(\psi)} : \mathbb{Q}(\chi)| \mid \chi(1)$. Since $\mathbb{Q}(\chi) \subseteq \mathbb{Q}_f \subseteq \mathbb{Q}_{f(\psi)}$, in fact $|\mathbb{Q}_f : \mathbb{Q}(\chi)| \mid \chi(1)$. We are done in this case. We may assume that χ is quasi-primitive.

Let $p \mid |G|$. By Theorem 21.7, $\chi = \alpha\beta$ where $\alpha \in X_p(G)$ and $\beta \in X_{p'}(G)$. Set $a = f(\alpha)$ and $b = f(\beta)$. Fix $P \in \mathrm{Syl}_p(G)$ and $H \in \mathrm{Hall}_{p'}(G)$. By Theorem 21.10 and Corollary 21.11, α_P and β_H are irreducible, $\mathbb{Q}(\alpha) = \mathbb{Q}(\alpha_P) \subseteq \mathbb{Q}_{\exp(P)}$ and $\mathbb{Q}(\beta) = \mathbb{Q}(\beta_H) \subseteq \mathbb{Q}_{\exp(H)}$. In particular, $a = f(\alpha_P)$,

$b = f(\beta_H)$ and $(a, b) = 1$. Since $H < G$, the inductive hypothesis implies that $|\mathbb{Q}_b : \mathbb{Q}(\beta)| \mid \beta(1)$. We also claim that $|\mathbb{Q}_a : \mathbb{Q}(\alpha)|$ divides $\alpha(1)$. This follows via the inductive hypothesis when $P < G$. Otherwise, $\alpha = \chi$ is a primitive character of $P = G$. In this case, α is linear (see [Is, 6.14]) and $|\mathbb{Q}_a : \mathbb{Q}(\alpha)| = 1 = \alpha(1)$. This establishes the claim.

We next observe that $\mathbb{Q}(\chi) = \mathbb{Q}(\alpha)\mathbb{Q}(\beta)$. One containment is trivial. If on the other hand $\tau \in \mathrm{Gal}\,(\mathbb{Q}(\alpha)\mathbb{Q}(\beta)/\mathbb{Q}(\chi))$, then $\alpha\beta = \chi = \chi^\tau = \alpha^\tau\beta^\tau$. Now α^τ is p-special and β^τ is p'-special. The uniqueness in Theorem 21.6 forces $\alpha^\tau = \alpha$ and $\beta^\tau = \beta$. Thus $\tau = 1$ and $\mathbb{Q}(\chi) = \mathbb{Q}(\alpha)\mathbb{Q}(\beta)$.

Since $(a, b) = 1$, we have that $\mathbb{Q} \subseteq \mathbb{Q}(\alpha) \cap \mathbb{Q}(\beta) \subseteq \mathbb{Q}_a \cap \mathbb{Q}_b = \mathbb{Q}$ and $\mathbb{Q}_a\mathbb{Q}_b = \mathbb{Q}_{ab}$. Since all these extensions of $\mathbb{Q}$ are Galois,

$$|\mathbb{Q}_a\mathbb{Q}_b : \mathbb{Q}(\alpha)\mathbb{Q}(\beta)| = |\mathbb{Q}_a : \mathbb{Q}(\alpha)| \cdot |\mathbb{Q}_b : \mathbb{Q}(\beta)|.$$

The right-hand side divides $\alpha(1)\beta(1) = \chi(1)$, and thus $|\mathbb{Q}_{ab} : \mathbb{Q}(\chi)| \mid \chi(1)$. As $\mathbb{Q}(\chi) \subseteq \mathbb{Q}_{ab}$, it follows that $\mathbb{Q}(\chi) \subseteq \mathbb{Q}_f \subseteq \mathbb{Q}_{ab}$, which in turn implies that $|\mathbb{Q}_f : \mathbb{Q}(\chi)| \mid \chi(1)$. $\qquad\square$

Now A_5 has two irreducible characters of degree 3, which are Galois-conjugate (see [Is, p. 288]). If χ is one of them, then $\mathbb{Q}(\chi) = \mathbb{Q}(\sqrt{5})$ and χ is rational-valued except on elements of order 5. Consequently, $f(\chi) = 5$ and $|\mathbb{Q}_{f(\chi)} : \mathbb{Q}(\chi)| = 2$. In particular, the above theorem does not extend to arbitrary G.

§23   Lifting Brauer Characters and Conjectures of Alperin and McKay

Let G be p-solvable. We begin this section by proving (Theorem 23.1) that $\chi \mapsto \chi^0$ (recall that ^{0} is restriction to p-regular elements) is a bijection from $X_{p'}(G)$ onto $\{\varphi \in \mathrm{IBr}_p(G) \mid p \nmid \varphi(1)\}$. It is an easy consequence of this theorem and Corollary 0.27 that every $\mu \in \mathrm{IBr}_p(G)$ has a p-rational lift ξ, i.e. $\xi^0 = \mu$ and $\mathbb{Q}(\xi) \subseteq \mathbb{Q}(\varepsilon)$ for a primitive n^{th} root of unity ε, $p \nmid n$. We use Theorem 23.1 to prove results about Brauer characters analogous to results in Section 15 on the McKay–Alperin conjecture for ordinary characters. For example, if B is a p-block of G with defect D and Brauer correspondent $b \in \text{bl}(N_G(D))$, then B and b have an equal number of height-zero Brauer characters, i.e. $\ell_0(B) = \ell_0(b)$. We also show that $|\text{IBr}_p(B)| \geq |\text{IBr}_p(b)|$. We close by giving Isaacs' canonical p-rational lift of Brauer character, $p \neq 2$, and discuss the case $p = 2$.

23.1 Theorem. *If G is p-solvable, then $\chi \mapsto \chi^0$ is a bijection from $X_{p'}(G)$ onto $\{\varphi \in \text{IBr}_p(G) \mid p \nmid \varphi(1)\}$.*

Proof. By induction on $|G|$. If G is a p'-group, then $X_{p'}(G) = \text{Irr}(G) = \text{IBr}_p(G)$ and the result is trivial. If G is a p-group, then $X_{p'}(G) = \{1_G\}$ and the result also follows.

Choose $N \lhd G$ such that G/N is either a p-group or p'-group. By the inductive hypothesis, the mapping $\theta \mapsto \theta^0$ is a bijection from $X_{p'}(N)$ onto $\{\mu \in \text{IBr}_p(N) \mid p \nmid \mu(1)\}$. Furthermore (see Remark 21.3 (iii)), conjugation by G commutes with this bijection. In particular, $I_G(\theta) = I_G(\theta^0)$ for all $\theta \in X_{p'}(N)$.

If $\chi \in X_{p'}(G)$, then $\chi \in \text{Irr}(G|\theta)$ for a unique (up to G-conjugacy) $\theta \in X_{p'}(N)$. If $\xi \in \text{IBr}_p(G)$ and $p \nmid \xi(1)$, then $\xi \in \text{IBr}_p(G|\mu)$ for a unique (up to G-conjugacy) $\mu \in \text{IBr}_p(N)$. Furthermore $p \nmid \mu(1)$. Hence, given the last paragraph, it suffices to fix $\theta \in X_{p'}(N)$ and show that $\chi \mapsto \chi^0$ is a bijection from $X_{p'}(G|\theta)$ onto $\{\varphi \in \text{IBr}_p(G|\theta^0) \mid p \nmid \varphi(1)\}$.

First assume that G/N is a p'-group. By Lemma 0.31, we have that $\chi \mapsto \chi^0$ is a bijection from $\text{Irr}(G|\theta)$ onto $\text{IBr}_p(G|\theta^0)$. Since $p \nmid |G/N|$, we have that $\text{Irr}(G|\theta) = X_{p'}(G|\theta)$ by Proposition 21.5. Now $p \nmid \chi(1)$ for all $\chi \in X_{p'}(G|\theta)$. Consequently $\chi \mapsto \chi^0$ is a bijection from $X_{p'}(G|\theta)$ onto $\{\varphi \in \text{IBr}_p(G|\theta^0) \mid p \nmid \varphi(1)\}$, as desired.

We now assume that G/N is a p-group. If $I_G(\theta) < G$, then $X_{p'}(G|\theta) = \varnothing$ because every $\xi \in X_{p'}(G)$ has p'-degree. Since $I_G(\theta) = I_G(\theta^0)$, we also have that $\{\varphi \in \mathrm{IBr}_p(G|\theta^0) \mid p \nmid \varphi(1)\} = \varnothing$ if $I_G(\theta) < G$. So we assume that $I_G(\theta) = G$. Applying Proposition 21.5 (d), $X_{p'}(G|\theta) = \{\chi\}$ for an extension χ of θ. By Corollary 0.27, $\mathrm{IBr}_p(G|\theta^0) = \{\tau\}$ for an extension τ of θ^0. Since $(\chi^0)_N = (\chi_N)^0 = \theta^0$ is irreducible, $\chi^0 \in \mathrm{IBr}_p(G|\theta^0)$, i.e. $\chi^0 = \tau$. $\square$

The following corollary has a slightly stronger statement than Theorem 23.1 does, but it is an immediate consequence of the theorem and Proposition 21.5 (a).

23.2 Corollary. *Suppose that $\mu \in \mathrm{IBr}_p(L)$, that $L \trianglelefteq G$ and G is p-solvable. If $p \nmid \mu(1)$, then there exists a unique $\theta \in X_{p'}(L)$ such that $\theta^0 = \mu$. Also $\chi \mapsto \chi^0$ is a bijection from $X_{p'}(G|\theta)$ onto $\{\psi \in \mathrm{IBr}_p(G|\mu) \mid p \nmid \psi(1)\}$.*

Let G be p-solvable. We use Theorem 23.1 to prove the next Theorem 23.5. Part (a) gives a result of Huppert [Hu 1] that states every $\varphi \in \mathrm{IBr}_p(G)$ is induced from an irreducible Brauer character of p'-degree. The proof here is somewhat different from Huppert's 1957 proof (which was only for solvable G). Part (b) is a result of Isaacs [Is 4] that says φ has a "p-rational" lift, thereby strengthening the Fong–Swan Theorem. (See Definition 23.4 of p-rational.) Isaacs [Is 4] does obtain Huppert's theorem for p-solvable G and also shows that when $p \neq 2$, there is a unique p-rational lift. We prove this uniqueness later. Part (c) gives a modular analogue of a well-known result for degrees of ordinary characters and those of normal subgroups. Proposition 23.3, a consequence of Theorem 23.1 and Corollary 0.27, gives a useful inductive tool.

23.3 Proposition. *Suppose that $\varphi \in \mathrm{IBr}_p(G)$ and $1 = M_0 \leq M_1 \leq \cdots \leq M_n = G$ is a normal series of G such that φ_{M_i} is homogeneous for each i. Assume that M_{i+1}/M_i is a p-group or a p'-group for each i. Then $p \nmid \varphi(1)$.*

Proof. By induction on $|G|$. We may assume that $M_{n-1} < G$ and we let

$M = M_{n-1}$. Let $\mu \in \mathrm{IBr}_p(M)$ be the irreducible constituent of φ_M so that $\varphi_M = f\mu$ for an integer f. By the inductive hypothesis, $p \nmid \mu(1)$.

First suppose that G/M is a p-group. By Corollary 0.27, φ extends μ. Thus $p \nmid \varphi(1)$. Hence we assume that G/M is a p'-group,

By Theorem 23.1, there exists a unique $\theta \in X_{p'}(M)$ such that $\theta^0 = \mu$. It follows from uniqueness that $I_G(\theta) = I_G(\mu)$. By Lemma 0.31, $\varphi = \chi^0$ for a (unique) $\chi \in \mathrm{Irr}(G|\theta)$. Since G/N is a p'-group, $p \nmid \chi(1)/\theta(1)$ (see [Is, Corollary 11.29]). Since p does not divide $\mu(1) = \theta(1)$, we also have that p does not divide $\chi(1) = \varphi(1)$. $\qquad \square$

23.4 Definition We say that $\chi \in \mathrm{Char}(G)$ is *p-rational* if $\mathbb{Q}(\chi) \subseteq \mathbb{Q}_r$ for an integer r such that $p \nmid r$ (i.e. $p \nmid f(\chi)$).

Recall that $\mathbb{Q}_r = \mathbb{Q}(\varepsilon)$ for a primitive r^{th}-root of unity ε. By Corollary 21.11, every $\chi \in X_{p'}(G)$ is p-rational. If $\psi \in \mathrm{Char}(H)$ is p-rational and $H \leq G$, then $\mathbb{Q}(\psi^G) \subseteq \mathbb{Q}(\psi)$ and so ψ^G is p-rational.

It is relatively easy to see that each $\varphi \in \mathrm{IBr}_p(G)$ is p-rational (extending the definition of p-rational to all complex-valued functions on subsets of G). Indeed, φ is only defined on p-regular elements $g \in G$ and $\varphi(g)$ is a sum of $o(g)^{\mathrm{th}}$ roots of unity. Consequently, φ is p-rational.

23.5 Theorem. *Let G be p-solvable and $\varphi \in \mathrm{IBr}_p(G)$. Then*

(a) *There exists $H \leq G$ and $\mu \in \mathrm{IBr}_p(H)$ such that $\mu^G = \varphi$ and $p \nmid \mu(1)$;*

(b) *There exists a p-rational $\chi \in \mathrm{Irr}(G)$ with $\chi^0 = \varphi$; and*

(c) *If $M \trianglelefteq G$ and α is a constituent of φ_M, then $\varphi(1)/\alpha(1) \mid |G/M|$.*

Proof. (a, b) By induction on $|G|$. If $p \nmid \varphi(1)$, then part (a) is trivial and Theorem 23.1 shows there exists $\chi \in X_{p'}(G)$ with $\chi^0 = \varphi$. By Corollary 21.11, χ is p-rational. Thus we assume that $p \mid \varphi(1)$.

By Proposition 23.3, there exist $K \lhd G$ and $\delta \in \mathrm{IBr}_p(K)$ such that $\varphi \in \mathrm{IBr}_p(G|\delta)$ and $I := I_G(\delta) < G$. Choose $\psi \in \mathrm{IBr}_p(I|\delta)$ with $\psi^G = \varphi$. By the inductive argument, there exists $H \leq I$ and $\mu \in \mathrm{IBr}_p(H)$ such that $\mu^I = \psi$ and $p \nmid \mu(1)$. Since $\mu^G = \varphi$, part (a) follows. The inductive hypothesis for (b) implies that there exists a p-rational $\lambda \in \mathrm{Irr}\,(I)$ such that $\lambda^0 = \psi$. Now $(\lambda^G)^0 = (\lambda^0)^G = \psi^G = \varphi$, hence $\lambda^G \in \mathrm{Irr}\,(G)$. Also λ^G is p-rational because λ is. This proves (b).

For (c), arguing by induction on $|G : M|\,|G|$, we may assume that M is a maximal normal subgroup of G. By Clifford's Theorem 0.8 and the inductive hypothesis, α is G-invariant. If $|G : M| = p$, then $\varphi(1) = \alpha(1)$ by Corollary 0.27. Hence we may assume that G/M is a p'-group. If $p \nmid \alpha(1)$, it follows from Corollary 23.2 that $\varphi(1)/\alpha(1) = \chi(1)/\theta(1)$ for some $\theta \in \mathrm{Irr}\,(M)$ and $\chi \in \mathrm{Irr}\,(G|\theta)$. Thus $\varphi(1)/\alpha(1) \mid |G/M|$ (see [Is, Corollary 11.29]). Hence we may assume that $p \mid \alpha(1)$.

By Proposition 23.3, we may choose $N \unlhd G$ with $N < M$ and $\gamma \in \mathrm{IBr}_p(N)$ such that γ is a constituent of α_N and $I_M(\gamma) < M$. We let $I = I_G(\gamma) < G$. Choose $\psi \in \mathrm{IBr}_p(I|\gamma)$ and $\tau \in \mathrm{IBr}_p(I \cap M|\gamma)$ such that $\psi^G = \varphi$ and $\tau^M = \alpha$. Since $(\psi^{MI})^G = \varphi$, it follows that ψ^{MI} is a constituent of φ_{MI} (see [HB, Theorem VII, 4.10]). Since α is G-invariant, α is a constituent of $\psi^{MI}{}_M$. But $\psi^{MI}{}_M = \psi_{M \cap I}{}^M$ (see [Is; Exercise 5.2]) and every irreducible constituent of $\psi_{M \cap I}$ lies in $\mathrm{Irr}\,(M \cap I|\gamma)$. By the uniqueness in Clifford's Theorem 0.8, τ must be a constituent of $\psi_{M \cap I}$. Applying the inductive hypothesis, $\psi(1)/\tau(1)$ divides $|I : M \cap I| = |MI/M|$. Now

$$\alpha(1) = |M : M \cap I|\tau(1) \text{ and } \psi^{MI}(1) = |MI : I|\psi(1) = |M : M \cap I|\psi(1).$$

Thus $\psi^{MI}(1)/\alpha(1)$ equals $\psi(1)/\tau(1)$ and divides $|MI/M|$. Since $(\psi^{MI})^G = \varphi$, $\varphi(1)/\alpha(1) \mid |G/M|$. $\square$

If, in Theorem 23.5 (c), G/M is a p'-group, then there exist p-rational $\beta \in \mathrm{Irr}\,(M)$ and $\chi \in \mathrm{Irr}\,(G|\beta)$ such that $\chi^0 = \varphi$ and $\beta^0 = \alpha$. Much of the motivation of Isaacs' work [Is 4, 5] on lifting Brauer characters was to develop a lift that works well with respect to normal subgroups, preferably a "canonical lift". We discuss this further below, but first present some modular versions of the Alperin–McKay conjecture.

We let $l(G) = |\mathrm{IBr}_p(G)|$ and $l_0(G) = |\{\varphi \in \mathrm{IBr}_p(G) \mid p \nmid \varphi(1)\}|$. If $N \trianglelefteq G$ and $\mu \in \mathrm{IBr}_p(N)$, then we let $l(G|\mu) = |\mathrm{IBr}_p(G|\mu)|$ and

$$l_0(G|\mu) = |\{\varphi \in \mathrm{IBr}_p(G|\mu) \mid p \nmid \varphi(1)/\mu(1)\}|.$$

Finally, if B is a p-block of G, we let $l(B) = |\mathrm{IBr}_p(B)|$ and $l_0(B)$ be the number of height-zero Brauer characters of B. Thus l counts modular characters analogously to how k counts ordinary characters.

The following proposition will often be used with Theorem 15.9.

23.6 Proposition. *Suppose that G/K has a normal Sylow p-subgroup M/K, that G is p-solvable and $\theta \in X_{p'}(K)$ is M-invariant. If $\hat{\theta}$ is the canonical extension of θ to M, then $l(G|\theta^0) = l_0(G|\theta^0) = k(G|\hat{\theta})$.*

Proof. By Corollary 23.2 and Corollary 0.27, $\{\hat{\theta}^0\} = \mathrm{IBr}_p(M|\theta^0)$ and so $\mathrm{IBr}_p(G|\theta^0) = \mathrm{IBr}_p(G|\hat{\theta}^0)$. Since $\hat{\theta}$ is the unique p-special lift of $\hat{\theta}^0$, we have that $I_G(\hat{\theta}) = I_G(\hat{\theta}^0)$. By Lemma 0.31, $\chi \mapsto \chi^0$ is a bijection from $\mathrm{Irr}\,(G|\hat{\theta})$ onto $\mathrm{IBr}_p(G|\hat{\theta}^0) = \mathrm{IBr}_p(G|\theta^0)$. Since $p \nmid \hat{\theta}(1)|G/M|$, $p \nmid \chi(1)$ for all $\chi \in \mathrm{Irr}\,(G|\hat{\theta})$ by [Is, Corollary 11.29]. Hence $k(G|\hat{\theta}) = l(G|\theta^0) = l_0(G|\theta^0)$. $\square$

23.7 Theorem. *Suppose that G is p-solvable, $N \trianglelefteq G$, and $\mu \in \mathrm{IBr}_p(N)$ is invariant in P, where $P/N \in \mathrm{Syl}_p(G/N)$. Let $H/N = \mathbf{N}_{G/N}(P/N)$. Then*

$$|\{\psi \in \mathrm{IBr}_p(G|\mu) \mid p \nmid \psi(1)\}| = |\{\sigma \in \mathrm{IBr}_p(H|\mu) \mid p \nmid \sigma(1)\}|.$$

Proof. We argue by induction on $|G : N|$. We assume that $p \nmid \mu(1)$. Choose $N \leq K \trianglelefteq G$ minimal such that G/K has a normal Sylow p-subgroup. If $K = N$, the result is trivial. Without loss of generality, choose $N \leq L < K$

such that K/L is a chief factor of G. Since G/K does have a normal Sylow p-subgroup and G/L does not, K/L is a p'-group.

Let $J/L = \mathbf{N}_{G/L}(LP/L)$. Then $J < G$. If $\eta \in \mathrm{IBr}_p(L)$ is P-invariant, then every J-conjugate of η is P-invariant. Furthermore, the Frattini argument shows that if η, $\alpha \in \mathrm{IBr}_p(L)$ and are P-invariant and G-conjugate, then η and α are indeed J-conjugate. So we may choose P-invariant $\eta_1,\ldots,\eta_t$ $\in \mathrm{IBr}_p(L)$ such that $p \nmid \eta_i(1)$ for each i and such that every P-invariant $\alpha \in \mathrm{IBr}_p(L|\mu)$ of p'-degree is G-conjugate (equivalently J-conjugate) to exactly one η_i. (We allow the possibility that there are no P-invariant $\eta \in \mathrm{IBr}_p(L|\mu)$ of p'-degree.)

If $\sigma \in \mathrm{IBr}_p(G|\mu) \cup \mathrm{IBr}_p(J|\mu)$ and $p \nmid \sigma(1)$, then σ lies over exactly one η_i. If $N < L$, the inductive hypothesis yields that $l_0(G|\eta_i) = l_0(J|\eta_i)$ for each i. Then

$$l_0(G|\mu) = \sum_{i=1}^{t} l_0(G|\eta_i) = \sum_{i=1}^{t} l_0(J|\eta_i) = l_0(J|\mu).$$

The inductive hypothesis also implies that $l_0(J|\mu) = l_0(H|\mu)$, because $J < G$. The result follows when $N < L$. So we assume that $N = L$ (and $H = J$).

Now $P \le I_G(\mu) \cap H = I_H(\mu)$. If $I_G(\mu) < G$, the inductive hypothesis and Clifford correspondence yields that

$$l_0(G|\mu) = l_0(I_G(\mu)|\mu) = l_0(I_H(\mu)|\mu) = l_0(H|\mu),$$

as desired. Hence we assume that μ is G-invariant.

Now $G = KH$ and we let $C = K \cap H$. Note that $C/N = \mathbf{C}_{K/N}(P)$. By Theorem 23.1, $\mu = \varphi^0$ for a G-invariant $\varphi \in X_{p'}(L)$. By Proposition 21.5 (c) and Lemma 0.31, $X_{p'}(K|\varphi) = \mathrm{Irr}(K|\varphi)$ and $\theta \mapsto \theta^0$ is a bijection from $\mathrm{Irr}(K|\varphi)$ onto $\mathrm{IBr}_p(K|\mu)$. Also $I_G(\theta) = I_G(\theta^0)$ for each $\theta \in \mathrm{Irr}(K|\varphi)$. Similarly $X_{p'}(C|\varphi) = \mathrm{Irr}(C|\varphi)$ and $\beta \mapsto \beta^0$ is a bijection from $\mathrm{Irr}(C|\varphi)$ onto $\mathrm{IBr}_p(C|\mu)$. Observe that φ extends to P by Lemma 21.1. By Theorem 15.9, there is a bijection from $\{\theta \in \mathrm{Irr}(K|\varphi) \mid \theta \text{ is } P\text{-invariant}\}$ onto $\mathrm{Irr}(C|\varphi)$ and this map commutes with conjugation by H. Now, if $\chi \in \mathrm{IBr}_p(G|\mu)$

and $p \nmid \chi(1)$, then $\chi \in \mathrm{IBr}_p(G|\theta^0)$ for a P-invariant $\theta \in \mathrm{Irr}\,(K|\varphi)$. Since $G = KH$, θ is unique up to H-conjugacy. Similarly, when $\tau \in \mathrm{IBr}_p(H|\mu)$, then $\tau \in \mathrm{IBr}_p(H|\beta^0)$ for some $\beta \in \mathrm{Irr}\,(C|\varphi)$. Of course β is unique up to H-conjugacy and note that β is P-invariant by Lemma 0.17. So it suffices to fix a P-invariant $\theta \in \mathrm{Irr}\,(K|\varphi)$ and $\beta \in \mathrm{Irr}\,(C|\varphi)$ such that $\theta \leftrightarrow \beta$ as in Theorem 15.9 and show that $l_0(G|\theta^0) = l_0(H|\beta^0)$. Theorem 15.9 (v) says that $k(G|\hat{\theta}) = k(C|\hat{\beta})$ where $\hat{\theta}$ and $\hat{\beta}$ are the canonical extensions of θ and β (respectively). Applying Proposition 23.6, $l_0(G|\theta^0) = l_0(H|\beta^0)$.   □

The conclusion of Theorem 23.7 says that $l_0(G|\mu) = l_0(H|\mu)$ whenever $p \nmid \mu(1)$. Is it true that $l_0(G|\mu) = l_0(H|\mu)$ even when $p|\mu(1)$? The answer is yes and we refer the reader to [Wo 7] and the discussion at the end of this section. In the meantime, we derive some consequences of Theorem 23.7.

23.8 Corollary. *If $P \in \mathrm{Syl}_p(G)$ and G is a p-solvable group, then $l_0(G) = l_0(\mathrm{N}_G(P))$.*

Proof. Set $N = 1$ in Theorem 23.7.   □

We next give the block-wise version of the last theorem. The proof is similar to that of Theorem 15.12.

23.9 Theorem. *Suppose that B is a p-block of a p-solvable group G, that D is a defect group of B, and $b \in \mathrm{bl}(\mathrm{N}_G(D))$ is the Brauer correspondent of B. Then $l_0(B) = l_0(b)$.*

Proof. Argue by induction on $|G|$. Let $K = \mathbf{O}_{p'}(G)$. We apply Corollary 0.30. We may choose $\theta \in \mathrm{Irr}\,(K)$ so that B covers θ and $D \le I_G(\theta)$. Let $\mu = \theta\rho(K,D) \in \mathrm{Irr}\,(\mathbf{C}_K(D))$ be the Glauberman correspondent of θ. Then $I \cap \mathrm{N}_G(D) = I_{\mathrm{N}_G(D)}(\mu)$. There exist blocks $B_0 \in \mathrm{bl}(I)$ and $b_0 \in \mathrm{bl}\,(I \cap \mathrm{N}_G(D))$ such that b_0 is the Brauer correspondent of B_0. Furthermore there is a height-preserving bijection (character induction) between B_0 and B. So $l_0(B_0) = l_0(B)$. Similarly, $l_0(b_0) = l_0(b)$. If $I < G$, the inductive hypothesis implies that $l_0(b_0) = l_0(B_0)$. Then $l_0(B) = l_0(b)$, as desired. We thus assume that $I = G$, i.e. θ is G-invariant.

By Theorems 0.28 and 0.29, $\mathrm{IBr}_p(B) = \mathrm{IBr}_p(G|\theta)$, $D \in \mathrm{Syl}_p(G)$, and $\mathrm{IBr}_p(b) = \mathrm{IBr}_p(\mathbf{N}_G(D)|\mu)$. Observe that $K\mathbf{N}_G(D)/K = \mathbf{N}_G(KD/K)$. By Theorem 23.7, we have that $l_0(B) = l_0(G|\theta) = l_0(K\mathbf{N}_G(D)|\theta)$. Trivially, $l_0(b) = l_0(\mathbf{N}_G(D)|\mu)$. It suffices to show that $l_0(K\mathbf{N}_G(D)|\theta) = l_0(\mathbf{N}_G(D)|\mu)$. Let $\hat{\theta} \in \mathrm{Irr}(KD)$ and $\hat{\mu} \in \mathrm{Irr}(CD)$ be the canonical extensions of θ and μ (respectively). By Proposition 23.6, $l_0(K\mathbf{N}_G(D)|\theta) = k(K\mathbf{N}_G(D)|\hat{\theta})$ and $l_0(\mathbf{N}_G(D)|\mu) = k(\mathbf{N}_G(D)|\hat{\mu})$. But on the other hand, $k(K\mathbf{N}_G(D)|\hat{\theta}) = k(\mathbf{N}_G(D)|\hat{\mu})$ by application of Theorem 15.9 (v) with $L = 1$. Hence $l_0(K\mathbf{N}_G(D)|\theta) = l_0(\mathbf{N}_G(D)|\mu)$, as desired.   □

A conjecture related to the Alperin–McKay conjecture is Alperin's *weight conjecture*. While we do not state the weight conjecture, we do mention that it would imply the Alperin–McKay conjecture. Another consequence would be that $l(B) \geq l(b)$ where B is a p-block with Brauer correspondent b. We next prove this inequality for p-solvable G. We do mention that it has been widely rumored for many years that Okuyama has verified the weight conjecture for p-solvable groups, but this has yet to appear in print.

23.10 Theorem. *Suppose that G is p-solvable, $P/L \in \mathrm{Syl}_p(G/L)$, and $\mu \in \mathrm{IBr}_p(L)$ is P-invariant. Assume that $p \nmid \mu(1)$. If $H/L = \mathbf{N}_G(P/L)$, then $l(G|\mu) \geq l(H|\mu)$.*

Proof. We argue by induction on $|G : L|$. Let $I = I_G(\mu)$. Then $P \leq I$ and $\mathbf{N}_I(P) = H \cap I = I_H(\mu)$. If $I < G$, the inductive argument and the Clifford correspondence (Theorem 0.8) yield that

$$l(G|\mu) = l(I|\mu) \geq l(I \cap H|\mu) = l(H|\mu).$$

Thus we assume that μ is G-invariant.

Let $M/L = \mathbf{O}_p(G/L)$. Since μ is invariant in G, it follows from Corollary 0.27 that $\mathrm{IBr}_p(M|\mu) = \{\sigma\}$ for a G-invariant σ and $p \nmid \sigma(1)$. Now $L \leq$

$M \leq P \leq H$ and $H/M = \mathbf{N}_G(P/M)$. If $L < M$, we employ the inductive hypothesis to conclude that

$$l(G|\mu) = l(G|\sigma) \geq l(H|\sigma) = l(H|\mu).$$

The conclusion of the theorem is satisfied in this case. We thus assume that $\mathbf{O}_p(G/L) = 1$ and $G > L$.

We now let $K/L = \mathbf{O}_{p'}(G/L)$ so that $K > L$. Observe that $KH/K = \mathbf{N}_{G/K}(KP/K)$. By Theorem 23.1, there exists a G-invariant $\varphi \in X_{p'}(L)$ with $\varphi^0 = \mu$. By Lemma 0.31, $\theta \mapsto \theta^0$ is a bijection from $\mathrm{Irr}\,(K|\varphi)$ onto $\mathrm{IBr}_p(K|\mu)$. In particular $I_G(\theta) = I_G(\theta^0)$ for $\theta \in \mathrm{Irr}\,(K|\varphi)$ and every element of $\mathrm{IBr}_p(K|\mu)$ has p'-degree.

If $\alpha, \gamma \in \mathrm{IBr}_p(K)$ are P-invariant and G-conjugate, the Frattini argument shows that α and γ are indeed H-conjugate. Thus there exist P-invariant $\theta_1, \ldots, \theta_t \in \mathrm{Irr}\,(K|\varphi)$ such that whenever $\alpha \in \mathrm{IBr}_p(K|\mu)$ is P-invariant, then α is G-conjugate (and H-conjugate) to exactly one θ_i^0. In particular, for $i \neq j$,

$$\mathrm{IBr}_p(G|\theta_i^0) \cap \mathrm{IBr}_p(G|\theta_j^0) = \varnothing \text{ and}$$

$$\mathrm{IBr}_p(KH|\theta_i^0) \cap \mathrm{IBr}_p(KH|\theta_j^0) = \varnothing.$$

By the inductive hypothesis, $l(G|\theta_i^0) \geq l(KH|\theta_i^0)$ for each i. Hence

$$l(G|\mu) \geq \sum_{i=1}^{t} l(G|\theta_i^0) \geq \sum_{i=1}^{t} l(KH|\theta_i^0).$$

We observe that φ extends to P by Lemma 21.1. We now apply Theorem 15.9 to KH to conclude there exist $\beta_1, \ldots, \beta_t \in \mathrm{Irr}\,(C|\varphi) = X_{p'}(C|\varphi)$ such that each $\beta \in \mathrm{Irr}\,(C|\varphi)$ is H-conjugate to exactly one β_i and such that $[(\theta_i)_C, \beta_i] \neq 0$. Furthermore part (v) of Theorem 15.9 (with $M = KP$ and $\lambda = 1$) and Proposition 23.6 imply for each i that $l(KH|\theta_i^0) = l(H|\beta_i^0)$. But each $\xi \in \mathrm{IBr}_p(C|\mu)$ is H-conjugate to exactly one β_i^0 and so $l(H|\mu) = \sum_{i=1}^{t} l(H|\beta_i^0)$. Combining with the last paragraph,

$$l(G|\mu) \geq \sum_{i=1}^{t} l(KH|\theta_i^0) = \sum_{i=1}^{t} l(H|\beta_i^0) = l(H|\mu).$$

This completes the proof.   □

23.11 Corollary. *Under the hypotheses of Theorem 23.10, $l(G|\mu) = l(H|\mu)$ if and only if $l(G|\mu) = l_0(G|\mu)$.*

Proof. Since $p \nmid \mu(1)$, we have that $\mu = \varphi^0$ for a unique $\varphi \in X_{p'}(L)$. Since μ and φ are P-invariant, it follows from Proposition 23.6 that $l(H|\mu) = l_0(H|\mu)$. By Theorems 23.7 and 23.10, we now have that

$$l_0(G|\mu) = l_0(H|\mu) = l(H|\mu) \leq l(G|\mu).$$

The corollary follows.   □

The hypothesis that $p \nmid \mu(1)$ in Theorem 23.10 and Corollary 23.11 is not really necessary. See [Wo 7] and the discussion at the end of this section.

23.12 Corollary. *Whenever G is p-solvable and $P \in \mathrm{Syl}_p(G)$, then $l(G) \geq l(\mathrm{N}_G(P))$. Equality holds if and only if $P \trianglelefteq G$.*

Proof. Set $N = 1$ in Theorem 23.10 to obtain $l(G) \geq l(\mathrm{N}_G(P))$. We trivially have equality if $P \trianglelefteq G$. If $l(G) = l(\mathrm{N}_G(P))$, then Corollary 23.11 yields that $l(G) = l_0(G)$. By Theorem 13.1 (c), $P \trianglelefteq G$.   □

23.13 Theorem. *Let B be a p-block of a p-solvable group. Suppose D is a defect group of B and $b \in \mathrm{bl}(\mathrm{N}_G(D))$ is the Brauer correspondent of B. Then $l(B) \geq l(b)$ with equality if and only if $l(B) = l_0(B)$.*

Proof. This theorem can be proved by Fong reduction, Theorem 23.10 and Corollary 23.11. Since the proof is essentially identical to those of Theorems 15.12 and 23.9, we omit the details.   □

23.14 Definition. We say $\chi \in \mathrm{Irr}\,(G)$ is *subnormally p-rational* if whenever S is subnormal in G and $\tau \in \mathrm{Irr}\,(S)$ is a constituent of χ_S, then τ is p-

rational. We let $\mathcal{S}_p(G)$ denote the set of subnormally p-rational irreducible characters of G.

We now proceed to give Isaacs' canonical lift of Brauer characters for p-solvable groups, $p \neq 2$. Indeed $\chi \mapsto \chi^0$ is a bijection from $\mathcal{S}_p(G)$ onto $\text{IBr}_p(G)$. This was originally done in [Is 4]. This theorem is not true for $p = 2$, as is evidenced by an elementary abelian 2-group. Indeed, part of the problem is that a linear character of order two is rational. For example, Theorem 6.30 of [Is] does not hold for $p = 2$. In a later paper, Isaacs [Is 5] does give a canonical lift when $p = 2$.

23.15 Lemma. *Suppose that G/N is a p-group, $\theta \in \text{Irr}(N)$ is p-rational and $p \neq 2$. Let $I = I_G(\theta)$. Then*

  (a) *There is a unique p-rational $\psi \in \text{Irr}(I|\theta)$. Furthermore $\psi_N = \theta$.*

  (b) *ψ^G is the unique p-rational constituent of θ^G.*

  (c) *If $\theta^0 \in \text{IBr}_p(N)$ and $I_G(\theta) = I_G(\theta^0)$, then $\text{IBr}_p(G|\theta^0) = \{(\psi^G)^0\}$.*

Proof. Part (a) is Theorem 6.30 of [Is]. Clearly $\psi^G \in \text{Irr}(G|\theta)$ is p-rational. If $\beta \in \text{Irr}(G|\theta)$ is p-rational, choose the unique $\eta \in \text{Irr}(I|\theta)$ with $\eta^G = \beta$. Since both β and θ are p-rational, a routine argument yields that η is p-rational. By part (a), $\eta = \psi$ and $\beta = \psi^G$.

Since $(\psi^0)_N = (\psi_N)^0 = \theta^0$ is irreducible, $\psi^0 \in \text{IBr}_p(I|\theta^0)$. Part (c) now follows from Clifford's Theorem 0.8 and Corollary 0.27.   □

23.16 Theorem. *Suppose G is p-solvable, $p \neq 2$. Then*

  (i) *$\chi \mapsto \chi^0$ is a bijection from $\mathcal{S}_p(G)$ onto $\text{IBr}_p(G)$.*

  (ii) *If $\beta \in \text{Irr}(G)$ is p-rational and $\beta^0 \in \text{IBr}_p(G)$, then $\beta \in \mathcal{S}_p(G)$.*

Proof. We argue by induction on $|G|$. The result is trivial if $p \nmid |G|$.

Step 1. If $L \lhd H \leq G$, then $\theta \mapsto \theta^0$ is a bijection from $\mathcal{S}_p(L)$ onto $\text{IBr}_p(L)$.

For each $\theta \in S_p(L)$, every H-conjugate of θ is in $S_p(L)$ and $I_H(\theta) = I_H(\theta^0)$.

Proof. The first statement follows from the inductive hypothesis as $L < G$. For $\theta \in S_p(L)$ and $h \in H$, $(\theta^h)^0 = (\theta^0)^h \in \mathrm{IBr}_p(L)$ and θ^h is p-rational. By the inductive hypothesis, $\theta^h \in S_p(L)$. If also $h \in I_H(\theta^0)$, then $(\theta^h)^0 = \theta^0$ and uniqueness in the inductive hypothesis yields that $\theta^h = \theta$. So $I_H(\theta^0) \leq I_H(\theta)$. The reverse inclusion is trivial.

Step 2. Suppose that $L \lhd H \leq G$ and H/L is a p'-group. If $\theta \in S_p(L)$, then

   (i) $S_p(H|\theta) = \mathrm{Irr}(H|\theta)$; and

   (ii) $\chi \mapsto \chi^0$ is a bijection from $S_p(H|\theta)$ onto $\mathrm{IBr}_p(H|\theta^0)$.

Proof. By Step 1, $\theta^0 \in \mathrm{IBr}_p(L)$ and $I_H(\theta) = I_H(\theta^0)$. By Lemma 0.31, $\chi \mapsto \chi^0$ is a bijection from $\mathrm{Irr}(H|\theta)$ onto $\mathrm{IBr}_p(H|\theta^0)$.

We next show that each $\chi \in \mathrm{Irr}(H|\theta)$ is p-rational. Now χ^0 is p-rational because it is a Brauer character (see discussion following Definition 23.4). Write $|G| = n = rp^l$ for integers $l \geq 0$, r and n, with $p \nmid r$. Whenever $\sigma \in \mathrm{Gal}(\mathbb{Q}_n/\mathbb{Q}_r)$, $(\chi^\sigma)^0 = (\chi^0)^\sigma = \chi^0$ and $\chi^\sigma \in \mathrm{Irr}(G|\theta)$, as θ is p-rational. By the uniqueness in the last paragraph, $\chi^\sigma = \chi$. Hence χ is p-rational.

To show that $\chi \in \mathrm{Irr}(H|\theta)$ is subnormally p-rational, it suffices to show that whenever $M \lhd H$, then the irreducible constituents of χ_M lie in $S_p(M)$. By Step 1, every H-conjugate of θ lies in $S_p(L)$. Hence every irreducible constituent of $\chi_{M\cap L}$ lies in $S_p(M \cap L)$. Since $M/M \cap L \cong LM/L$ is a p'-group, the argument of the last paragraph shows that whenever $\alpha \in \mathrm{Irr}(M)$ is a constituent of χ_M, then $\alpha^0 \in \mathrm{IBr}_p(M)$ and α is p-rational. The inductive hypothesis implies that $\alpha \in S_p(M)$, as desired. So $\chi \in S_p(H)$. Thus $\mathrm{Irr}(H|\theta) = S_p(H|\theta)$.

Step 3. Suppose that $L \lhd H \leq G$ and H/L is a p-group. Let $\theta \in S_p(L)$. Then

   (i) There is a unique $\tau \in S_p(H|\theta)$;

(ii) τ is the unique p-rational constituent of θ^H; and

(iii) $\mathrm{IBr}_p(H|\theta^0) = \{\tau^0\}$.

Proof. By Step 1, $I_H(\theta) = I_H(\theta^0)$. By Lemma 23.15, θ^H has a unique p-rational constituent $\tau \in \mathrm{Irr}\,(H)$ and $\mathrm{IBr}_p(H|\theta) = \{\tau^0\}$. It suffices to show $\tau \in \mathcal{S}_p(H)$. To this end, it suffices to show that the irreducible constituents of τ_M are in $\mathcal{S}_p(M)$ for all maximal normal subgroups M of H. By Step 1, it suffices to show that some irreducible constituent of τ_M is in $\mathcal{S}_p(M)$.

Every irreducible constituent γ of $\theta_{L \cap M}$ lies in $\mathcal{S}_p(L \cap M)$. The last paragraph shows some irreducible constituent ν of γ^M is p-rational and $\nu^0 \in \mathrm{IBr}_p(M)$. By the inductive hypothesis $\nu \in \mathcal{S}_p(M)$.

First assume that $|H/M| = p$, so that $H/L \cap M$ is a p-group. By Lemma 23.15, θ, τ, and ν are the unique p-rational irreducible constituents of γ^L, γ^H and γ^M, respectively. The lemma also implies that ν^H has a p-rational constituent ξ. Then ξ is a constituent of γ^H. By uniqueness, $\xi = \tau$ and thus $[\tau_M, \nu] \neq 0$. Since $\nu \in \mathcal{S}_p(M)$, we are done when $|H/M| = p$.

We now assume that H/M is a p'-group. By Step 2, every irreducible constituent of ν^H lies in $\mathcal{S}_p(H)$. Now $H = ML$, and ν is a constituent of $\theta_{L \cap M}{}^M = (\theta^H)_M$. Hence some irreducible constituent of θ^H is in $\mathcal{S}_p(H)$. But τ is the unique p-rational constituent of θ^H and so $\tau \in \mathcal{S}_p(H)$. This completes Step 3.

Step 4. $\chi \mapsto \chi^0$ is a bijection from $\mathcal{S}_p(G)$ onto $\mathrm{IBr}_p(G)$, i.e. conclusion (a) of the theorem holds.

Proof. Since G is p-solvable, we may choose $L \lhd G$ such that G/L is a p'-group or p-group. By Step 1, $\theta \mapsto \theta^0$ is a bijection from $\mathcal{S}_p(L)$ onto $\mathrm{IBr}_p(L)$. By Steps 2 and 3, whenever $\theta \in \mathcal{S}_p(L)$, then $\chi \mapsto \chi^0$ is a bijection from $\mathcal{S}_p(G|\theta)$ onto $\mathrm{IBr}_p(G|\theta^0)$. Thus $\chi \mapsto \chi^0$ maps $\mathcal{S}_p(G)$ onto $\mathrm{IBr}_p(G)$.

Suppose $\chi, \eta \in \mathcal{S}_p(G)$ and $\chi^0 = \eta^0$. We may choose $\delta \in \mathcal{S}_p(L)$ such that

δ^0 is a constituent of both χ_L^0 and η_L^0. It follows from uniqueness that χ, $\eta \in \mathcal{S}_p(G|\delta)$. By the last paragraph, $\chi = \eta$.

Step 5. Conclusion

Proof. To complete the proof, we assume that $\beta \in \mathrm{Irr}\,(G)$ is p-rational and $\beta^0 \in \mathrm{IBr}_p(G)$. We need to show that $\beta \in \mathcal{S}_p(G)$. Choose $K \trianglelefteq G$ maximal such that the irreducible constituents of β_K lie in $\mathcal{S}_p(K)$. We may assume that $K < G$. Fix $\alpha \in \mathcal{S}_p(K)$ with $\beta \in \mathrm{Irr}\,(G|\alpha)$.

Let N/K be a chief factor of G. If $p \nmid |N/K|$, then every irreducible constituent of α^N is in $\mathcal{S}_p(N)$. Then some (and hence all) irreducible constituents of β_N are subnormally p-rational, contradicting the choice of K. Hence N/K is a p-group.

Choose $\gamma \in \mathrm{Irr}\,(N|\alpha)$ such that $[\beta_N, \gamma] \neq 0$. Then $\gamma \notin \mathcal{S}_p(N)$. By Step 3, α^N has a unique p-rational irreducible constituent and that lies in $\mathcal{S}_p(N)$. Hence γ is not p-rational.

Let $I = I_G(\gamma)$ and $T = \{g \in G \mid \gamma^g \text{ is Galois-conjugate to } \gamma\}$. Let $\psi \in \mathrm{Irr}\,(I|\gamma)$ with $\psi^G = \beta$. Since $(\psi^G)^0 = ((\psi^T)^G)^0 = \beta^0 \in \mathrm{IBr}_p(G)$, we have that $\psi^0 \in \mathrm{IBr}_p(I)$ and $(\psi^T)^0 \in \mathrm{IBr}_p(T)$. By Proposition 22.2, $\mathbb{Q}(\psi^T) = \mathbb{Q}(\beta)$ and thus ψ^T is p-rational. If $T < G$, the inductive hypothesis implies that $\psi^T \in \mathcal{S}_p(T)$. But then $\gamma \in \mathcal{S}_p(N)$, a contradiction. Hence $T = G$. By Proposition 22.2, $I \trianglelefteq G$.

Now let $S = \{g \in G \mid \gamma^g = \gamma^\sigma \text{ for some } \sigma \in \mathrm{Gal}\,(\mathbb{Q}_n/\mathbb{Q}_r)\}$. Recall $|G| = n = rp^\ell$ with $p \nmid r$. Then $I \leq S \leq T$ and $\psi^S \in \mathrm{Irr}\,(S|\gamma)$. Let $\tau \in \mathrm{Gal}\,(\mathbb{Q}_n/\mathbb{Q}_r)$. Since $(\psi^S)^G = \beta$ is p-rational, $((\psi^S)^\tau)^G = (\psi^G)^\tau = \beta$. Now γ^τ and γ are constituents of β_N and thus G-conjugate. Hence γ^τ and γ are S-conjugate. Thus $(\psi^S)^\tau \in \mathrm{Irr}\,(S|\gamma)$. Since $((\psi^S)^\tau)^G = \beta = (\psi^S)^G$, it follows from uniqueness in Clifford's Theorem 0.8 that $(\psi^S)^\tau = \psi^S$. Hence ψ^S is p-rational. Since $((\psi^S)^G)^0 \in \mathrm{IBr}_p(G)$, indeed $(\psi^S)^0 \in \mathrm{IBr}_p(S)$. If $S < G$, the inductive hypothesis implies that $\psi^S \in \mathcal{S}_p(S)$, whence $\gamma \in \mathcal{S}_p(N)$, a contradiction. Thus $S = G$.

Now $I \trianglelefteq G$ and $\psi^0 \in \mathrm{IBr}_p(I)$ because $(\psi^0)^G = \beta \in \mathrm{IBr}_p(G)$. We claim that ψ^0 is G-invariant. If $g \in G$, then $\psi^g \in \mathrm{Irr}\,(I|\gamma^g)$ and $(\psi^g)^G = \beta$. On the other hand, $\gamma^g = \gamma^\sigma$ for some $\sigma \in \mathrm{Gal}\,(\mathbb{Q}_n/\mathbb{Q}_r)$. Then $I = I_G(\gamma^g)$ and $\psi^\sigma \in \mathrm{Irr}\,(I|\gamma^g)$. Since $(\psi^\sigma)^G = \beta^\sigma = \beta$, the uniqueness in Clifford's Theorem 0.8 yields that $\psi^g = \psi^\sigma$. Then $(\psi^0)^g = (\psi^0)^\sigma = \psi^0$ because every Brauer character is p-rational. Hence ψ^0 is G-invariant.

Now $I \trianglelefteq G$, $\psi^0 \in \mathrm{IBr}_p(I)$ is G-invariant, and $(\psi^0)^G = \beta^0 \in \mathrm{IBr}_p(G)$. If $I < G$, we may apply Step 1 to conclude there exists a G-invariant $\eta \in \mathrm{Irr}\,(I)$ such that $\eta^0 = \psi^0$. Also η^G is irreducible because $(\eta^G)^0 = \beta^0 \in \mathrm{IBr}_p(G)$. We have both $\eta \in \mathrm{Irr}\,(I)$ is G-invariant and $\eta^G \in \mathrm{Irr}\,(G)$. By Frobenius reciprocity, $1 = [\eta^G, \eta^G] = [\eta^G{}_I, \eta] = |G : I|$. Thus $I = G$ and γ is G-invariant. Because β is p-rational and $\beta_N = e\gamma$ for an integer e, γ is p-rational. This contradiction completes the proof. $\qquad\square$

One can derive from Theorem 23.16 and its proof a number of corollaries about $\mathcal{S}_p$-characters, $p \neq 2$. For example, Steps 1 to 3 are valid and the Clifford correspondence works as one would hope.

We can now remove the hypothesis that $p \nmid \mu(1)$ for Theorems 23.7 and 23.10, at least for $p \neq 2$, by using subnormally p-rational lifts of Brauer characters instead of p'-special lifts. Observe that when $p \neq 2$, Theorem 15.9 (v) may be stated in terms of $\mathcal{S}_p$-characters instead of p-rational characters. Otherwise, the details of the proof, which we leave to the reader, are identical. An alternative method to remove the hypothesis $p \nmid \mu(1)$ from Theorem 23.7 (and $p \neq 2$ below) is the use of projective representations over fields of characteristic p.

23.17 Theorem. *Suppose that G is p-solvable, that $L \trianglelefteq G$ and $P/L \in \mathrm{Syl}_p(G/L)$. If $\mu \in \mathrm{IBr}_p(L)$ is invariant in P, if $p \neq 2$ and $H/L = \mathbf{N}_{G/L}(P)$, then*

  (i) $l_0(G|\mu) = l_0(H|\mu)$; *and*

(ii) $l(G|\mu) \geq l(H|\mu)$ with equality if and only if $l(G|\mu) = l_0(G|\mu)$.

Again $p \neq 2$ is not really necessary. Isaacs [Is 5] developed a lift that works for all p and is the same lift (when $p \neq 2$) as Theorem 23.16. This lift is slightly more tricky. While Clifford correspondence works smoothly for $\mathcal{S}_p$-characters, when $p \neq 2$, it does not work quite as expected when $p = 2$. We will give a brief description of Isaacs' lift, but first mention one result related to Theorem 23.17 (i). The proof is essentially the same, but one must extract a little more information from Theorem 15.9. A proof is given in [Wo 7].

23.18 Theorem. Let G be p-solvable and q-solvable for not necessarily distinct primes p and q. Suppose that $L \trianglelefteq G$, $Q/L \in \mathrm{Syl}_q(G/L)$ and $H/L = \mathbf{N}_{G/L}(Q/L)$. If $\mu \in \mathrm{IBr}_p(L)$ is invariant in Q, then

$$|\{\chi \in \mathrm{IBr}_p(G|\mu) \mid q \nmid \chi(1)/\mu(1)\}| = |\{\psi \in \mathrm{IBr}_p(H|\mu) \mid q \nmid \psi(1)/\mu(1)\}|.$$

Let G be π-separable. Isaacs [Is 6] has shown there is a uniquely defined subset $B_\pi(G) \subseteq \mathrm{Irr}(G)$ such that the following hold whenever $N \trianglelefteq G$:

   (i) If $\chi \in B_\pi(G)$, every irreducible constituent of χ_N is in $B_\pi(N)$;

   (ii) If $\theta \in B_\pi(N)$ and G/N is a π-group, then $B_\pi(G|\theta) = \mathrm{Irr}(G|\theta)$;

   (iii) If $\theta \in B_\pi(N)$ and G/N is a π'-group, then there is a unique $\psi \in B_\pi(G|\theta)$. Also $\psi = \eta^G$ for the unique $\eta \in B_\pi(I_G(\theta)|\theta)$ and $\eta_N = \theta$;

   (iv) Each $\chi \in B_\pi(G)$ is "π'-rational";

   (v) $X_\pi(G) = \{\chi \in B_\pi(G) \mid \chi(1) \text{ is a } \pi\text{-number}\}$;

   (vi) If $2 \in \pi$, then $B_\pi(G) = \mathcal{S}_{\pi'}(G)$; and

   (vii) If $\pi = \{p\}$, then $\chi \mapsto \chi^0$ is a bijection from $B_{p'}(G)$ onto $\mathrm{IBr}_p(G)$.

The definitions of π-rational and $\mathcal{S}_\pi(G)$ are the obvious generalizations of Definitions 23.4 and 23.14. The notion of B_π-characters has also been used to develop Brauer theorems in "characteristic π". For this, we refer the reader to Isaacs [Is 6, 8], Slattery [Sl 1, 2] and Wolf [Wo 6, 7].

REFERENCES

AB 1] D. Alvis and M. Barry, *Character degrees of simple groups*, J. Algebra 140 (1991), 116–123.

AC 1] G. Amit and D. Chillag, *On a question of Feit concerning character values of finite solvable groups*, Pacific J. Math. 122 (1986), 257–261.

AOT] K. Asano, M. Osima, and M. Takahasi, *Über die Darstellung von Gruppen durch Kollineationen in Körpen der Charactistik p*, Proc. Phys. Math. Soc. Japan 19 (1937), 199–209.

[Be 1] T. Berger, *Characters and derived length in groups of odd order*, J. Algebra 39 (1976), 199–207.

[Be 2] T. Berger, *Hall–Higman type theorems VI*, J. Algebra 51 (1978), 416–424.

[BG] T. Berger and F. Gross, *2-length and the derived length of a Sylow 2-subgroup*, Proc. London Math. Soc. 34 (1977), 520–534.

[Br 1] E. Bryukhanova, *Connections between the 2-length and the derived length of a Sylow 2-subgroup of a solvable group*, Math. Notes 29 (1981), 85–90 (translated from Mat. Zametki 29 (1981), 161–170).

Cm 1] A. Camina, *Some conditions which almost characterize Frobenius groups*, Israel J. Math. 31 (1978), 153–160.

Ca 1] W. Carlip, *Regular orbits of nilpotent groups in non-coprime representations*, Ph. D. Thesis, U. of Chicago (1989).

Ca 2] W. Carlip, *Regular orbits of nilpotent subgroups of solvable groups*, Illinois J. Math., to appear.

CDR] M. Coates, M. Dwan, and J.S. Rose, *A note on Burnside's other $p^\alpha q^\beta$-theorem*, J. London Math. Soc. (1976), 160–166.

JPW] J. Conway, R. Curtis, S. Norton, R. Parker and R. Wilson, "The Atlas of Finite Groups," Oxford University Press, New York (1985).

Cr 1] G.-M. Cram, *Lokal isomorphe Algebren und Anwendungen auf Gruppenalgebren*, Dissertation, Universität Augsburg, 1985.

Cr 2] G.-M. Cram, *On the field of character values of finite solvable groups*, Archiv der Math. 51 (1988), 294–296.

[Da 1] E. Dade, *Characters of groups with normal extra-special subgroups*, Math. Z. 152 (1976), 1–31.

[Da 2] E. Dade, *A correspondence of characters*, Proc. Symposia Pure Math. 37 (1979), 401–404.

[Di 1] J. Dixon, *The solvable length of a solvable linear group*, Math. Z. 107 (1968), 152–158.

[Di 2] J. Dixon, "The Structure of Linear Groups", Van Nostrand-Reinhold, Princeton, NJ, 1971.

[Es 1] A. Espuelas, *The existence of regular orbits*, J. Algebra 127 (1989), 259–268.

[Es 2] A. Espuelas, *Regular orbits on symplectic modules*, J. Algebra 138 (1991), 524–527.

[Es 3] A. Espuelas, *Large character degrees of groups of odd order*, Illinois J. Math. 35 (1991), 499–505.

[Fe 1] W. Feit, "Characters of Finite Groups", Benjamin, New York, 1967.

[Fe 2] W. Feit, "The Representation Theory of Finite Groups", North-Holland, Amsterdam, 1982.

[FT 1] P. Ferguson and A. Turull, *On a question of Feit*, Proc. Amer. Math. Soc. 97 (1986), 21–22.

[Fl 1] P. Fleischmann, *Finite groups with regular orbits on vector spaces*, J. Algebra 103 (1986), 211–215.

[Fo 1] P. Fong, *On the characters of p-solvable groups*, Trans. Amer. Math. Soc. 98 (1961), 263–284.

[Fo 2] P. Fong, *A note on a conjecture of Brauer*, Nagoya Mathematical J. 22 (1963), 1–13.

[Ga 1] D. Gajendragadkar, *A characteristic class of characters of finite π-separable groups*, J. Algebra 59 (1979), 237–259.

[GH] S. Glasby and R. Howlett, *Extraspecial towers and Weil representations*, J. Algebra 151 (1992), 236–260.

[Gl 1] D. Gluck, *Trivial set stabilizers in finite permutation groups*, Can. J. Math. 35 (1983), 59–67.

[Gl 2] D. Gluck, *Bounding the number of character degrees of a solvable group*, J. London Math. Soc. 31 (1985), 457–462.

[Gl 3] D. Gluck, *Primes dividing character degrees and character orbit sizes*, Proc. Amer. Math. Soc. 101 (1987), 219–225.

[GM 1] D. Gluck and O. Manz, *Prime factors of character degrees of solvable groups*, Bull. London Math. Soc. 19 (1987), 431–437.

[GW 1] D. Gluck and T.R. Wolf, *Defect groups and character heights in solvable groups II*, J. Algebra 87 (1984), 222–246.

[GW 2] D. Gluck and T.R. Wolf, *Brauer's height conjecture for p-solvable groups*, Trans. Amer. Math. Soc. 282 (1984), 137–152.

[Go 1] R. Gow, *On the number of characters in a p-block of a p-soluble group*, J. Algebra 65 (1980), 421–426.

[Go 2] R. Gow, "The p-subgroups of Classical Groups and Some Topics in the Representation Theory of Solvable Groups", Ph. D. Thesis, University of Liverpool, 1973.

[Go 3] R. Gow, *Character values of groups of odd order and a conjecture of Feit*, J. Algebra 68 (1981), 75–78.

[HH] P. Hall and G. Higman, *The p-length of a p-soluble group, and reduction theorems for Burnside's problems*, Proc. London Math. Soc. 6 (1956) 1–42.

[Hu] B. Huppert, "Endliche Gruppen I", Springer-Verlag, Berlin, 1967.

[Hu 1] B. Huppert, *Lineare auflösbare Gruppen*, Math. Z. 67 (1957), 479–518.

[Hu 2] B. Huppert, *Zweifach transitive auflösbare Permutationsgruppen*, Math. Z. 68 (1957), 126–150.

[Hu 3] B. Huppert, *Singer-Zyklen in klassischen Gruppen*, Math. Z. 117 (1970), 141–150.

[HB] B. Huppert and N. Blackburn, "Finite Groups II, III", Springer-Verlag, Berlin, 1982.

[HM 1] B. Huppert and O. Manz, *Degree problems: squarefree character degrees*, Archiv der Math. 45 (1985), 125–132.

[HM 2] B. Huppert and O. Manz, *Orbit sizes of p-groups*, Archiv der Math. 54 (1990), 105–110.

[Is] I.M. Isaacs, "Character Theory of Finite Groups", Academic Press, New York, 1976.

[Is 1] I.M. Isaacs, *The p-parts of character degrees in p-solvable groups*, Pacific J. Math. 36 (1971), 677–691.

[Is 2] I.M. Isaacs, *Characters of solvable and symplectic groups*, Amer. J. Math. 95 (1973), 594–635.

[Is 3] I.M. Isaacs, *Character degrees and derived length of a solvable group*, Can. J. Math. 27 (1975), 146–151.

[Is 4] I.M. Isaacs, *Lifting Brauer characters of p-solvable groups*, Pacific J. Math. 53 (1974), 171–188.

[Is 5] I.M. Isaacs, *Lifting Brauer characters of p-solvable groups II*, J. Algebra 51 (1978), 476–490.

[Is 6] I.M. Isaacs, *Characters of π-separable groups*, J. Algebra 86 (1984), 98–128.

[Is 7] I.M. Isaacs, *Solvable group character degrees and sets of primes*, J. Algebra 104 (1986), 209–219.

[Is 8] I.M. Issacs, *Counting objects that behave like Brauer characters of finite groups*, J. Algebra 117 (1988), 419–433.

[Is 9] I.M. Isaacs, *Coprime group actions fixing all non-linear irreducible characters*, Can. J. Math. 41 (1989), 68–88.

[Ja 1] N. Jacobson, "Basic Algebra I", W.H. Freeman and Co., San Francisco, 1974.

[LM 1] U. Leisering and O. Manz, *A note on character degrees of solvable groups*, Archiv der Math. 48 (1987), 32–35.

[Lü 1] H. Lüneberg, *Ein einfacher Beweis für den Satz von Zsigmondy über primitive Primteiler von $a^N - 1$*, in "Geometries and Groups, proceedings", edited by M. Aigner and D. Jungnickel, Springer Lecture Notes 893 (1981), 219–222.

[Ma 1] A. Mann, *Solvable subgroups of symmetric and linear groups*, Israel J. Math. 55 (1986), 162–172.

[Mz 1] O. Manz, *Endliche auflösbare Gruppen, deren sämtliche Charaktergrade Primzahlpotenzen sind*, J. Algebra 94 (1985), 211–255.

[Mz 2] O. Manz, *Endliche nicht-auflösbare Gruppen, deren sämtliche Charaktergrade Primzahlpotenzen sind*, J. Algebra 96 (1985), 114–119.

[Mz 3] O. Manz, *Degree problems: the p-rank in p-solvable groups*, Bull. London Math. Soc. 17 (1985), 545–548.

[Mz 4] O. Manz, *Degree problems II: π-separable character degrees*, Communications in Algebra 13 (1985), 2421–2431.

[Mz 5] O. Manz, *On the modular version of Ito's Theorem for groups of odd order*, Nagoya Math. J. 105 (1987), 121–128.

[MS 1] O. Manz and R. Staszewski, *Some applications of a fundamental theorem by Gluck and Wolf in the character theory of finite groups*, Math. Z. 192 (1986), 383–389.

MSW] O. Manz, R. Staszewski and W. Willems, *On the number of components of a graph related to character degrees*, Proc. Amer. Math. Soc. 103 (1988), 31–37.

MWW] O. Manz, W. Willems and T.R. Wolf, *The diameter of the character degree graph*, J. reine angew. Math. 402 (1989), 181–198.

MW 1] O. Manz and T.R. Wolf, *Brauer characters of q'-degree in p-solvable groups*, J. Algebra 115 (1988), 75–91.

MW 2] O. Manz and T.R. Wolf, *The q-parts of degrees of Brauer characters*, Illinois J. Math. 33 (1989), 583–591.

MW 3] O. Manz and T.R. Wolf, *Arithmetically long orbits of solvable linear groups*, Illinois J. Math., to appear.

[Mi 1] G. Michler, *Brauer's conjectures and the classification of finite simple groups*, in Representation Theory II: Groups and Orders, Proc. Ottawa 1984, Springer Lecture Notes 1178, 129–142.

[Mi 2] G. Michler, *A finite simple Lie group has p-blocks with different defects, $p \neq 2$*, J. Algebra 104 (1986), 220–230.

[NT] H. Nagao and Y. Tsushima, "Representations of Finite Groups", Academic Press, San Diego CA, 1987.

[Ok 1] T. Okuyama, *On a problem of Wallace*, preprint.

OW 1] T. Okuyama and M. Wajima, *Irreducible characters of p-solvable groups*, Proceedings of the Japan Academy 55 (1979), 309–312.

OW 2] T. Okuyama and M. Wajima, *Character correspondence and p-blocks of p-solvable groups*, Osaka J. Math. 17 (1980), 801–806.

[Pl 1] P.P. Palfy, *A polynomial bound for the orders of primitive solvable groups*, J. Algebra 77 (1982), 127–137.

[Pl 2] P.P. Palfy, *On the character degree graph of solvable groups I*, preprint.

[Pa 1] D. Passman, *Groups with normal Hall p'-subgroups*, Trans. Amer. Math. Soc. 123 (1966), 99–111.

[Pa 2] D. Passman, "Permutation Groups", W.A. Benjamin Inc., New York, 1968.

[Pa 3] D. Passman, *Blocks and normal subgroups*, J. Algebra 12 (1969), 569–575.

[Sa 1] S. Saeger, *The rank of a finite primitive solvable permutation group*, J. Algebra 105 (1987), 389–394.

[Sc 1] I. Schur, *Über die Darstellung der endlichen Gruppen durch gebrochene lineare Substitutionen*, J. reine angew. Math. 127 (1904), 20–50.

[Sl 1] M. Slattery, *Pi-blocks of pi-separable groups I*, J. Algebra 102 (1986), 60–77 .

[Sl 2] M. Slattery, *Pi-blocks of pi-separable groups II*, J. Algebra 124 (1989), 236–269.

[Su] D. Suprunenko, "Solvable and Nilpotent Linear Groups", Trans. Math. Monographs, vol. 9, Amer. Math. Soc., Providence, RI, 1963.

[Tu 1] A. Turull, *Supersolvable automorphism groups of solvable groups*, Math. Z. 183 (1983), 47–73.

[Un 1] K. Uno, *Character correspondences in p-solvable groups*, Osaka J. Math. 20 (1983), 713–725.

[Wa 1] Y.-Q. Wang, *A note on Brauer characters of solvable groups*, Can. Math. Bull. 34 (1991), 423–425.

[Wo 1] T.R. Wolf, *Characters of p'-degree in solvable groups*, Pacific J. Math. 74 (1978), 267–271.

[Wo 2] T.R. Wolf, *Character correspondences in solvable groups*, Illinois J. Math. 22 (1978), 327–340.

[Wo 3] T.R. Wolf, *Defect groups and character heights in solvable groups*, J. Algebra 72 (1981), 183–209.

[Wo 4] T.R. Wolf, *Solvable and nilpotent subgroups of $GL(n, q^m)$*, Can. J. Math. 34 (1982), 1097–1111.

[Wo 5] T.R. Wolf, *Sylow-p-subgroups of p-solvable subgroups of $GL(n, p)$*, Archiv der Math. 43 (1984), 1–10.

[Wo 6] T.R. Wolf, *Character correspondences and π-special characters in π-separable groups*, Can. J. Math. 39 (1987), 920–937.

[Wo 7] T.R. Wolf, *Variations on McKay's character degree conjecture*, J. Algebra 135 (1990), 123–138.

INDEX